Nutrition and Obsessive-Compulsive Disorder

This reference book discusses the role of nutrition in obsessive-compulsive disorders (OCD). It covers the investigational therapies in OCD, use of animal models, and biomarkers for better diagnosis, monitoring, and possible therapeutic options with nutrients for patients with OCD. This book highlights the role of probiotics, amino acids, vitamins, micronutrients, and salts in the management of OCD. It includes such topics as applications of computational modeling and nano-biotechnology in OCD, the effects of environmental pollutants, and the regulation of antidepressants in OCD.

KEY FEATURES

- Covers emerging therapeutic and nutritional approaches for the treatment and management of OCD
- Reviews the role of various micronutrients in OCD
- Discusses the prenatal genetic diagnosis and application of computational modelling in OCD
- Includes the nutritional and dietary roles in such neuropsychiatric disorders as depression and obsessive-compulsive disorder (OCD)
- Explores the application of nano-biotechnology in OCD

This book is meant for researchers and professionals in the field of nutritional neuroscience and psychology.

Nutrition and Obsessive-Compulsive Disorder

The Interplay

Edited by

Senthilkumar Rajagopal
REVA University, Karnataka, India

Musthafa M Essa
Sultan Qaboos University, Muscat, Oman

Saravanan Ramachandran
Chettinad Academy of Research and Education, Tamil Nadu, India

Geethalakshmi Sundararaman
Sree Narayana Guru College, Tamil Nadu, India

Hanan Fahad Alharbi
Princess Nourah bint Abdulrahman University, Riyadh, Saudi Arabia

CRC Press
Taylor & Francis Group
Boca Raton London New York

CRC Press is an imprint of the
Taylor & Francis Group, an **informa** business

First edition published 2024
by CRC Press
2385 Executive Center Drive, Suite 320, Boca Raton, FL 33431

and by CRC Press
4 Park Square, Milton Park, Abingdon, Oxon, OX14 4RN

CRC Press is an imprint of Taylor & Francis Group, LLC

ISBN: 978-1-032-47089-4 (hbk)
ISBN: 978-1-032-48208-8 (pbk)
ISBN: 978-1-003-38789-3 (ebk)

DOI: 10.1201/9781003387893

Typeset in Times
by Apex CoVantage, LLC

Contents

SECTION 2 Diagnosis of OCD

SECTION 3 Treatment and Intervention of OCD

Foreword

Prof. Min Li MD, PhD

It is a distinct honor to have been invited by my colleagues in the neurosciences arena to write the foreword to this very articulate and scientifically state-of-the-art book titled *Nutrition and Obsessive-Compulsive Disorder: The Interplay.* This book covers all the areas required to create a robust category and perform a read-across. I am certain that the readers, including faculties, researchers, and students will find this book extremely informative, interesting, and inspiring. Hence, I hope that you will find this book possesses sufficient disclosure and adequate utility. This book aims to simplify the revolution and to fortify the researcher with the information needed to use nutrition and their supplementations for neurological disorders with complete confidence and the best compound can be applied for therapy of the individual. A book differs in important ways from a journal. This book is an important publication, and the time for it has come. The extraordinary effort required to create a book that comprehensively and authoritatively presents the state of knowledge in an arena as broad and ever-expanding as nutrition influences in neuronal malfunctioning is a daunting task.

As a consequence, a large number of books, thick and thin, have been and continue to be published on various aspects of beneficial usage of nutrition. The central theme of this book is giving fundamental mechanisms of nutrition and their supplementations for involvement of brain pathophysiology. They have made an effort to unify all the content of scattered research literature in this area of research and tried to provide in-depth contents about the topics they have chosen. Overall, this book is not just a collection of chapters, but it is an essence of the diverse nutritional aspects on OCD. In all chapters, they have provided the basic information relevant to the topics, and at the same time, they have described the perspective knowledge about using nutritional approaches. This believes that this book could be an informative resource, in the form of condensed handbook for research students as well as advanced researchers.

Consequently, this book covers our information of the field by focusing on the overview of OCD, diagnosis and treatment of OCD with different therapeutics approaches of nutritional values. The editors, Senthilkumar Rajagopal, Musthafa M Essa, Saravanan Ramachandran, Geethalakshmi Sundararaman, and Hanan Fahad Alharbi, bring understanding to the milieu of clinical research, experimental research, and pharmacology, which, revenue collected, focus the most active areas of brain research.

Prof. Min Li MD, PhD
Associate Dean
Program Director, BCM and BSc. BIOMED
Director, Mr. and Mrs. Ko Chi-Ming Centre for Parkinson's Disease Research

School of Chinese Medicine, Hong Kong Baptist University
Kowloon Tong, Hong Kong
TEL: (852) 3411 2919
FAX: (852) 3411 2461
E-mail: limin@hkbu.edu.hk

Foreword

Dr. N. R. Jagannathan FNA, FASc, FNASc, FAMS, FISC, FISMRM (USA)

It is a great privilege to write this foreword to *Nutrition and Obsessive-Compulsive Disorder*: *The Interplay* by Drs. Senthilkumar Rajagopal, Musthafa M Essa, Saravanan Ramachandran, Geethalakshmi Sundararaman, and Hanan Fahad Alharbi.

This edited volume covers the area of neurosciences that is very well articulated by various authors, and no doubt this volume definitely would serve as a cutting-edge source material in this fascinating area of research. The volume aims to simplify the current trends and advances in the area of nutrition and their supplements used for neurological disorders and for individual therapy. The central theme of this volume is to provide fundamental mechanisms of nutrition, their supplements, and their role in brain development and pathophysiology. Overall, this volume is more than just a collection of number of chapters on different topics; it is a careful compilation of articles that cover a vast area of the various nutritional aspects of OCD. The authors have provided necessary basic information relevant to the topic covered during the early part of this book so as to enable the readers to understand the advances that are covered in the later part of this book.

It is tough to produce a volume that comprehensively describes the various aspects of the state of knowledge in a concise manner and the editors along with the authors have accomplished it in a very simple way and would like to congratulate them for the same. I am confident that the readers, that includes faculty, researchers, and students, will find this book to be extremely informative, useful, interesting, and inspiring.

Dr. N. R. Jagannathan FNA, FASc, FNASc, FAMS, FISC, FISMRM (USA)
Distinguished Visiting Professor, Koita Centre for Digital Health, IIT, Mumbai 400076
Adjunct Professor, Dept. of Electrical Engineering, IIT Madras, Chennai
Advisor, Chettinad Hospital and Research Institute, Kelambakkam, TN 603 103
Visiting Professor, Dept. of Radiology, Sri Ramachandra Inst. Of Higher Education and Research, Chennai—600116
Former Professor and Head, Dept. of NMR and MRI Facility, All India Institute of Medical Sciences, New Delhi-110029

Preface

Obsessive-compulsive disorder (OCD) is a condition of the brain in which unwanted thoughts or obsessions get generated, which leads to compulsive behaviors that last for a longer time. Such disorders do not have much cure and can occur in people of all age group and gender. If left untreated, the conditions are likely to worsen, which leads to other fatal nervous system disorders like Alzheimer's, dementia, etc. Usually, such disorders are treated with a combination of cognitive-behavioral therapy and pharmacotherapy. But these therapies do not provide significant improvement in patient's conditions. Moreover, their pharmaceutical preparations tend to induce severe side effects, and hence many hesitate to use the pharmacotherapy. Due to these drawbacks, the current day treatment has started to focus on remedy through food and food supplements.

Certain components in food are found to associate with neurotransmitters and enhance their activity in a positive manner. Some supplements are found to possess antioxidant with synergistic effect, helping the nervous system to rejuvenate. Gene level interaction of certain food components have also been studied, which has the efficacy to be considered as a potent drug for OCD treatment. Some minerals are found to activate the non-functional hormones and enzymes involved in brain functioning and helps in reducing stress and anxiety.

This book has a collection of chapters that describes the histopathology of OCD, effect of anti-depressant drug on OCD, effect of herbal drugs and micronutrients in reducing OCD, role of microbiota in OCD, and other possible nutritional treatments to decrease the symptoms and effects of this neurodegenerative disorder. There are also chapters which explain how nanoparticles can be used for treating the brain damage and computational modelling of OCD through neural networks. This book also highlights up-to-date information of beneficial effects of dietary intake and uniquely covers diversity of micronutrients and its roles on OCD. It has easily understandable illustrations on prophylactic activities of dietary supplementation in various CNS disorders, and research-oriented topics are discussed with results and evidences.

Overall, this book provides the role of nutrition and nutritional supplement in suppressing the fatal and dreadful effects of brain and nervous system damage.

Senthilkumar Rajagopal
Musthafa M Essa
Saravanan Ramachandran
Geethalakshmi Sundararaman
Hanan Fahad Alharbi

Editors

Dr. Senthilkumar Rajagopal is an associate professor/Ramalingaswami Re-entry fellow [DBT] of Department of Biotechnology, REVA University, Bangalore, India. He secured his PhD in biochemistry from Annamalai University, India. He did his three postdoctoral researches in anesthesiology in University of Virginia, Charlottesville, Virginia, followed by Department of Cancer Immunology and AIDS, Dana Farber Cancer Institute, Harvard University, Boston, Massachusetts, and Department of Physiology and Biophysics, Virginia Commonwealth University, Richmond, Virginia, USA. Dr. Senthilkumar is a recipient of Rameshwardasji Birla Smarak Kosh Endowment award from Mumbai Medical Trust, Mumbai, India, and Ramalingaswami Re-entry Fellowship from Department of Biotechnology, Ministry of Science and Technology, government of India. He has more than 18 years of research experiences in various fields such as alcoholic hepatotoxicity, voltage gated calcium channels, protein kinase C phosphorylation, ADAM family, smooth muscle physiology, G-protein coupled receptor signaling pathways, neurotransmitters transporters to central nervous system. He has published nearly 67 research articles in various peer-reviewed international journals, 32 book chapters, 10 abstracts, 65 conferences proceedings/abstracts and attended various workshops/seminars in India, Canada, and USA. He conducts research in the area of G-protein coupled receptor mediated intracellular mechanisms by alcohol-induced digestive disorders in the gastrointestinal tract. Currently, he is mentoring graduate students and having research projects from various funding agencies.

Prof. Musthafa M Essa, PhD, is a professor of nutrition at Sultan Qaboos University, Oman. He is an editor-in-chief for the *International Journal of Nutrition, Pharmacology, and Neurological Diseases*. He is also acting as an associate editor for BMC complementary and alternative medicine and is an editor for Frontiers in Neuroscience and Frontiers in Biosciences. Prof. Essa is an expert in the field of nutritional neuroscience/neuropharmacology and has published more than 200 papers, 64 book chapters, and 16 books. Prof. Essa is the recipient of several awards, including the best book in the world, National Research Award, Outstanding Neuroscientist, and Distinguished Researcher. He has over 15 years of research and teaching experience in the field of nutritional neuroscience and under his guidance, five PhD and ten masters students completed their research. He is the editor for a unique and first in its kind book series titled *Nutritional Neurosciences* published by Springer Nature.

Dr. Saravanan Ramachandran is an associate professor of Marine Pharmacology at Chettinad Academy of Research and Education in Kelabmakkam, Tamil Nadu, India. At Annamalai University in India, he completed his PhD in marine biotechnology. His scientific interests are around the effects of pharmaceuticals on marine species. He has 50 research publications published in international and national journals and has received two research awards from the Indian Council of Medical Research and the Centre for Marine Living Resources and Ecology in Kochi, India. He possesses two patents and has submitted 65 gene sequences to the NCBI, including three entire genomes of halophilic bacteria. He is a member of the Society for Biotechnology (India) and the International Neurotoxin Association. He is currently the Principal Investigator of a research project financed by the by the Department of Biotechnology, Government of India.

Dr. Geethalakshmi Sundararaman is working as an associate professor and head in the Department of Biotechnology, Sree Narayana Guru College, Coimbatore, Tamilnadu. She has 15 years of teaching and research experience. Her research interest is in bioinformatics and biotechnology with specialization in plant biotechnology, functional genomics, and drug designing. She was the university rank holder in her postgraduation from University of Madras. She has published more than 25 research articles in international and peer reviewed journals and has co-authored many book chapters and books. To add to her credit, she has completed one project funded by UGC-SERO. She is a life time member of many international scientific societies and organizations.

Dr. Hanan Fahad Alharbi is an Assistant Professor in the Maternity and Child Health Nursing Department, College of Nursing, Princess Nourah bint Abdulrahman University, Riyadh, Saudi Arabia. Her specialty is Pediatric Nursing, and she is classified as a consultant in Pediatric nursing by the Saudi Commission for Health Specialties. She has Seven years of clinical experience in the Pediatric Intensive Care Unit and as a quality analyst. She was awarded an Outstanding Clinician by the American Society of Science, Engineering and Technology (ASSET) in 2022. Dr. Hanan has over nine years of academic experience. She published many articles and supervised several undergraduate research projects. She also worked as an external examiner for several master students' research projects. She reviewed many papers for different journals. Her research interest includes Pediatric Nursing, Healthcare, and Mental Health.

Contributors

Anitha Arumugam
Women Scientist, Department of Botany, Bharathiar University
Coimbatore, Tamil Nadu, India

Hanan Fahad Alharbi
Maternity and Child Health Nursing Department, College of Nursing, Princess Nourah bint Abdulrahman University

Thabitha Aavula
Chettinad Hospital and Research Institute, Chettinad Academy of Research and Education
Chettinad Health City, Kelambakkam, Chennai

Anu Augustine
Department of Biotechnology
Sree Narayana Guru College
Coimbatore, TN, India

Kartheek. R. Balapala
Physiological Sciences Unit, Department of Basic Medical Sciences
MCS School of Medicine, Copperbelt University
Zambia

Vishal Chanda
Department of Biotechnology
School of Applied Sciences, REVA University
Bengaluru, KA, India

Sabyasachi Choudhury
Navigators software Pvt Ltd
Kolkata, WB, India

Arunkumar Dhailappan
Department of Chemistry and Biosciences
SrinivasaRamanujan Centre, SASTRA Deemed to be University
TN, India

Thilagar Gobinath
Centre of Advanced Study in Marine Biology, Annamalai University
Parangipettai, TamilNadu, India

Shoumi Haldar
Department of Biotechnology
School of Applied Sciences, REVA University
Bengaluru, KA, India

Neelakanta Sarvashiva Kiran
Department of Biotechnology, School of Applied Sciences, REVA University
Bengaluru, KA, India

Kumar Ebenezar Kesavarao
Chettinad Hospital and Research Institute, Chettinad Academy of Research and Education
Chettinad Health City, Kelambakkam, Chennai

K. Kalaiarasi
Department of Biotechnology
Sree Narayana Guru College
Coimbatore, TN, India

Sowmya Kumar
Department of Life Science
Mount Carmel College
Bengaluru, KA, India

Ramya Kumari
Department of Chemistry and Biochemistry, M S Ramaiah College of Arts, Science and Commerce
Bengaluru, KA, India

Neelakanta Sarvashiva Kiran
Department of Biotechnology, School of Applied Sciences, REVA University
Bengaluru, KA, India

Gaonkar Lowkesh
Department of Biotechnology, School of Applied Sciences, REVA University
Bengaluru, KA, India

Krishnavignesh Lakshmanan
Academics and Research, Chennai Fertility Centre and Research Institute
Chennai, Tamil Nadu, India

Sudha Devi Manickam
Department of Biochemistry, Biotechnology and Bioinformatics, School of Biosciences, Avinashilingam Institute for Home Science and Higher Education for Women
Coimbatore, Tamil Nadu, India

Diana Miguez
Latitud—Fundación LATU Laboratorio Tecnológico del Uruguay (LATU) Edificio Los Abetos
Avenida Italia 6201, C.P. 11500, Montevideo, Uruguay

Prabhakar Mishra
Department of Biotechnology
School of Applied Sciences, REVA University
Bengaluru, KA, India

Krusha Moradiya
Department of Biotechnology
School of Applied Sciences, REVA University
Bengaluru, KA, India

Yuvashree Muralidaran
Department of Biotechnology
School of Applied Sciences, REVA University
Bengaluru, KA, India

Samuthirapandian Ravichandran
Government Thirumagal Mills College
Gudiyatham, TN, India

Saravanan Ramachandran
Chettinad Hospital and Research Institute, Chettinad Academy of Research and Education
Chettinad Health City, Kelambakkam, Chennai

Senthilkumar Rajagopal
Department of Biotechnology
School of Applied Sciences, REVA University
Bengaluru, KA, India

Subitha R.
Department of Biotechnology, Sree Narayana Guru College
Coimbatore—Tamilnadu, India

Geetika Pant
Department of Biotechnology and Genetics
M S Ramaiah College of Arts, Science and Commerce
Bengaluru, KA, India

Aparna K.S.
Department of Biotechnology, Sree Narayana Guru College
Coimbatore—Tamilnadu, India

Srushti Amarsingh Rajput
Department of Biotechnology School of Applied Sciences, REVA University
Bengaluru, KA, India

Supta Sarkar
Department of Foods and Nutrition, College of Community Science, Professor Jayashankar Telangana State Agricultural University
Hyderabad, Telangana, India

Santhi Priya Sobha
Chettinad Hospital and Research Institute, Chettinad Academy of Research and Education
Chettinad Health City, Kelambakkam, Chennai

Sumathi C Samiappan
Department of Chemistry and Biosciences SrinivasaRamanujan Centre, SASTRA Deemed to be University
TN, India

Varun Varma Sangaraju
Department of Information Technology University of the Cumberlands
Williamsburg, KY, USA

Parthana Sarma
Department of Biotechnology School of Applied Sciences, REVA University
Bengaluru, KA, India

Upasana Sarma
Department of Biotechnology School of Applied Sciences, REVA University
Bengaluru, KA, India

Dhasarathdev Srinivasan
Chettinad Hospital and Research Institute, Chettinad Academy of Research and Education
Chettinad Health City, Kelambakkam, Chennai

Balakrishnan Subramanian
Physiological Sciences Unit, Department of Basic Medical Sciences MCS School of Medicine, Copperbelt University
Zambia

Gowtham Kumar Subbaraj
Chettinad Hospital and Research Institute, Chettinad Academy of Research and Education
Chettinad Health City, Kelambakkam, Chennai

Shubham Singh
Department of Molecular Metabolism, Harvard T.H. Chan School of Public Health, Harvard University
Boston, MA, USA

Dhanalakshmi Thiyagarajan
Nehru Institute of Health Sciences
Coimbatore, Tamil Nadu, India

Aswani T
Department of Biotechnology School of Applied Sciences, REVA University
Bengaluru, KA, India

C. Thangam
Chettinad Hospital and Research Institute, Chettinad Academy of Research and Education
Chettinad Health City, Kelambakkam, Chennai

Mrinalini UG
Dept of Biotechnology and Microbiology
Bangalore University
Bengaluru, KA, India

Gorthi Vaishnavi
Department of Biotechnology, School of Applied Sciences, REVA University
Bengaluru, KA, India

Rohini Venugobal
Department of Science and Humanities, Nehru Institute of Engineering and Technology
Coimbatore, Tamil Nadu, India

Nethravathy V
Department of Biotechnology School of Applied Sciences, REVA University
Bengaluru, KA, India

Santhosh Kumar Yasam
Chettinad Hospital and Research Institute, Chettinad Academy of Research and Education
Chettinad Health City, Kelambakkam, Chennai

Chandrashekar Yashaswini
Department of Biotechnology, School of Applied Sciences, REVA University
Bengaluru, KA, India

Section 1

Background of OCD

1 Obsessive-Compulsive Disorder

A Perspective Review

Vishal Chanda, Senthilkumar Rajagopal, and Yuvashree Muralidaran
Department of Biotechnology, School of Applied Sciences, REVA University, Bengaluru, KA, India

1.1 INTRODUCTION

Obsessive-compulsive disorder (OCD) is an anxiety-related, chronic, costly condition where a person experiences intrusive thoughts, commonly referred to as obsessions (Purdon, 2022; Stein, 2002; Drummond and Edwards, 2018). These obsessive thoughts certainly lead the person to do uncontrolled repetitive behavioral activities, which is called compulsion (Saboowala, 2022; Jenike, 2004). Compulsions or repetitive behavior brings relief to the anxiety caused by obsessive thoughts, but the relief is temporary and relapses each time the person's obsessive thoughts are triggered (Salkovskis et al., 1998; Marks, 1990). Some of the common obsessional issues are worrying about hurting someone, being a bad example to the society, or contamination on the places touched by oneself or others (Barnhill, 2022). The purpose of compulsion is to reduce anxiety and to impede any feared event. Over time, these compulsive actions become a habit, where the actual obsessive episodes and worries are forgotten (Marincowitz et al., 2023; Berman, 1942). These instances provide evidence that obsession and compulsion to complete that obsession is interlinked. For instance, an obsessive fright of contamination leads to compulsive action of washing and cleaning surfaces very often. This habit serves a short-term relief to the stress brought by the obsession. However, this obsessive thought reactivates the compulsion cycle, and the cycle of obsession-compulsion continues (Gans, 1998).

1.1.1 The Science Behind OCD

Many medical researchers have proven that OCD, the brain disorder, is caused by processing and passing of incorrect information in the brain (Ruchsow et al., 2005). Usually, the brain-body coordination happens by passing of neural chemical message (neurotransmitters—serotonin in this case). OCD is caused due to the problem in communication between the frontal part of the brain (the frontal cortex) and the deeper thalamus, by incorrect processing or passing of this neurotransmitter (Burton et al., 2019; Abramowitz et al., 2018). OCD presents heterogenous nature (Olatunji et al., 2019), that is, there are categorical subtypes in obsessive-compulsion, and this varies person to person. For example, if a person is obsessed with a contamination-free environment, they are categorized under (1) "checkers and washers" subtype. Similarly, there are three more categories: (2) symmetry, repeating; (3) forbidden/taboo thoughts, sexual, aggressive, religious; and (4) doubt/checking (Williams et al., 2013). Among these subtypes, symmetry is the most typical sign/symptom presented by many individuals expressing OCD. However, patterns of such symptoms depend on varying factors like age, gender, occasionally environmental factors, and there are instances where people stack their

DOI: 10.1201/9781003387893-2

wares systematically, for the sake of proper ordering, and storing huge amounts of goods in smaller area possible. This task cannot be considered as symmetry subtype of OCD.

The first instance of OCD in psychiatric literature appears as early as 1838 by Jean-Etienne Esquirol (Grant and Kim, 2006). Ever since, the hereditary role of OCD is considered to be one of the factors for developing OCD, but inheritance pattern of this disorder is unclear. However, in a study conducted, it was reported that the risk of developing this condition is higher in first-degree relatives, and according to familial aggregation studies, OCD is reported to be familial (Pampaloni et al., 2022). And the same study concluded that OCD is hereditary due to some genetic factors.

1.1.2 Epidemiology: The Spread of OCD

The enduring occurrence of OCD is approximately 1–3% (Nestadt et al., 2010), with constant percentages of people with this disorder across different nations (Fawcett et al., 2020; Shapiro, 2020; Weissman et al., 1994; Ruscio et al., 2010). Males are more frequently affected in childhood despite the fact that females are affected at a slightly higher rate in adolescence and adulthood than males, this was a hypothetical conclusion by Mathes, Morabito et al., 2019, without any genome-wide study to prove the same. The age of onset is distributed in a bimodal fashion, peaking in early adulthood (13–18 years) and adolescence (Albert et al., 2015; Anholt et al., 2014). This bimodal pattern leads to the conclusion that childhood-onset OCD is a unique subtype of the condition with unique clinical characteristics and putative etiological causes (Geller et al., 1998). Early onset OCD has actually been linked to a larger hereditary component, a higher genetic risk of developing the disorder and more severe than late-onset OCD (Fontenelle and Hasler, 2008; Nicolini et al., 2009).

OCD has a major detrimental impact on many facets of quality of life, especially on social connections, and can be profoundly debilitating illness (Stein et al., 2019; Subramaniam et al., 2013). Compared to the general population, OCD patients have a significantly higher chance of dying from natural or unnatural causes (mortality risk ratio ~ 1.68 and 2.61, respectively) (Meier et al., 2016). Similar findings were reported in a recent review, which found that significant percentages of OCD patients had suicidal thoughts and that the mean lifetime suicide attempt rate is ~14.2% (Mattera et al., 2022; Albert et al., 2019; Fernández et al., 2017).

Majority of the OCD patients present at least one comorbid depressive disorder; anxiety disorder, social nervousness, specific phobia, or any psychiatric disorder are most frequent co-occurring problems in such patients. Autism spectrum disorder (ASD), anorexia nervosa, and attention-deficit hyperactivity disorder (ADHD) are some more prevalent comorbidities (Brakoulias et al., 2017). The most common lifetime comorbidity of OCD is major depression, which has a frequency of 50% worldwide (Brakoulias et al., 2017). The neurological disease Tourette syndrome, stroke, traumatic brain injury, progressive supranuclear palsy, Huntington disease, Parkinson's disease, and different dementias can be related to OCD (Richter and Ramos, 2018; Drubach, 2015).

1.2 CAUSES OF OCD

The cause of OCD is still unclear. OCD with an early onset (childhood-onset) is thought to be 45–65% heritable, whereas OCD with an adolescent or adult onset is thought to be 27–47% heritable (Adams et al., 2018). Although through genome association study, or also known as genome-wide association studies (GWAS), some candidate genes have been reported, but the results are highly variable, and many have not been repeated or have not borne the test of rigorous statistics. Numerous brain regions and processes have been linked to OCD. Studies have repeatedly shown caudate and orbitofrontal brain activation in sufferer with this disorder. The anterior cingulate cortex, thalamus, amygdala, and partial cortex are other significant implicated regions (suggesting

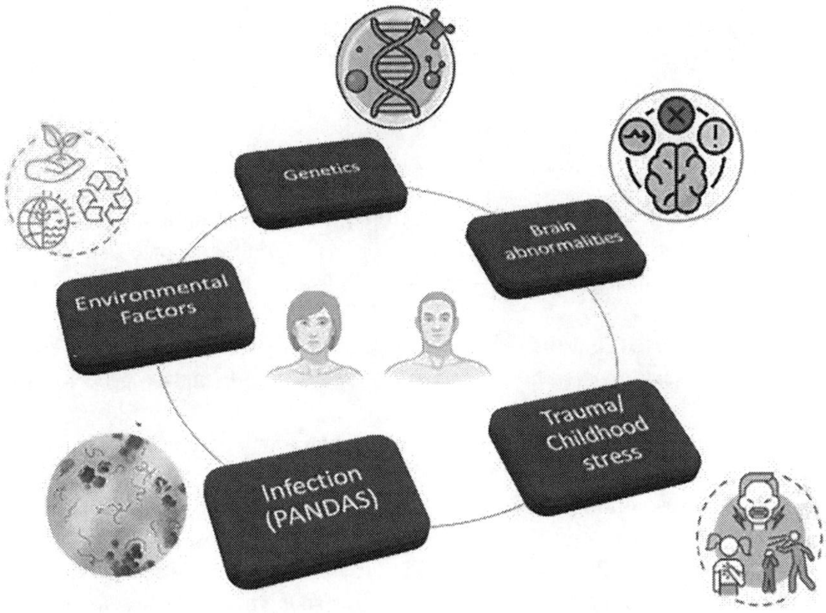

FIGURE 1.1 Causes of obsessive-compulsive disorder.

anomalies in functional or structural connections) (Perrotta, 2019). Patients with OCD experience cognitive abnormalities in neuropsychological testing, which are related to the frontal lobes and related front-subcortical structure functioning (Drummond and Edwards, 2022). Figure 1.1 shows pictorial representation of major causes.

1.2.1 Genetics

OCD is an extremely difficult condition. It is exceedingly improbable that one gene out of ~30,000 genes we have could ever be the cause of the intricate obsessions and compulsions that characterize OCD (Grünblatt, 2021; Mahjani et al., 2021). It is considerably more likely to be caused by a variety of genes interacting to raise vulnerability (Purty et al., 2019). With human genome finally being completely mapped out, the search for the responsible genes of various psychotic disorders is underway. Same is the case with obsessive-compulsive disorder (Thalamuthu et al., 2022). Although, OCD's genetic basis is still not clear, that which gene is important under what circumstances. But study of identical twins and closely related people with OCD concluded that genetics is the greatest contributor for developing OCD in a person, along with effects from environmental factors (Sidorchuk et al., 2021; Alemany-Navarro et al., 2020). Some of the preliminary evidence shows that certain alleles or versions of genes that control production of serotonin, a monoamine neurotransmitter in the brain that may be important for OCD, brain-derived neurotrophic factor, and serotonin may indicate a vulnerability to develop OCD. These chemical transmitters play important roles in controlling brain development. However, it is not yet obvious on how these genes affect the onset of OCD, and further study is still under process.

1.2.2 Brain Abnormalities

According to recent studies, OCD individuals frequently have reduced gray matter in specific regions of the brain. In general, the brain's gray-matter regions help us with controlling impulses; managing senses; processing information, thoughts, and feelings; and also developing, regulating, and exhibiting motor skills like talking, writing, reaction time, balance, coordination, and drawing. Low quantities of gray matter might also alter how a person receives information, increasing the likelihood of unintentional negative thoughts (Maziero et al., 2021; Mukhopadhyay and Tarafder, 2018).

1.2.3 Environmental

OCD may also be brought on by environmental factors like stress, trauma, and even microbes. It is well-known that stress can make OCD symptoms worse. OCD, for instance, is frequently associated with significant life transitions like divorce, losing a loved one, existential crisis, mental stress (Kracker et al., 2020).

Previous research reported that the degree of disorder and childhood trauma were related, but not directly. Instead, it was impacted by issues with connection and emotion processing (Carpenter et al., 2011).

Pediatric autoimmune neuropsychiatric disorder associated with strep (PANDAS) is yet another environmental element that puts youngsters at the risk. This is an uncommon form of childhood OCD where symptoms resembling OCD are brought on by an immunological response to streptococcus infection. PANDAS symptoms resemble those of childhood-onset OCD. However, onset can appear very differently. Typically, early onset OCD often manifests between the ages of 8 and 12, and symptoms worsen over time, whereas PANDAS can strike children between the ages of 4 to 14 (Porth and Geller, 2018), and its symptoms start more abruptly, and the onset is much more intense from initial stages itself (Alsheikh and Alsheikh, 2021; Plana et al., 2019).

1.3 CLASSIFICATION OF OCD: THE FOUR TYPES OF OCD

A variety of subjects can be covered by obsessions and compulsions. However, the majority of them fit into one of the four categories discussed in the following section. More than one category of obsessive behaviors and thoughts are possible. Figure 1.2 will show conceptual illustration of various categories of OCD.

1.3.1 Contamination/Washing

Obsessive worries about becoming sick and spreading germs. Their obsessions with cleanliness go well beyond the norm for hygiene practices and housekeeping. According to research, those who have this particular sort of OCD could respond to revulsion too easily. Thus, individuals experience disgust more frequently than the majority of people (Mancini, 2018). For instance, someone exhibiting contamination OCD could establish routines that entail repeatedly washing their hands until they bleed. After that, they struggle more to let the emotion pass. People with contamination OCD may also acquire phobias of blood, spoiled food, body fluids, broken glass, etc. in addition to dirt and germs (Yu et al., 2022).

1.3.2 Doubt/Checking

OCD sufferers frequently lack trust in their own judgement. Every kind of OCD has the central component of doubt. However, some people have a hard time questioning their perspective of reality. They could also question their recollection of the recent occurrences (Cole and Tubbs, 2022). When someone with this sort of OCD leaves the house, locks the door, and then gets in the automobile, they could start doubting if they really did lock the door. Also, an OCD sufferer need to check back several times before they feel certain that the work has been finished (Moreira et al., 2020). Another characteristic of OCD is incompleteness, or the perception that something is not "quite

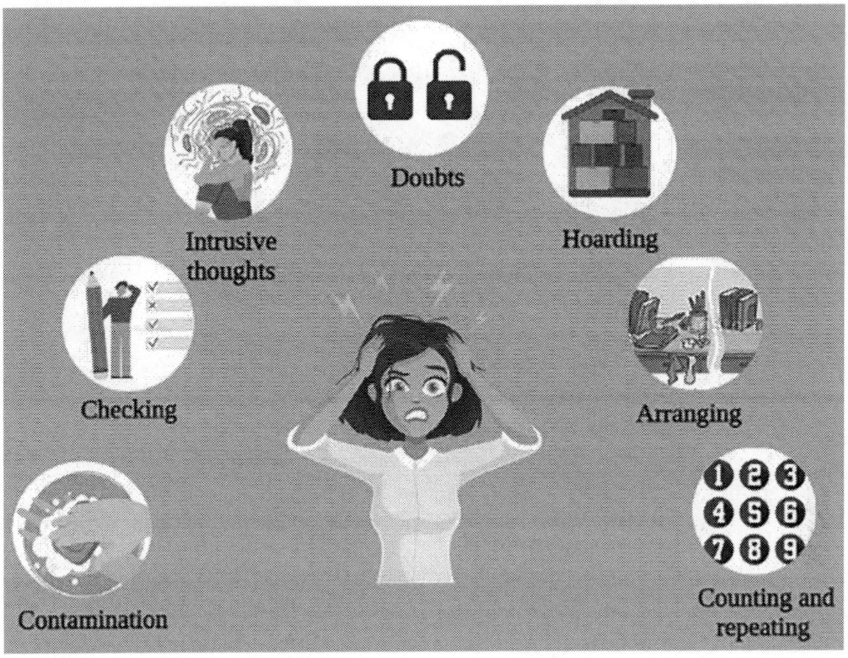

FIGURE 1.2 Types of obsessive-compulsive disorder.

right" (Praptomojati, 2019). This incompleteness might cause individuals to repeat the action until it seems "just right," even though the door was locked securely the first time.

1.3.3 Ordering/Arranging

This kind of OCD can cause a person to get obsessed on the symmetry and order in their surroundings (Radomsky and Rachman, 2004). This is distinct from taking pleasure in keeping the home tidy. The repetitive arrangement of the same things until they match a given standard can take hours off of a person's day who suffers from this kind of OCD (Pitman, 1987). Rearranging objects in your home until they feel "correct," moving objects a predetermined number of times (for instance, putting the same book on a shelf eight times because that number has special significance), and repeatedly counting objects are examples of compulsive behaviors that fall into this category.

1.3.4 Unacceptable/Taboo Thoughts

Taboo, usually referred to as "pure obsessions," include irrational sexual, violent, and religious obsessions (Moulding et al., 2014). Unacceptable or forbidden ideas are unmistakably ego-dystonic and have a repulsive character that is less noticeable in other OCD symptoms. As their names imply, these obsessions frequently include inappropriate, taboo, or prohibited elements at their core, such as blasphemy, incest, or stabbing a family member (Baer, 2002). In addition to the standard behavioral treatment for OCD in the form of ERP, psychological therapies that target underlying beliefs and cognitive processes are supported by the ego-dystonic nature of unacceptable/forbidden ideas, thoughts, and their association with the idea that it is crucial to control one's thoughts. Unacceptable or taboo ideas may also relate to substance misuse and animosity (Williams and Wetterneck, 2019; Freeman and Leonard, 2000).

TABLE 1.1
Detailed Description of Additional Types of Obsessive-Compulsive Disorder

Type	Details	References
Checking	Reassurance, past memories, door locks/ windows, housc/office alarm, gas or electric stove knobs, water taps, valuable items.	Leonhart and Radomsky, 2019
Contamination	Keep away from public toilets, shaking hands, chemicals, door handles, eating in public, crowds, outside air.	Mathes et al., 2020
Mental Contamination	The feeling of mental contamination is related to contact contamination; the person feels that touching something after washing hands will again contaminate their hand.	Mathes, McDermott et al., 2019
Hoarding	Inability to discard unwanted or impractical possessions.	Samuels et al., 2002
Ruminations	A deep or considered thought about something that is useless.	Raines et al., 2017
Intrusive Thoughts	Overthinking impossible instances. For example, thoughts of causing harm to loved ones.	Julien, O'Connor and Aardema et al., 2007
Symmetry Orderliness	Keeping everything neat, clean, and in order always.	Sasson et al., 1997
HOCD (Homosexual OCD)/SOOCD (Sexual orientation OCD)	Fear of homosexuality (homophobia).	Bhatia et al., 2015
POCD (Pedophile OCD)	Prenatal or postpartum OCD "Pure O."	Munslow-Davies and Anderson, 2022
ROCD (Relationship OCD)	Fear of being in a wrong relationship or doing something wrong in relationships.	Doron, Derby and Szepsenwol, 2014

Additionally, one can experience obsessions and compulsions unrelated to those listed, as shown in Table 1.1.

1.4 EFFECTS OF OCD

OCD sufferers may feel isolated. A person's worries could prevent them from engaging in their favorite activities with the people they care about. Their free time could be limited by rituals. And shame may make it challenging for someone to express their most private feelings. Sometimes it seems like the safest way of dealing is to tackle all problems alone. But in doing so, they are also changing how they might create and nurture helpful relationships. Figure 1.3 represents the daily challenges faced by a person with OCD.

1.4.1 Personal

These obsessive triggers and compulsive repetitive behavior can consume a large portion of a person's day. "Avoidance" may become a bigger issue when OCD grows more serious. Anything that can make their irrational thoughts come true may be avoided by the person. People with OCD may find it challenging to carry out simple tasks like reading, drinking, shopping, or eating. Many OCD sufferers could end up being housebound (Gokhale and Chakole, 2022). Depression and related anxiety disorders such as social anxiety, panic disorder, and separation anxiety can intensify OCD (Fennell, 2022).

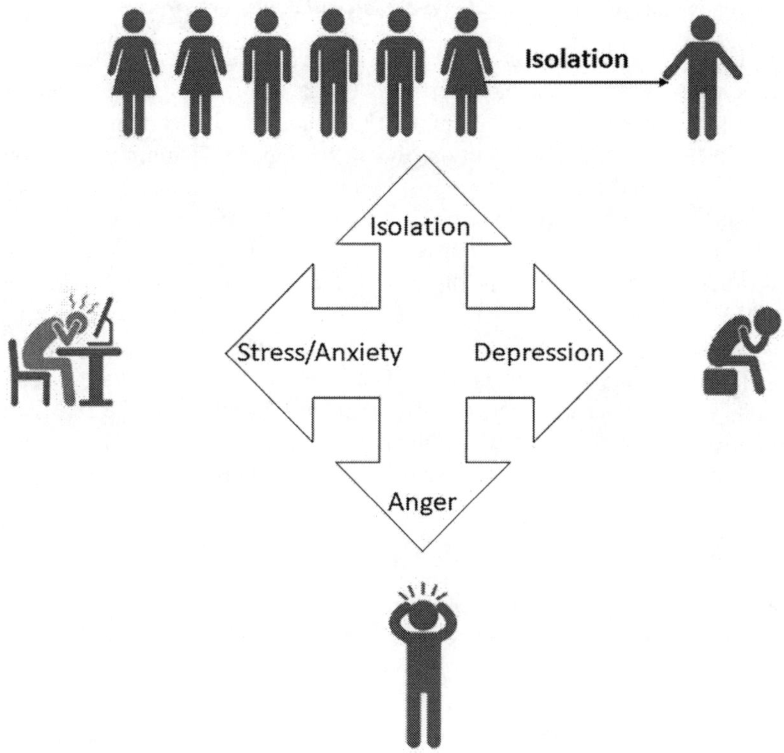

FIGURE 1.3 Effects of obsessive-compulsive disorder.

1.4.2 Social

In majority of cases, anxiety-related issues and personal problems can be managed through therapy and medications. However, the illness could have a significant detrimental effect on social relationships, resulting in frequent marital and family conflict, separation, or divorce (Mariaskin, 2022). Additionally, it hinders a person's capacity to study or work, interferes with their leisure time activities.

1.4.3 Day-to-Day Challenges in Person With OCD

Repetitive compulsions to lower the obsessive anxiety attacks could take up a lot of time in a person's life. This may prevent the person from going to work, visiting loved ones, eating out, or even going outside. The patient may find it challenging to concentrate and feel weary as a result of obsessions. As discussed previously, a person with OCD also finds it difficult to make new relationships, complete easy daily tasks.

1.5 TREATMENT AND MANAGEMENT OF OCD

There are no 100% treatments for any mental or psychotic disorders, but multiple management types of psychotherapy have been examined in the treatment of such disorders, including OCD (Krzyszkowiak et al., 2019; Mukhopadhyay and Tarafder, 2018). However, evidence from randomized trials significantly favors the use of cognitive treatment or exposure-response therapy for OCD (Berman et al., 2021; Fineberg et al., 2020).

1.5.1 Cognitive Behavior Therapy—Exposure Therapy Prevention

Treatments like cognitive-behavior therapy assist patients in managing and altering negative emotions, actions, and ideas (Clark, 2019; Gagné et al., 2018). Exposure and ritual prevention is a particular type of cognitive-behavioral therapy employed to treat obsessive-compulsive disorder (Reddy et al., 2020). Two different sorts of connections that develop in OCD are intended to be broken by this therapy. Firstly, the connection between distressing items, situations, or ideas and distressing sensations is broken. The second association is that ritualistic conduct reduces discomfort when it is practiced (Leahy et al., 2022). The therapy provided will serve the ingrained link between anxious sensations and ritualistic actions. Additionally, it will teach you to avoid ritualizing when one feels anxious (Foa, 2022).

Exposure also teaches you that dealing with these circumstances won't make you feel unbearably anxious. Ritual prevention aims to dispel myths like the one that performing rituals can shield you from negative outcomes like harm to you or others. You will discover that the anxiety and discomfort go away on their own by deliberately getting anxious during exposure practice and restraining the impulse to ritualize. The need to ritualize will consequently lessen.

1.5.2 Psychoeducation

Psychotherapy or educating the person about their OCD is a therapy strategy to get a thorough history of the client's OCD symptoms and concerns, which is followed by two types of therapy techniques, imaginal exposure and in vivo exposure (Del et al., 2019; Arden, 2019).

1.5.2.1 In Vivo Exposure

Exposure is the foundation of exposure and ritual prevention (EX/RP) treatment. It has been demonstrated that in vivo exposure lessens obsessions and associate suffering. This method entails lengthy, repeated confrontation with anxiety-inducing circumstances. The length of an exposure session might range from 30 minutes to two hours. The client must stay in the scenario long enough to experience some anxiety relief and to understand that the feared disaster results do not materialize. With more exposures, both the peak and the general degree of distress decrease over the course of sessions. As a result, the client becomes accustomed to disturbing stimuli in two ways: during sessions and between them (Dua et al., 2021).

1.5.2.2 Imaginal Exposure

In rare cases, construction of in vivo exposure in impossible, and in these circumstances, the exposure can be done in the imagination (similar to a deep meditation sessions). Situations where the client worries that they would change fundamentally (sexual orientation, mental stability, or becoming a serial killer) cause a distal catastrophe (start a chain reaction that results in harming others) (Berman et al., 2019).

Imaginal exposure is effective when it causes the same level of distress in the subject as the fixation itself. Usually, an OCD sufferer fights the preoccupation because they think that if they entertain the thoughts, the dreaded event is more likely to materialize. But resisting the obsession simply makes it stronger. The OCD sufferer becomes accustomed to the anxieties by telling themselves the same story over and over again, and they also learn that worrying about the troubling thoughts does not make them come true.

1.5.3 Habit Reversal Training

For OCD patients of any age, a psychological training called habit reversal training (HRT) is an evidence-based behavioral therapy (Pilcher et al., 2020). For those who engage in undesired repeated habits like nail-biting, skin picking, and hair pulling, to mention a few, HRT is a very effective treatment. It consists of methods for calming down, instruction in awareness, encouragement, and social support (Lee et al., 2019). Practice the habit or tic (involuntary compulsions) as part of awareness training, pay attention to the patient's bodily sensations and certain muscles, and note when the habit or tic happens. Such an approach encourages awareness of when and how cravings arise, enabling the person to take action and redirect their focus. The five elements of this training are as follows: (1) training for awareness, (2) training for competing responses, (3) training for motivation and compliance, (4) training for relaxation, and (5) training for generalization (Moritz et al., 2022).

1.6 CONCLUSION

OCD is a polygenic, hereditary condition that is influenced by both common and unusual variations, as well as risk factors from the environment. There is solid evidence from numerous research that there is a sizable additive genetic component to the risk of developing OCD and other obsessive-compulsive symptoms.

Due to insufficient studies and date sizes, no significant genome-wide loci for OCD have yet been found. The results of genetic correlation, polygenic risk, and gSEM (general linear structural model) studies to date must also be interpreted considering these relatively smaller sizes, as numerous studies like these are underpowered. However, researchers are currently working on a substantially larger meta-analysis with promising early findings. In order to understand the genetic makeup of OCD and its genetic linkages with other mental health illness and features, larger samples will be essential for identifying strong genetic variations for OCD.

Future research has a number of critical directions. Most GWAS samples till now have originated from populations of European cohort studies. It will be crucial to investigate the function of environmental factors and how they interact with genetic factors, especially when considering people who have different genetic backgrounds. It will also be crucial to comprehend whether the genetic makeup and risk for OCD vary depending on the age at which the symptoms first appear (early onset or late onset), since this may help with nosology and risk prediction. Further study is also needed about the gender variances. Although twin, family, and GWAS research indicate that there is probably sex-specific genetic overlap in the risk for OCD, it is yet unknown how this caries by age and symptom dimensions.

REFERENCES

Abramowitz, Jonathan S., Shannon M. Blakey, Lillian Reuman, and Jennifer L. Buchholz. "New directions in the cognitive-behavioral treatment of OCD: Theory, research, and practice." *Behavior Therapy* 49, no. 3 (2018): 311–322.

Adams, Thomas G., Benjamin Kelmendi, C. Alex Brake, Patricia Gruner, Christal L. Badour, and Christopher Pittenger. "The role of stress in the pathogenesis and maintenance of obsessive-compulsive disorder." *Chronic Stress* 2 (2018): 2470547018758043.

Albert, Umberto, Diana De Ronchi, Giuseppe Maina, and Maurizio Pompili. "Suicide risk in obsessive-compulsive disorder and exploration of risk factors: A systematic review." *Current neuropharmacology* 17, no. 8 (2019): 681–696.

Albert, U., M. Manchia, A. Tortorella, U. Volpe, G. Rosso, B. Carpiniello, and G. Maina. "Admixture analysis of age at symptom onset and age at disorder onset in a large sample of patients with obsessive–compulsive disorder." *Journal of Affective Disorders* 187 (2015): 188–196.

Alemany-Navarro, María, Raquel Cruz, Eva Real, Cinto Segalàs, Sara Bertolín, Raquel Rabionet, Ángel Carracedo, Jose M. Menchón, and Pino Alonso. "Looking into the genetic bases of OCD dimensions: A pilot genome-wide association study." *Translational psychiatry* 10, no. 1 (2020): 1–16.

Alsheikh, Ahmed M., and Maram M. Alsheikh. "Obsessive-compulsive disorder with rheumatological and inflammatory diseases: A systematic review." *Cureus* 13, no. 5 (2021).

Anholt, G. E., I. M. Aderka, A. J. L. M. Van Balkom, J. H. Smit, K. Schruers, N. J. A. Van Der Wee, M. Eikelenboom, V. De Luca, and P. Van Oppen. "Age of onset in obsessive–compulsive disorder: Admixture analysis with a large sample." *Psychological Medicine* 44, no. 1 (2014): 185–194.

Arden, John. *Mind-brain-gene: Toward psychotherapy integration*. WW Norton & Company, 2019.

Baer, Lee. *The imp of the mind: Exploring the silent epidemic of obsessive bad thoughts*. Penguin, 2002.

Barnhill, Jarrett. "Obsessive-Compulsive and Related Disorders." In *Textbook of psychiatry for intellectual disability and autism spectrum disorder*, pp. 625–654. Springer, 2022.

Berman, Leo. "The obsessive-compulsive neurosis in children." *The Journal of Nervous and Mental Disease* 95, no. 1 (1942): 26–39.

Berman, Noah Chase, Dianne M. Hezel, and Sabine Wilhelm. "Is my patient too sad to approach their fear? Depression severity and imaginal exposure outcomes for patients with OCD." *Journal of Behavior Therapy and Experimental Psychiatry* 70 (2021): 101615.

Berman, Noah Chase, Berta J. Summers, Hilary Weingarden, and Sabine Wilhelm. "Positive affect and imaginal exposure processes in patients with taboo obsessions." *Journal of Obsessive-Compulsive and Related Disorders* 23 (2019): 100474.

Bhatia, Manjeet S., and Jaswinder Kaur. "Homosexual obsessive compulsive disorder (HOCD): A rare case report." *Journal of Clinical and Diagnostic Research: JCDR* 9, no. 1 (2015): VD01.

Brakoulias, Vlasios, Vladan Starcevic, Amparo Belloch, Christopher Brown, Ygor Arzeno Ferrao, Leonardo F. Fontenelle, Christine Lochner et al. "Comorbidity, age of onset and suicidality in obsessive–compulsive disorder (OCD): An international collaboration." *Comprehensive Psychiatry* 76 (2017): 79–86.

Burton, Christie, Mehdi Zarrei, Worrawat Enghuan, Daniele Merico, Jeff MacDonald, Bowei Xiao, Andrew Paterson et al. "Copy number variants in brain-related genes are associated with neuropsychiatric traits in childhood." *European Neuropsychopharmacology* 29 (2019): S825–S826.

Carpenter, Leanne, and Man Cheung Chung. "Childhood trauma in obsessive compulsive disorder: The roles of alexithymia and attachment." *Psychology and Psychotherapy: Theory, Research and Practice* 84, no. 4 (2011): 367–388.

Clark, David A. *Cognitive-behavioral therapy for OCD and its subtypes*. Guilford Publications, 2019.

Cole, Scott N., and Peter Tubbs. "Predictors of obsessive–compulsive symptomology: Mind wandering about the past and future." *Psychological Research* 86, no. 5 (2022): 1518–1534.

Del Casale, Antonio, Serena Sorice, Alessio Padovano, Maurizio Simmaco, Stefano Ferracuti, Dorian A. Lamis, Chiara Rapinesi et al. "Psychopharmacological treatment of obsessive-compulsive disorder (OCD)." *Current Neuropharmacology* 17, no. 8 (2019): 710–736.

Doron, Guy, Danny S. Derby, and Ohad Szepsenwol. "Relationship obsessive compulsive disorder (ROCD): A conceptual framework." *Journal of Obsessive-Compulsive and Related Disorders* 3, no. 2 (2014): 169–180.

Drubach, Daniel A. "Obsessive-compulsive disorder." *Continuum: Lifelong Learning in Neurology* 21, no. 3 (2015): 783–788.

Drummond, Lynne M., and Laura J. Edwards. *Obsessive compulsive disorder: All you want to know about OCD for People living with OCD, carers, and clinicians*. Cambridge University Press, 2018.

Drummond, Lynne M., and Laura J. Edwards. *Everything you need to know about OCD*. Cambridge University Press, 2022.

Dua, Devakshi, Gopika Jagota, and Sandeep Grover. "Management of obsessive–compulsive disorder with virtual reality-based exposure." *Industrial Psychiatry Journal* 30, no. 1 (2021): 179.

Fawcett, Emily J., Hilary Power, and Jonathan M. Fawcett. "Women are at greater risk of OCD than men: A meta-analytic review of OCD prevalence worldwide." *The Journal of Clinical Psychiatry* 81, no. 4 (2020): 13075.

Fennell, Dana. *The world of obsessive-compulsive disorder: The experiences of living with OCD*. NYU Press, 2022.

Fernández de la Cruz, Lorena, Mina Rydell, B. Runeson, Brian M. D'Onofrio, Gustaf Brander, Christian Rück, P. Lichtenstein, Henrik Larsson, and David Mataix-Cols. "Suicide in obsessive–compulsive disorder: A population-based study of 36 788 Swedish patients." *Molecular Psychiatry* 22, no. 11 (2017): 1626–1632.

Fineberg, Naomi A., Michael Van Ameringen, L. Drummond, Eric Hollander, D. J. Stein, Daniel Geller, Susanne Walitza et al. "How to manage obsessive-compulsive disorder (OCD) under COVID-19: A clinician's guide from the International College of Obsessive Compulsive Spectrum Disorders (ICOCS) and the Obsessive-Compulsive and Related Disorders Research Network (OCRN) of the European college of neuropsychopharmacology." *Comprehensive Psychiatry* 100 (2020): 152174.

Foa, Edna B. "Cognitive behavioral therapy of obsessive-compulsive disorder." *Dialogues in Clinical Neuroscience* 12, no. 2 (2022): 199–207.

Fontenelle, Leonardo F., and Gregor Hasler. "The analytical epidemiology of obsessive–compulsive disorder: Risk factors and correlates." *Progress in Neuro-Psychopharmacology and Biological Psychiatry* 32, no. 1 (2008): 1–15.

Freeman, Jennifer B., and Henrietta L. Leonard. "Sexual obsessions in obsessive-compulsive disorder." *Journal of the American Academy of Child and Adolescent Psychiatry* 39, no. 2 (2000): 141–142.

Gagné, Jean-Philippe, Kenneth Kelly-Turner, and Adam S. Radomsky. "From the laboratory to the clinic (and back again): How experiments have informed cognitive–behavior therapy for obsessive–compulsive disorder." *Journal of Experimental Psychopathology* 9, no. 4 (2018): 2043808718810030.

Gans, Steven. "Doing it again: Obsessions." *The European Journal of Psychotherapy, Counselling & Health* 1, no. 3 (1998): 405–420.

Geller, Daniel, Joseph Biederman, Janice Jones, Kenneth Park, Suzanne Schwartz, Stephanie Shapiro, and Barbara Coffey. "Is juvenile obsessive-compulsive disorder a developmental subtype of the disorder? A review of the pediatric literature." *Journal of the American Academy of Child & Adolescent Psychiatry* 37, no. 4 (1998): 420–427.

Gokhale, Mihika V., and Swarupa Chakole. "A review of effects of pandemic on the patients of obsessive-compulsive disorder." *Cureus* 14, no. 10 (2022).

Grant, Jon, and Daniel Kim. "Present for centuries, shoplifting seems to have become a serious problem in the 19th century with the rise of the department store (p. 4). With easy." *Mental Disorders of the New Millennium* 3 (2006): 99.

Grünblatt, Edna. "Genetics of OCD and related disorders; Searching for shared factors." In *The Neurobiology and Treatment of OCD: Accelerating Progress*, pp. 1–16. Springer, 2021.

Jenike, Michael A. "Obsessive–compulsive disorder." *New England Journal of Medicine* 350, no. 3 (2004): 259–265.

Julien, Dominic, Kieron P. O'Connor, and Frederick Aardema. "Intrusive thoughts, obsessions, and appraisals in obsessive–compulsive disorder: A critical review." *Clinical Psychology Review* 27, no. 3 (2007): 366–383.

Kracker Imthon, André, César Antônio Caldart, Maria Conceição Do Rosário, Leonardo F. Fontenelle, Euripedes Constantino Miguel, and Ygor Arzeno Ferrão. "Stressful life events and the clinical expression of Obsessive–Compulsive Disorder (OCD): An exploratory study." *Journal of Clinical Medicine* 9, no. 10 (2020): 3371.

Krzyszkowiak, Waldemar, Małgorzata Kuleta-Krzyszkowiak, and Eliza Krzanowska. "Treatment of obsessive-compulsive disorders (OCD) and obsessive-compulsive-related disorders (OCRD)." *Psychiatria Polska* 53, no. 4 (2019): 825–843.

Leahy, Robert L., David A. Clark, and David J. A. Dozois. "Cognitive-behavioral theories." *Gabbard's Textbook of Psychotherapeutic Treatments* (2022): 151.

Lee, Melissa T., Davis N. Mpavaenda, and Naomi A. Fineberg. "Habit reversal therapy in obsessive compulsive related disorders: A systematic review of the evidence and CONSORT evaluation of randomized controlled trials." *Frontiers in Behavioral Neuroscience* 13, no. 79 (2019): 1–15.

Leonhart, Mark W., and Adam S. Radomsky. "Responsibility causes reassurance seeking, too: An experimental investigation." *Journal of Obsessive-Compulsive and Related Disorders* 20 (2019): 66–74.

Mahjani, Behrang, Katharina Bey, Julia Boberg, and Christie Burton. "Genetics of obsessive-compulsive disorder." *Psychological Medicine* 51, no. 13 (2021): 2247–2259.

Mancini, Francesco, ed. *The obsessive mind: Understanding and treating obsessive-compulsive disorder*. Routledge, 2018.

Mariaskin, Amy. *Thriving in relationships when you have OCD: How to keep obsessions and compulsions from sabotaging love, friendship, and family connections*. New Harbinger Publications, 2022.

Marincowitz, Clara, Nastassja Koen, Christine Lochner, and Dan J. Stein. "Obsessive–compulsive disorder." In *Neurobiology of Brain Disorders*, pp. 825–842. Academic Press, 2023.

Marks, Isaac. "Behavioural (non-chemical) addictions." *British Journal of Addiction* 85, no. 11 (1990): 1389–1394.

Mathes, Brittany M., Katherine A. McDermott, Sarah A. Okey, Ana Vazquez, Ashleigh M. Harvey, and Jesse R. Cougle. "Mental contamination in obsessive-compulsive disorder: Associations with contamination symptoms and treatment response." *Behavior Therapy* 50, no. 1 (2019): 15–24.

Mathes, Brittany M., Danielle M. Morabito, and Norman B. Schmidt. "Epidemiological and clinical gender differences in OCD." *Current Psychiatry Reports* 21, no. 5 (2019): 1–7.

Mathes, Brittany M., et al. "Indices of change in exposure and response prevention for contamination-based OCD." *Behaviour Research and Therapy* 133 (2020): 103707.

Mattera, Elizabeth F., Terence HW Ching, Brian A. Zaboski, and Stephen A. Kichuk. "Suicidal obsessions or suicidal ideation? A case report and practical guide for differential assessment." *Cognitive and Behavioral Practice* (2022). Doi: 10.1016/j.cbpra.2022.09.002.

Maziero, Maria Paula, Johanna Seitz-Holland, Kang Ik K. Cho, Joshua E. Goldenberg, Taís W. Tanamatis, Juliana B. Diniz, Carolina Cappi et al. "Cellular and extracellular white matter abnormalities in obsessive-compulsive disorder: A diffusion magnetic resonance imaging study." *Biological Psychiatry: Cognitive Neuroscience and Neuroimaging* 6, no. 10 (2021): 983–991.

Meier, Sandra M., Manuel Mattheisen, Ole Mors, Diana E. Schendel, Preben B. Mortensen, and Kerstin J. Plessen. "Mortality among persons with obsessive-compulsive disorder in Denmark." *JAMA Psychiatry* 73, no. 3 (2016): 268–274.

Moreira, Pedro Silva, Julian Macoveanu, Paulo Marques, Ana Coelho, Ricardo Magalhães, Hartwig R. Siebner, José Miguel Soares, Nuno Sousa, and Pedro Morgado. "Altered response to risky decisions and reward in patients with obsessive–compulsive disorder." *Journal of Psychiatry and Neuroscience* 45, no. 2 (2020): 98–107.

Moritz, Steffen, Danielle Penney, Alina Bruhns, Sarah Weidinger, and Stella Schmotz. "Habit reversal training and variants of decoupling for use in body-focused repetitive behaviors. A randomized controlled trial." *Cognitive Therapy and Research* (2022): 1–14.

Moulding, Richard, Frederick Aardema, and Kieron P. O'Connor. "Repugnant obsessions: A review of the phenomenology, theoretical models, and treatment of sexual and aggressive obsessional themes in OCD." *Journal of Obsessive-Compulsive and Related Disorders* 3, no. 2 (2014): 161–168.

Mukhopadhyay, P., and S. Tarafder, eds. *Obsessive compulsive disorder: A neuropsychological approach.* SAGE Publishing India, 2018.

Munslow-Davies, Samantha, and Rebecca A. Anderson. "Effects of a brief education on psychologists' attitudes and diagnostic impressions of paedophilic obsessive-compulsive intrusive thoughts." *Journal of Obsessive-Compulsive and Related Disorders* 32 (2022): 100707.

Nestadt, Gerald, Marco Grados, and Jack F. Samuels. "Genetics of obsessive-compulsive disorder." *Psychiatric Clinics* 33, no. 1 (2010): 141–158.

Nicolini, Humberto, Paul Arnold, Gerald Nestadt, Nuria Lanzagorta, and James L. Kennedy. "Overview of genetics and obsessive–compulsive disorder." *Psychiatry Research* 170, no. 1 (2009): 7–14.

Olatunji, Bunmi O., Caroline Christian, Leigh Brosof, David F. Tolin, and Cheri A. Levinson. "What is at the core of OCD? A network analysis of selected obsessive-compulsive symptoms and beliefs." *Journal of Affective Disorders* 257 (2019): 45–54.

Pampaloni, Ilenia, Sabina Marriott, Enrico Pessina, Claire Fisher, Anusha Govender, Heba Mohammed, Augusta Chandler, Himanshu Tyagi, Lucy Morris, and Stefano Pallanti. "The global assessment of OCD." *Comprehensive Psychiatry* (2022): 152342.

Perrotta, Giulio. "Anxiety disorders: Definitions, contexts, neural correlates and strategic therapy." Journal of Neuroscience 6, no. 1 (2019): 042.

Pilcher, Felicity, Stefano Seri, and Andrea E. Cavanna. "Initial treatment retention for habit reversal training in adults with Tourette syndrome." *Cited in: EMBASE-Excerpta Medica Database• Index Copernicus• PsycINFO• SCOPUS• Google Scholar• Emerging Sources Citation Index (ESCI), a new edition of Web of Science* (2020): 290.

Pitman, Roger K. "A cybernetic model of obsessive-compulsive psychopathology." *Comprehensive Psychiatry* 28, no. 4 (1987): 334–343.

Plana, Maria Teresa, Teresa Torres, Natalia Rodríguez, Daniel Boloc, Patricia Gassó, Elena Moreno, Amalia Lafuente, Josefina Castro-Fornieles, Sergi Mas, and Luisa Lazaro. "Genetic variability in the serotoninergic system and age of onset in anorexia nervosa and obsessive-compulsive disorder." *Psychiatry Research* 271 (2019): 554–558.

Porth, Rachel, and Daniel Geller. "Atypical symptom presentations in children and adolescents with obsessive compulsive disorder." *Comprehensive Psychiatry* 86 (2018): 25–30.

Praptomojati, Ardian. ""How do I stop checking things?" Understanding obsessive-compulsive disorder from neuropsychological perspective." *Buletin Psikologi* 27, no. 1 (2019): 15–29.

Purdon, Christine. "The internal logic of OCD: A review." *Journal of Behavior Therapy and Experimental Psychiatry* (2022): 101773.

Purty, Abhishek, Gerald Nestadt, Jack F. Samuels, and Biju Viswanath. "Genetics of obsessive-compulsive disorder." *Indian Journal of Psychiatry* 61, no. Suppl 1 (2019): S37.

Radomsky, Adam S., and S. Rachman. "Symmetry, ordering and arranging compulsive behaviour." *Behaviour Research and Therapy* 42, no. 8 (2004): 893–913.

Raines, Amanda M., Desirae N. Vidaurri, Amberly K. Portero, and Norman B. Schmidt. "Associations between rumination and obsessive-compulsive symptom dimensions." *Personality and Individual Differences* 113 (2017): 63–67.

Reddy, YC Janardhan, Paulomi M. Sudhir, M. Manjula, Shyam Sundar Arumugham, and Janardhanan C. Narayanaswamy. "Clinical practice guidelines for cognitive-behavioral therapies in anxiety disorders and obsessive-compulsive and related disorders." *Indian Journal of Psychiatry* 62, no. Suppl 2 (2020): S230.

Richter, Peggy M. A., and Renato T. Ramos. "Obsessive-compulsive disorder." *CONTINUUM: Lifelong Learning in Neurology* 24, no. 3 (2018): 828–844.

Ruchsow, Martin, Georg Grön, Kathleen Reuter, Manfred Spitzer, Leopold Hermle, and Markus Kiefer. "Error-related brain activity in patients with obsessive-compulsive disorder and in healthy controls." *Journal of Psychophysiology* 19, no. 4 (2005): 298.

Ruscio, Ayelet M., Dan J. Stein, Wai Tat Chiu, and Ronald C. Kessler. "The epidemiology of obsessive-compulsive disorder in the national comorbidity survey replication." *Molecular Psychiatry* 15, no. 1 (2010): 53–63.

Saboowala, H. K. *What is Obsessive Compulsive Disorder (OCD)? A complete pedagogy.* Dr. Hakim Saboowala, 2022.

Salkovskis, Paul M., Elizabeth Forrester, and Candida Richards. "Cognitive–behavioural approach to understanding obsessional thinking." *The British Journal of Psychiatry* 173, no. S35 (1998): 53–63.

Samuels, J., O. Joseph Bienvenu III, M. A. Riddle, B. A. M. Cullen, M. A. Grados, K-Y. Liang, R. Hoehn-Saric, and G. Nestadt. "Hoarding in obsessive compulsive disorder: Results from a case-control study." *Behaviour Research and Therapy* 40, no. 5 (2002): 517–528.

Sasson, Yehuda, Joseph Zohar, Miriam Chopra, Michal Lustig, Iulian Iancu, and Talma Hendler. "Epidemiology of obsessive-compulsive disorder: A world view." *Journal of Clinical Psychiatry* 58, no. 12 (1997): 7–10.

Shapiro, Leslie J. *Obsessive compulsive disorder: Elements, history, treatments, and research.* Abc-Clio, 2020.

Sidorchuk, Anna, Ralf Kuja-Halkola, Bo Runeson, Paul Lichtenstein, Henrik Larsson, Christian Rück, Brian M. D'Onofrio, David Mataix-Cols, and Lorena Fernández de la Cruz. "Genetic and environmental sources of familial coaggregation of obsessive– compulsive disorder and suicidal behavior: A population-based birth cohort and family study." *Molecular Psychiatry* 26, no. 3 (2021): 974–985.

Stein, Dan J. "Obsessive-compulsive disorder." *The Lancet* 360, no. 9330 (2002): 397–405.

Stein, Dan J., Daniel L. C. Costa, Christine Lochner, Euripedes C. Miguel, Y. C. Reddy, Roseli G. Shavitt, Odile A. van den Heuvel, and H. Blair Simpson. "Obsessive–compulsive disorder." *Nature reviews Disease Primers* 5, no. 1 (2019): 1–21.

Subramaniam, Mythily, Pauline Soh, Janhavi Ajit Vaingankar, Louisa Picco, and Siow Ann Chong. "Quality of life in obsessive-compulsive disorder: Impact of the disorder and of treatment." *CNS Drugs* 27, no. 5 (2013): 367–383.

Thalamuthu, Anbupalam, Natalie T. Mills, Klaus Berger, Heike Minnerup, Dominik Grotegerd, Udo Dannlowski, Susanne Meinert et al. "Genome-wide interaction study with major depression identifies novel variants associated with cognitive function." *Molecular Psychiatry* 27, no. 2 (2022): 1111–1119.

Weissman, Myrna M., Roger C. Bland, Glorisa J. Canino, Steven Greenwald, Hai-Gwo Hwu, CHUNG KYOON Lee, Stephen C. Newman, Mark A. Oakley-Browne, Maritza Rubio-Stipec, and Priya J. Wickramaratne. "The cross national epidemiology of obsessive compulsive disorder. The Cross National Collaborative Group." *The Journal of Clinical Psychiatry* 55 (1994): 5–10, PMID: 8077177.

Williams, Monnica T., Beth Mugno, Martin Franklin, and Sonya Faber. "Symptom dimensions in obsessive-compulsive disorder: Phenomenology and treatment outcomes with exposure and ritual prevention." *Psychopathology* 46, no. 6 (2013): 365–376.

Williams, Monnica T., and Chad T. Wetterneck. *Sexual obsessions in obsessive-compulsive disorder: A step-by-step, definitive guide to understanding, diagnosis, and treatment.* Oxford University Press, 2019.

Yu, Jianping, Minyao Xie, Shasha Song, Ping Zhou, Fangzheng Yuan, Mengyuan Ouyang, Chun Wang, Na Liu, and Ning Zhang. "Functional connectivity within the frontal–striatal network differentiates checkers from washers of obsessive-compulsive disorder." *Brain Sciences* 12, no. 8 (2022): 998.

2 Toxicological Impact of Environmental Pollutant in Occurrence of OCD

Diana Miguez,[a] *Shoumi Haldar,*[b] *Krusha Moradiya,*[b] *Srushti Amarsingh Rajput,*[b] *and Prabhakar Mishra*[b]

[a] Latitud—Fundación LATU, Laboratorio Tecnológico del Uruguay (LATU), Edificio Los Abetos, Avenida Italia 6201, C.P. 11500, Montevideo, Uruguay

[b] Department of Biotechnology, School of Applied Sciences, REVA University, Bengaluru, Karnataka, India

2.1 INTRODUCTION

This disorder includes obsessions that are repetitive, intrusive thoughts that are unwanted and generally originate from anxiety (Prasad and Manna, 2022). Compulsions are repetitive behaviors or mental acts that the individuals feel a responsibility towards, to perform in response to an obsession to achieve a sense of satisfaction or completeness (Purdon, 2018). Examples of obsessions include unwanted thoughts or thoughts about harming someone or our loved ones, persisting thoughts of whether the light switch is off or not, repugnant thoughts, etc. (Jassi et al., 2020; Philpot et al., 2022). Compulsions include cleaning hands, repetitive praying, checking if the light switch is turned off by turning it on and off exactly ten times, and other behaviors that result due to a certain obsession (Summers et al., 2020; Smith, 2022). It is aimed as said to achieve satisfaction and to reduce anxiety and distress to the related thoughts but are excessive and are not realistically related to what the individual wants to prevent (Kumar and Rai, 2020). Obsessive-compulsive disorder, also abbreviated as OCD, is a disorder that involves certain obsessions and/or compulsive rituals (Pinciotti et al., 2021; McKay et al., 2021). Obsessive-compulsive disorder has two major features, repetitive actions and recurrence of intrusive thoughts or obsessions typically performed by specific regions of the brain (Saboowala et al., 2022). Prominent involved regions of the brain seen in OCD are dorsolateral prefrontal cortex (DLPC), anterior cingulate cortex (ACC), orbitofrontal cortex (OFC), and cortico-striatal circuitry (Srivastava, 2018; Dragoi et al., 2020; Park et al., 2022). The prefrontal cortex is involved in various cognitive functions, including reasoning, problem solving, comprehension, impulse-control, inhibition, and various complex cognitive functions (Fuster, 2019; Van Roessel et al., 2022). Dysfunction and dysregulation of neurotransmitters and specific brain circuits in the prefrontal cortex can lead to OCD and various disorders (Robbins et al., 2019). Prefrontal cortex has two major subdivisions: The lateral PFC, which is also known as dorsolateral prefrontal cortex and ventrolateral prefrontal cortex, and the ventromedial PFC, also known as the orbitofrontal PFC (Hathaway and Newton, 2023). The anterior cingulate cortex (ACC) is majorly responsible for the host of error detection and problem solving (Guglielmo et al., 2021). However, the hyperactivity of anterior cingulate cortex (ACC) is observed to be increased along with symptoms in patients with OCD even though the functional involvement of anterior cingulate cortex (ACC) is unknown in case of OCD and electrophysiological evidence has linked this region to error-processing abnormalities (Carmi et al., 2018; Xiao et al., 2021; Ahmari and Rauch, 2022).

 DOI: 10.1201/9781003387893-3

The dorsolateral prefrontal cortex (DCPL) plays a crucial role in higher cognitive functions, working memory, switching attention, and decision-making (Ngetich et al., 2020). However, the functional neuroimaging data have shown hypoactivity of dorsolateral prefrontal cortex (DCPL) for OCD patients and various other disorders, including depression (Badhe and De Sousa, 2022). Further orbitofrontal cortex (OFC) plays a crucial role in response inhibition, behavior regulation along with thalamus and striatum, which are part of cortico-striato-thalamic circuit (Tumati et al., 2019; Conn et al., 2020). Cortico-striato-thalamic circuits consist of two pathways, "direct" and "indirect," though they both have completely opposite roles in managing the circuit (Dougherty et al., 2020). Direct pathway involves direct projections from striatum to the globus pallidus internalis (Gpi) and thalamus, whereas indirect pathway involves projections from striatum to globus pallidus internalis (Gpi) via the globus pallidus pars externalis (Gpe) (Young et al., 2021). The striatum is the determiner for activity of the CSTC pathway through the invigoration of two different types of spiny neurons (MSNs) which express D1 or D2 dopamine receptors (Biria et al., 2021; Foa, 2010). The exact mechanism by which balanced excitation/inhibition (E/I) of these cells is mastered is unknown (Rădulescu et al., 2017). However, changes in excitation or inhibition in the activity of D1 and D2 can generate a different pattern of activity such as increase in activity of D1-MSNs and reduction in activity of D2-MSNs can lead motor cortex activity and performance of movement, which is significant to the motor symptoms of OCD (McCarthy et al., 2011). Clinical studies in patients with OCD have shown hyperactivation in the direct pathway of CSTC. Furthermore, the findings have suggested that glutamatergic synaptic dysfunction leads to changes in GABAergic inhibition, and that can lead to the onset of OCD (Rădulescu et al., 2017). OCD is also associated with misfunctioning of serotonin and availability of serotonin transporter (SERT) (Politis and Niccolini, 2015). Although the role of serotonin in symptoms of OCD remains unclear as results were inconsistent while studying the OCD subject using tracers. Multiple studies have shown low availability of serotonin transporter (SERT) in prefrontal circuits in patients with OCD as compared to healthy subjects (Kuo CY et al., 2022).

2.2 ENVIRONMENTAL TOXIC RESPONSE

2.2.1 Pollutants Causing Neurological Disorders

A higher risk of neurological disorders, such as autistic spectrum disorder, etc., have been linked to many pollutants, and the major contribution is through air pollution (Costa et al., 2020). In addition to behavioral disorders, the most prominent effects of air pollution can be oxidative stress and neuro-inflammation (Block and Lilian, 2009). The composition in the air, especially the ultrafine particulate matter (UFPM), is one of the most concerning and alarming pollutants. It is because these particles can easily enter and distribute throughout the body to different organs and the brain (Ali et al., 2022). There also is the fact that UFPM can gain direct excess to the brain through nasal olfactory mucosa reaching first the olfactory bulb (Costa et al., 2014). DE (diesel exhaust) is also a major contributor to air pollution and induces a heavy deal of damage to CNS (central nervous system). The National Ambient Air Quality Standards set limits for six primary air pollutants: carbon monoxide (CO), lead, nitrogen dioxide (NO_2), ozone (O_3), sulfur dioxide (SO_2), and particulate matter (PM) (Costa et al., 2020). Decreased cognitive function, olfactory dysfunction, auditory deficits, depressive symptoms, and other adverse neuropsychological effects have been reported in humans, and one of the main causes for these are reported to be $PM_{2.5}$ and UFPM, respectively (Costa et al., 2020). Among the various environmental pollutants, particulate matter (PM), ultrafine particulate matter (UFPM), nanoparticles, lipophilic vaporized toxicant (acrolein), they can easily cross the blood-brain barrier and induce the innate immune response in astrocytes, microglia, and neurons and cause neurotoxicity (Iqubal et al., 2020).

Some of the pollutants that may cause neurological disorders such as autism spectrum disorder (ASD), Alzheimer's disease (AD), Parkinson's disease (PD), etc., in various age groups are the following: Air pollutants like TRAP (Traffic Related Air Pollution), PM exposures ($PM_{2.5}$, PM_{10}, etc.), maternal exposure to NO_x, long-term exposure to black carbon and nickel, increasing NO_2 and CO, occupational exposure to DE (diesel exhaust), association with O_3 and $PM_{2.5}$, etc. (Costa et al., 2020). Heavy metals like zinc, copper, lithium, manganese, magnesium, iron, nickel, and such are some essential metals that are required by cofactors and enzymes in our body in trace amounts (Asif, 2017). Deficiency or elevated (activation of apoptosis, oxidative stress, etc.) levels of these metals may lead to neurodegenerative diseases like autism, Alzheimer's, Parkinson's, Huntington's disease, etc. (Wojtunik et al., 2019). For example, elevated levels of copper (Cu) may result in Alzheimer's disease (AD), amyotrophic lateral sclerosis (ALS), Huntington's disease (HD), Parkinson's disease (PD), etc. (Khan et al., 2020). Similarly elevated levels of iron (Fe) may also result in AD, ALS, HD, PD, and other neurological disorders. Manganese (Mn) may cause PD. Aluminum is associated with AD, ALS, multiple sclerosis, etc. Arsenic, cadmium, lead, etc., all are some heavy metals that may cause neurological disorders, respectively (Chen et al., 2016). Prenatal exposure to persistent organic pollutants, such as perfluorooctanoate (PFOA) and perfluorononanoate (PFNA), have been associated with increased risk of ASD (Oh et al., 2021) and PFOA to negative mental development in six months age female infants (Goudarzi et al., 2016). Although still requiring further validation, prenatal exposure to PFOA has also been associated with increased risk of attention-deficit/hyperactivity disorder (ADHD) (Skogheim et al., 2021) and to lower visual motor abilities in children (Harris et al., 2018).

Various studies on nanoplastics (NPs) and their effects on living things suggest that the major threat posed by plastic exposure is toxic, especially in terms of the neurological and gut microbiome (Prüst et al., 2020; Yong et al., 2020). The GI tract's sensitivity can be explained by the fact that it serves as one of the main entrance points for NPs into the body because of its smaller size and is subjected to high plastic load (Carbery et al., 2018; Cox et al., 2019; Hernandez et al., 2019). There is a resemblance between the effects caused by NPs and other nano-sized particles on the CNS-related structures. The vagus nerve, the center of the GB axis, is able to store nano-particulate contaminants, such as combustion-derived nanoparticles, and this can lead to neurodegeneration, according to findings from human studies (Calderón et al., 2017). As a result, similar-sized NPs can be anticipated to give similar results. To the best of our knowledge, no research has specifically examined the effect of NPs on the GB axis, but the theory suggests that exposure to NPs may have negative effects on the communication between the GI tract and the brain, leading to neurological diseases such as depression, anxiety, OCD, etc. due to the disturbance in the gut homeostasis.

2.2.2 Biochemistry/Molecular Understanding of OCD

2.2.2.1 Serotonin as the Possible Solution to OCD

Serotonin (5-hydroxytryptamine or 5-HT) is a neurotransmitter that modulates the neural activity and neuropsychological processes in our day-to-day life (Berger et al., 2009; Wang et al., 2022). Vast majority of serotonin content is found outside the central nervous system (CNS), and the 15 serotonin receptors are expressed both outside and within the brain (Lv and Feng Liu, 2017). Serotonin controls several biological and metabolic processes such as cardiovascular functions, bladder control, platelet aggregation, pulmonary functions, gastrointestinal functions, genitourinary systems, as well as central nervous system (Kanova and Pavel Kohout, 2021). CNS serotonin are ideally positioned to regulate the activity of a wide variety of human brain circuits which can explain the pleiotropic behavioral effects of brain serotonin, that is, individual neurons may express multiple serotonin receptors (Giorgetti and Laurence, 2004; Berger et al., 2009). For example, Layer V pyramidal neurons express 5-HT_{1A} and 5-HT_{2A} receptors which exert opposing effects on pyramidal neuron firing, respectively (Amargós-Bosch et al., 2004). The behavioral and neuropsychological processes controlled by serotonin include mood, reward, memory, appetite, sexuality, etc. (Berger et al., 2009). As behavior is regulated by multiple serotonin receptors, each serotonin receptor is expressed in multiple brain regions and likely modulates the multiple behavioral processes, such as, anxiety like behavior is regulated by 5-HT$_{1A}$ and 5-HT$_{2C}$ receptors but 5-HT$_{2C}$ not only regulates anxiety but also locomotion, appetite, and energy balance, etc. (Alex, K. D., and EA2562467 Pehek, 2007; Rueda-Ruzafa et al., 2019).

The previously mentioned paragraph was about serotonin and its functions. Now let's see how serotonin works in case of obsessive-compulsive disorder (OCD). OCD is said to be linked to the dysregulation of the brain's serotonin system, but since serotonin is also involved in other psychological disorders, the claim doesn't prove to be entirely confirmed (Abramowitz and Jacoby, 2015). But the contributions and involvement of serotonin cannot be denied. Obsessive-compulsive disorder has been associated with hypersensitivity of postsynaptic serotonin receptors (El Mansari and Pierre Blier, 2006). Individuals with the disorder might have a particular dysfunction in the genes encoding for the serotonin transporter (5-HTT) and serotonin receptor (5-HT$_{2A}$), but these have not been consistently identified (Abramowitz et al., 2009). The usually given prescription for OCD are serotonin reuptake inhibitors (SRIs), and these inhibitors target the serotonin transporters (SERT) which remain in reduced amounts in prefrontal-basal ganglia-thalamic-prefrontal circuits in OCD patients as compared to healthy individuals, respectively (Pauls, 2010). Other prescriptions given for OCD can be clomipramine, SSRIs (selective serotonin reuptake inhibitors), etc. (Sinopoli et al., 2017). Clomipramine is a potent inhibitor of serotonin transporter (SERT). These inhibitors (SRIs, clomipramine, SSRIs) are used to prevent the SERT to transport serotonin to other parts of brain and make it available for the elevation of the OCD symptoms, respectively (Baldessarini, 2006; Edinoff et al., 2021).

2.3 TOXICITY-BASED GUT-MICROBIOTA DYSBIOSIS AND OCD

2.3.1 Gut Microbiome–Brain Association with OCD

Earlier OCD was termed as an anxiety disorder until in 2013 when DSM-5 was published declaring OCD as an independent disorder and no longer an anxiety disorder, but the fact cannot be ignored that the OCD's predominant symptom is anxiety, and multiple studies have led to the conclusion where microbiome of gut is dramatically affected due to depression, anxiety, schizophrenia, and other disorders (Turna et al., 2016; Cryan et al., 2019). Etiology of OCD is unknown on a large scale, although microbiota has shown a causative role, but mechanistic understanding is still unclear. The one pathway which is well-known is bidirectional signaling between gut microbiota and brain functioning has been related to disorders. Gut is estimated to have different distributions of α- and β-diversity microbes. In a recent study, they investigated stool samples by 16S ribosomal RNA gene-based metagenomic sequencing, where patients with OCD showed relative increase in abundance of genus *Alistipes* associated with the *Rikenellaceae* family and low presence of bacterial α-diversity and other two genera within *Lachnospiraceae: Agathobacer* and *Coprococcus* was observed. *Coprococcus* is associated with synthesis of DOPAC, which is a dopamine neurotransmitter. Although a series of studies on OCD using probiotics were performed to check the influence of gut microbiome. In this experiment for two to four weeks, mice received probiotic pretreatment. After inducing OCD-like behaviors, locomotor behavior was dramatically diminished in the probiotic-treated mice when compared to the control group. Furthermore, it has been demonstrated that probiotic formulations like *L. helveticus* and *B. longum*, when taken for 30 days, lower "obsessive-compulsive" subscores on the Hopkins symptoms checklist in healthy humans (Turna et al., 2016). Given these results, there is reason to believe that the dysregulation of the microbiome may contribute to OCD. The gut influence is seen in direct or indirect pathways (Doménech et al., 2022; Van Bruggen et al., 2021). One pathway among them is production of neurotransmitters by gut bacteria for maintaining the microbiota-gut-brain axis. γ-aminobutyric acid (GABA) is produced by strains *Lactobacillus* and *Bifidobacterium*, as GABA is an important inhibitory neurotransmitter of brain circuits (Turna et al., 2016). Hence, we can hypothesize that imbalance in gut microbiota can lead changes in GABAergic inhibition, which can lead to onset of OCD. Another pathway for gut-brain communication is hypothalamic-pituitary-adrenal axis (HPA or HTPA axis), where significant elevation is observed in patients with OCD, specifically an increment in basal activity is seen (Turna et al., 2016; Rude et al., 2019). A range of human diseases are brought on by dysbiosis in the gut microbiome, which is induced by numerous environmental variables (Figure 2.1). We have seen that arsenic modifies the composition of the gut microbiota. However, especially at environmentally relevant doses, the functional influence of arsenic on the gut microbiome has not been sufficiently evaluated. In a study, mice were exposed to 100 ppb arsenic for 13 weeks, which altered the capacity and functioning of the gut microbiome. Arsenic exposure reconstructed the alpha and beta diversities along with the overall composition of the gut microbiota. Results stated, subjects with OCD were observed with low presence of bacterial α-diversity, hence with such findings in mind, there is reason to hypothesize that arsenic exposure can lead to dysregulation in gut microbiome (Dempsey et al., 2019). Although we can only postulate that, it may lead to the onset of OCD, which can be explained by communication between the pollutant, gut microbiome, and OCD.

Another research stated, the Val66Met polymorphism decreases brain-derived neurotrophic factor (BDNF) production and activity, which is linked to attention deficit hyperactivity disorder, hyperactivity, and eating disorders, as well as obsessive-compulsive disorder and eating disorders (Rosas-Vargas et al., 2011; Bonaccorso et al., 2015). Another neurodevelopmental issue is tic disorder, which includes temporary and persistent verbal or physical tics. Many of these individuals have coexisting mental disorders like depression or obsessive-compulsive disorder, which may affect how well they operate on a daily basis (Wanderer et al., 2012; Roessner et al., 2013). Although the etiology is not fully understood, environmental and genetic factors may both contribute to disease (Hoekstra et al., 2013). The variations in diversity of patient samples were not found in two investigations that have yet to be published (Xi et al., 2021; Wang et al., 2022). The study discovered significant diversity disparities. It was discovered that TD patients have reduced concentrations of *Prevotella stercorea* and *Streptococcus lutetiensis* and greater concentrations of other bacterial species, including *Bacteroides plebeius* and *Ruminococcus lactaris* (Xi et al., 2021). The presence of *Klebsiella pneumoniae, Akkermansia muciniphila, Bacteroides spp., Bifidobacterium spp.*, and *Eubacterium*

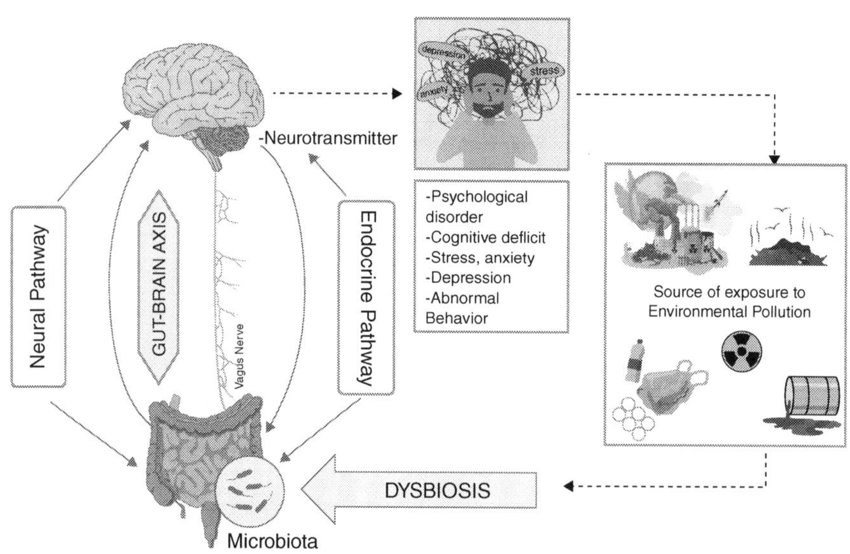

FIGURE 2.1 Neural pathway and gut dysbiosis.

spp. was also linked with tic severity and these microorganisms (Xi et al., 2021). In the other study, it was shown that TD patients had a substantial decrease in *Bifidobacterium* genus. The release of neurotransmitters may play a role in TD, and authors hypothesize that *Bifidobacterium* may be used to treat mild to moderate tic disorder symptoms. Children with TD were shown to have significantly higher levels of *Prevotella* and *Odoribacter* (Wang et al., 2022). *Prevotella* is thought to have a pro-inflammatory impact, and TD was shown to have higher levels of inflammatory markers (Bernard et al., 2014; Yeon et al., 2017). Increased *Odoribacter* was also discovered in neuropsychiatric conditions like pediatric acute onset neuropsychiatric syndrome (PANS) and pediatric autoimmune neuropsychiatric disorders associated with streptococcal infections (PANDAS), so this discovery may also point to a connection between the etiopathogenesis of tic disorders (Quagliariello et al., 2018).

In recent years, there has been a lot of interest in the possibility of altering the makeup of the bacteria in the gut using certain probiotics, prebiotics, or fecal microbiota transplants in order to enhance health. Although there is some evidence supporting the use of gut microbiota manipulation, the existing body of information is far from definitive due to the difficulty of the study and other restrictions. As a result, it is also necessary to take into account the geographic region, the diet that corresponds to it, the use of different psychotropic medications, and dietary supplements. It is necessary to recognize and take into consideration a bacterial signature associated with ageing.

2.4 CONCLUSION

Obsessive-compulsive disorder is a neurological ailment which relates to outline of unwanted obsessions related to thoughts and fear which results in repetitive compulsions. These compulsions and obsessions make the interference in daily life, making it stressed and miserable. These defects till date have no profound pathophysiology which could relate to any therapeutics formulation. Currently the two main incidences which relates to obsessions and compulsions could be the side effects of certain neurological disorders and dysbiosis of the gut microbiome which has an imminent impact on the regulation of neurobiology. In this context, there are many factors which trigger the phenomenon of the neurological defects and gut microbiome dysbiosis, one of them are the toxicological impact of environmental pollutants. These xenobiotics, upon interaction with the human body, dysregulates metabolic pathways, making the neurobiology to get disturbed, while on the other hand these hazardous pollutants tend to impact the gastrointestinal tract, resulting in dysbiosis and further disorientation of the metabolism regulating the neurological system. At the outset, the current chapter relates to all profound phenomenon which relates the toxicity of hazardous xenobiotic pollutants resulting in OCD via neurotoxicity and gut microbiome dysbiosis.

CONFLICT OF INTEREST

The authors of this chapter declare no conflict of interest.

ACKNOWLEDGMENT

The authors thank REVA University for providing the support and encouragement towards completion of the manuscript.

REFERENCES

Abramowitz, Jonathan S., and Ryan J. Jacoby. "Obsessive-compulsive and related disorders: A critical review of the new diagnostic class." *Annual Review of Clinical Psychology* 11 (2015): 165–186.

Ahmari, Susanne E., and Scott L. Rauch. "The prefrontal cortex and OCD." *Neuropsychopharmacology* 47, no. 1 (2022): 211–224.

Alex, K. D., and EA2562467 Pehek. "Pharmacologic mechanisms of serotonergic regulation of dopamine neurotransmission." *Pharmacology & Therapeutics* 113, no. 2 (2007): 296–320.

Ali, Muhammad Ubaid, Siyi Lin, Balal Yousaf, Qumber Abbas, Mehr Ahmed Mujtaba Munir, Audil Rashid, Chunmiao Zheng, Xingxing Kuang, and Ming Hung Wong. "Pollution characteristics, mechanism of toxicity and health effects of the ultrafine particles in the indoor environment: Current status and future perspectives." *Critical Reviews in Environmental Science and Technology* 52, no. 3 (2022): 436–473.

Amargós-Bosch, Mercè, Analía Bortolozzi, M. Victoria Puig, Jordi Serrats, Albert Adell, Pau Celada, Miklos Toth, Guadalupe Mengod, and Francesc Artigas. "Co-expression and in vivo interaction of serotonin1A and serotonin2A receptors in pyramidal neurons of prefrontal cortex." *Cerebral Cortex* 14, no. 3 (2004): 281–299.

Asif, Mohammad. "Role of heavy metals in human health and particularly in respect to diabetic patients." *CELLMED* 7, no. 1 (2017): 1–1.

Badhe, Suvarna, and Avinash De Sousa. "Optimizing Psychiatric Care Using Functional Neuroimaging." In *Handbook on optimizing patient care in psychiatry*, pp. 515–523. Routledge, 2022.

Baldessarini, Ross J. "Drug therapy of depression and anxiety disorders." *Goodman and Gilman's The Pharmacological Basis of Therapeutics. Edited by Brunton LL, Lazo JS, Parker KL. New York, McGraw-Hill* (2006): 429–460.

Berger, Miles, John A. Gray, and Bryan L. Roth. "The expanded biology of serotonin." *Annual Review of Medicine* 60 (2009): 355.

Bernard, N. J. "Rheumatoid arthritis: Prevotella copri associated with new-onset untreated RA." *Nature Reviews Rheumatology* 10 (2014): 2.

Biria, Marjan, Lucia-Manuela Cantonas, and Paula Banca. "Magnetic Resonance Spectroscopy (MRS) and Positron Emission Tomography (PET) imaging in obsessive-compulsive disorder." In *The neurobiology and treatment of OCD: Accelerating progress*, pp. 231–268. Springer, 2021.

Block, Michelle L., and Lilian Calderón-Garcidueñas. "Air pollution: Mechanisms of neuroinflammation and CNS disease." *Trends in Neurosciences* 32, no. 9 (2009): 506–516.

Bonaccorso, S., M. Sodhi, J. Li, W. V. Bobo, Y. Chen, M. Tumuklu, C. Theleritis, K. Jayathilake, and H. Y. Meltzer. "The brain-derived neurotrophic factor (BDNF) Val66Met polymorphism is associated with increased body mass index and insulin resistance measures in bipolar disorder and schizophrenia." *Bipolar Disorder* 17 (2015): 528–535.

Calderón-Garcidueñas, Lilian, Rafael Reynoso-Robles, Beatriz Pérez-Guillé, Partha S. Mukherjee, and Angélica Gónzalez-Maciel. "Combustion-derived nanoparticles, the neuroenteric system, cervical vagus, hyperphosphorylated alpha synuclein and tau in young Mexico City residents." *Environmental Research* 159 (2017): 186–201.

Carbery, Maddison, Wayne O'Connor, and Thavamani Palanisami. "Trophic transfer of microplastics and mixed contaminants in the marine food web and implications for human health." *Environment International* 115 (2018): 400–409.

Carmi, Lior, Uri Alyagon, Noam Barnea-Ygael, Joseph Zohar, Reuven Dar, and Abraham Zangen. "Clinical and electrophysiological outcomes of deep TMS over the medial prefrontal and anterior cingulate cortices in OCD patients." *Brain Stimulation* 11, no. 1 (2018): 158–165.

Chen, Pan, Mahfuzur Rahman Miah, and Michael Aschner. "Metals and neurodegeneration." *F1000Research* 5 (2016).

Conn, Kyna-Anne, Thomas HJ Burne, and James P. Kesby. "Subcortical dopamine and cognition in schizophrenia: Looking beyond psychosis in preclinical models." *Frontiers in Neuroscience* 14 (2020): 542.

Costa LG, Cole TB, Coburn J, Chang YC, Dao K, Roque P. Neurotoxicants are in the air: convergence of human, animal, and in vitro studies on the effects of air pollution on the brain. Biomed Res Int. 2014;2014:736385. doi: 10.1155/2014/736385. Epub 2014 Jan 12. PMID: 24524086; PMCID: PMC3912642.

Costa, Lucio G., Toby B. Cole, Khoi Dao, Yu-Chi Chang, Jacki Coburn, and Jacqueline M. Garrick. "Effects of air pollution on the nervous system and its possible role in neurodevelopmental and neurodegenerative disorders." *Pharmacology & Therapeutics* 210 (2020): 107523.

Cox, Kieran D., Garth A. Covernton, Hailey L. Davies, John F. Dower, Francis Juanes, and Sarah E. Dudas. "Human consumption of microplastics." *Environmental Science & Technology* 53, no. 12 (2019): 7068–7074.

Cryan, J. F., K. J. O'Riordan, C. S. M. Cowan, K. V. Sandhu, T. F. S. Bastiaanssen, M. Boehme, M. G. Codagnone, S. Cussotto, C. Fulling, A. V. Golubeva, K. E. Guzzetta, M. Jaggar, C. M. Long-Smith, J. M. Lyte, J. A. Martin, A. Molinero-Perez, G. Moloney, E. Morelli, E. Morillas, R. O'Connor, J. S. Cruz-Pereira, V. L. Peterson, K. Rea, N. L. Ritz, E. Sherwin, S. Spichak, E. M. Teichman, M. van de Wouw, A. P. Ventura-Silva, S. E. Wallace-Fitzsimons, N. Hyland, G. Clarke, and T. G. Dinan. "The microbiota-gut-brain axis. Physiological Reviews 99, no. 4 (2019, Oct 1): 1877–2013. doi: 10.1152/physrev.00018.2018. PMID: 31460832.

Dempsey, Joseph L., Mallory Little, and Julia Yue Cui. "Gut microbiome: An intermediary to neurotoxicity." *Neurotoxicology* 75 (2019): 41–69.

Domènech, Laura, Jesse Willis, Maria Alemany-Navarro, Marta Morell, Eva Real, Geòrgia Escaramís, Sara Bertolín et al. "Changes in the stool and oropharyngeal microbiome in obsessive-compulsive disorder." *Scientific Reports* 12, no. 1 (2022): 1–9.

Dougherty, D., S. Rauch, and Benjamin D. Greenberg. "Pathophysiology of obsessive-compulsive and related disorders." In *The American Psychiatric Association Publishing Textbook of Anxiety, Trauma and OCD-Related Disorders*. The American Psychiatric Association Publishing, 2020.

Drăgoi, Ana Miruna, Loredana-Georgiana Pecie, Bogdan-Eduard Patrichi, and Maria Ladea. "Morphopathological changes in obsessive-compulsive disorder." *Romanian Journal of Morphology and Embryology* 61, no. 1 (2020): 51.

Edinoff, Amber N., Juliana M. Fort, Joshua J. Woo, Christopher D. Causey, Caroline R. Burroughs, Elyse M. Cornett, Adam M. Kaye, and Alan D. Kaye. "Selective serotonin reuptake inhibitors and clozapine: Clinically relevant interactions and considerations." *Neurology International* 13, no. 3 (2021): 445–463.

El Mansari, Mostafa, and Pierre Blier. "Mechanisms of action of current and potential pharmacotherapies of obsessive-compulsive disorder." *Progress in Neuro-Psychopharmacology and Biological Psychiatry* 30, no. 3 (2006): 362–373.

Foa, E. B. "Cognitive behavioral therapy of obsessive-compulsive disorder." *Dialogues in Clinical Neuroscience* 12, no. 2 (2010): 199–207. doi: 10.31887/DCNS.2010.12.2/efoa. PMID: 20623924; PMCID: PMC3181959.

Fuster, Joaquín M. "The prefrontal cortex in the neurology clinic." *Handbook of Clinical Neurology* 163 (2019): 3–15.

Giorgetti, Marco, and Laurence H. Tecott. "Contributions of 5-HT2C receptors to multiple actions of central serotonin systems." *European Journal of Pharmacology* 488, no. 1–3 (2004): 1–9.

Goudarzi, Houman, Sonomi Nakajima, Tamiko Ikeno, Seiko Sasaki, Sachiko Kobayashi, Chihiro Miyashita, Sachiko Ito, Atsuko Araki, Hiroyuki Nakazawa, and Reiko Kishi. "Prenatal exposure to perfluorinated chemicals and neurodevelopment in early infancy: The Hokkaido Study." *Science of the Total Environment* 541 (2016): 1002–1010.

Guglielmo, Riccardo, Kamilla Woznica Miskowiak, and Gregor Hasler. "Evaluating endophenotypes for bipolar disorder." *International Journal of Bipolar Disorders* 9, no. 1 (2021): 1–20.

Harris, Maria H., Emily Oken, Sheryl L. Rifas-Shiman, Antonia M. Calafat, Xiaoyun Ye, David C. Bellinger, Thomas F. Webster, Roberta F. White, and Sharon K. Sagiv. "Prenatal and childhood exposure to per-and polyfluoroalkyl substances (PFASs) and child cognition." *Environment International* 115 (2018): 358–369.

Hathaway, William R., and Bruce W. Newton. "Neuroanatomy, prefrontal cortex." In *StatPearls [Internet].* StatPearls Publishing, 2023, May 29. PMID: 29763094

Hernandez, Laura M., Elvis Genbo Xu, Hans CE Larsson, Rui Tahara, Vimal B. Maisuria, and Nathalie Tufenkji. "Plastic teabags release billions of microparticles and nanoparticles into tea." *Environmental Science & Technology* 53, no. 21 (2019): 12300–12310.

Hoekstra, P. J., A. Dietrich, M. J. Edwards, I. Elamin, and D. Martino. "Environmental factors in Tourette syndrome." *Neuroscience & Biobehavioral Reviews* 37 (2013): 1040–1049.

Iqubal, Ashif, Musheer Ahmed, Shahnawaz Ahmad, Chita Ranjan Sahoo, Mohammad Kashif Iqubal, and Syed Ehtaishamul Haque. "Environmental neurotoxic pollutants." *Environmental Science and Pollution Research* 27, no. 33 (2020): 41175–41198.

Jassi, Amita, Khodayar Shahriyarmolki, Tracey Taylor, Lauren Peile, Fiona Challacombe, Bruce Clark, and David Veale. "OCD and COVID-19: A new frontier." *The Cognitive Behaviour Therapist* 13 (2020).

Kanova, Marcela, and Pavel Kohout. "Serotonin—Its synthesis and roles in the healthy and the critically ill." *International Journal of Molecular Sciences* 22, no. 9 (2021): 4837.

Khan, Andleeb, Sadaf Jahan, Zuha Imtiyaz, Saeed Alshahrani, Hafiz Antar Makeen, Bader Mohammed Alshehri, Ajay Kumar, Azher Arafah, and Muneeb U. Rehman. "Neuroprotection: Targeting multiple pathways by naturally occurring phytochemicals." *Biomedicines* 8, no. 8 (2020): 284.

Kumar, P. and V. Rai. "Catechol-O-methyltransferase gene Val158Met polymorphism and obsessive compulsive disorder susceptibility: A meta-analysis." *Metabolic Brain Disease* 35, no. 2 (2020): 241–251.

Kuo, Che Yu, Kao Chin Chen, I. Hui Lee, Huai-Hsuan Tseng, Nan Tsing Chiu, Po See Chen, Yen Kuang Yang, and Wei Hung Chang. "Serotonin modulates the correlations between obsessive-compulsive trait and heart rate variability in normal healthy subjects: A Spect study with [123I] ADAM and heart rate variability measurement." *Clinical Psychopharmacology and Neuroscience* 20, no. 2 (2022): 271.

Lv, Junhua, and Feng Liu. "The role of serotonin beyond the central nervous system during embryogenesis." *Frontiers in Cellular Neuroscience* 11 (2017): 74.

McCarthy, M. M., C. Moore-Kochlacs, X. Gu, E. S. Boyden, X. Han, and N. Kopell. "Striatal origin of the pathologic beta oscillations in Parkinson's disease." *Proceedings of the National Academy of Sciences* 108, no. 28 (2011): 11620–11625.

McKay, Dean, Jonathan S. Abramowitz, and Eric A. Storch. "Mechanisms of harmful treatments for obsessive–compulsive disorder." *Clinical Psychology: Science and Practice* 28, no. 1 (2021): 52.

Ngetich, Ronald, Jing Zhou, Junjun Zhang, Zhenlan Jin, and Ling Li. "Assessing the effects of continuous theta burst stimulation over the dorsolateral prefrontal cortex on human cognition: A systematic review." *Frontiers in Integrative Neuroscience* 14 (2020): 35.

Oh, Jiwon, Deborah H. Bennett, Antonia M. Calafat, Daniel Tancredi, Dorcas L. Roa, Rebecca J. Schmidt, Irva Hertz-Picciotto, and Hyeong-Moo Shin. "Prenatal exposure to per-and polyfluoroalkyl substances in association with autism spectrum disorder in the MARBLES study." *Environment International* 147 (2021): 106328.

Park, Hyungyou, Minah Kim, Yoo Bin Kwak, Kang Ik K. Cho, Junhee Lee, Sun-Young Moon, Silvia Kyungjin Lho, and Jun Soo Kwon. "Aberrant cortico-striatal white matter connectivity and associated subregional microstructure of the striatum in obsessive-compulsive disorder." *Molecular Psychiatry* (2022): 1–8.

Pauls, D. L. "The genetics of obsessive-compulsive disorder: A review." *Dialogues in Clinical Neuroscience* 12, no. 2 (2010): 149–163. doi: 10.31887/DCNS.2010.12.2/dpauls. PMID: 20623920; PMCID: PMC3181951.

Philpot, Nicola, Richard Thwaites, and Mark Freeston. "Understanding why people with OCD do what they do, and why other people get involved: Supporting people with OCD and loved ones to move from safety-seeking behaviours to approach-supporting behaviours." *The Cognitive Behaviour Therapist* 15 (2022).

Pinciotti, Caitlin M., Bradley C. Riemann, and Jonathan S. Abramowitz. "Intolerance of uncertainty and obsessive-compulsive disorder dimensions." *Journal of Anxiety Disorders* 81 (2021): 102417.

Politis, Marios, and Flavia Niccolini. "Serotonin in Parkinson's disease." *Behavioural Brain Research* 277 (2015): 136–145.

Prasad, Sambhu, and Chayan Manna. "Clinical description of rare phenomenology of obsessive–Compulsive disorder." *Indian Journal of Social Psychiatry* 38, no. 3: 304 (2022).

Prüst, Minne, Jonelle Meijer, and Remco HS Westerink. "The plastic brain: Neurotoxicity of micro-and nanoplastics." *Particle and Fibre Toxicology* 17, no. 1 (2020): 1–16.

Purdon, Christine. "There is a lot more to compulsions than meets the eye." *Clinical Neuropsychiatry* 15, no. 5 (2018).

Quagliariello, A., F. Del Chierico, A. Russo, S. Reddel, G. Conte, L. R. Lopetuso, G. Ianiro, B. Dallapiccola, F. Cardona, A. Gasbarrini et al. "Gut microbiota profiling and gut-brain crosstalk in children affected by pediatric acute-onset neuropsychiatric syndrome and pediatric autoimmune neuropsychiatric disorders associated with streptococcal infections." *Frontiers in Microbiology* 9 (2018): 675.

Rădulescu, Anca, Joanna Herron, Caitlin Kennedy, and Annalisa Scimemi. "Global and local excitation and inhibition shape the dynamics of the cortico-striatal-thalamo-cortical pathway." *Scientific Reports* 7, no. 1 (2017): 1–21.

Robbins, Trevor W., Matilde M. Vaghi, and Paula Banca. "Obsessive-compulsive disorder: Puzzles and prospects." *Neuron* 102, no. 1 (2019): 27–47.

Roessner, V., K. Schoenefeld, J. Buse, S. Bender, S. Ehrlich, A. Münchau. "Pharmacological treatment of tic disorders and Tourette Syndrome." *Neuropharmacology* 68 (2013): 143–149.

Rosas-Vargas, H., J. D. Martínez-Ezquerro, and T. Bienvenu. "Brain-derived neurotrophic factor, food intake regulation, and obesity." *Archives of Medical Research* 42 (2011): 482–494.

Rude, Kavi M., Matteo M. Pusceddu, Ciara E. Keogh, Jessica A. Sladek, Gonzalo Rabasa, Elaine N. Miller, Sunjay Sethi et al. "Developmental exposure to polychlorinated biphenyls (PCBs) in the maternal diet causes host-microbe defects in weanling offspring mice." *Environmental Pollution* 253 (2019): 708–721.

Rueda-Ruzafa, Lola, Francisco Cruz, Pablo Roman, and Diana Cardona. "Gut microbiota and neurological effects of glyphosate." *Neurotoxicology* 75 (2019): 1–8.

Saboowala, H. K. *What is Obsessive Compulsive Disorder (OCD)? A complete pedagogy*. Dr. Hakim Saboowala, 2022.

Sinopoli, Vanessa M., Christie L. Burton, Sefi Kronenberg, and Paul D. Arnold. "A review of the role of serotonin system genes in obsessive-compulsive disorder." *Neuroscience & Biobehavioral Reviews* 80 (2017): 372–381.

Skogheim, T. S., K. V. F. Weyde, H. Aase, S. M. Engel, P. Surén, M. G. Øie, G. Biele, T. Reichborn-Kjennerud, A. L. Brantsæter, L. S. Haug, and A. Sabaredzovic. "Prenatal exposure to per- and polyfluoroalkyl substances (PFAS) and associations with attention-deficit/hyperactivity disorder and autism spectrum disorder in children." *Environmental Research* 202 (2021): 111692.

Smith, Jared Nathaniel. "The moral psychology of obsessive-compulsive disorder." PhD diss., UC Riverside, 2022.

Srivastava, Abhinav. "Efficacy of continuous theta-burst stimulation of orbito-frontal cortex in patients with alcohol dependence with craving: A sham-controlled study." PhD diss., Central Institute of Psychiatry (India), 2018.

Summers, Jesse S., and Walter Sinnott-Armstrong. *Clean hands: Philosophical lessons from scrupulosity.* Oxford University Press, 2019.

Summers, B. J., G. Aalbers, P. J. Jones, R. J. McNally, K. A. Phillips, and S. Wilhelm. "A network perspective on body dysmorphic disorder and major depressive disorder." *Journal of Affective Disorders* 262 (2020): 165–173.

Tumati, S., S. Martens, B. M. de Jong, and A. Aleman. "Lateral parietal cortex in the generation of behavior: Implications for apathy." *Progress in Neurobiology* 175 (2019): 20–34.

Turna, Jasmine, Keren Grosman Kaplan, Rebecca Anglin, and Michael Van Ameringen. " 'What's bugging the gut in OCD?' A review of the gut microbiome in obsessive–compulsive disorder." *Depression and Anxiety* 33, no. 3 (2016): 171–178.

Van Bruggen, A. H. C., M. R. Finckh, M. He, C. J. Ritsema, P. Harkes, D. Knuth, and V. Geissen. "Indirect effects of the herbicide glyphosate on plant, animal and human health through its effects on microbial communities." *Frontiers in Environmental Science* (2021): 464.

Van Roessel, Peter, Giacomo Grassi, Elias N. Aboujaoude, Jose M. Menchon, Michael Van Ameringen, and Carolyn I. Rodriguez. "Treatment-resistant OCD: Pharmacotherapies in adults." *Comprehensive Psychiatry* (2022): 152352.

Wanderer, S., V. Roessner, R. Freeman, N. Bock, A. Rothenberger, and A. Becker. "Relationship of obsessive-compulsive disorder to age-related comorbidity in children and adolescents with Tourette syndrome." *Journal of Developmental and Behavioral Pediatrics* 33 (2012): 124–133.

Wang, Y., H. Xu, M. Jing, X. Hu, J. Wang, and Y. Hua. "Gut microbiome composition abnormalities determined using high-throughput sequencing in children with tic disorder." *Frontiers in Pediatrics* 10 (2022): 831944.

Wojtunik-Kulesza, Karolina, Anna Oniszczuk, and Monika Waksmundzka-Hajnos. "An attempt to elucidate the role of iron and zinc ions in development of Alzheimer's and Parkinson's diseases." *Biomedicine & Pharmacotherapy* 111 (2019): 1277–1289.

Xi, W., X. Gao, H. Zhao, X. Luo, J. Li, X. Tan, L. Wang, J. B. Zhao, J. Wang, G. Yang, and L. Y. Liu. Depicting the composition of gut microbiota in children with tic disorders: An exploratory study. *Journal of Child Psychology and Psychiatry* 62, no. 10 (2021): 1246–1254.

Xiao, Xiao, Ming Ding, and Yu-Qiu Zhang. "Role of the anterior cingulate cortex in translational pain research." *Neuroscience Bulletin* 37, no. 3 (2021): 405–422.

Yeon, S. M., J. H. Lee, D. Kang, H. Bae, K. Y. Lee, S. Jin, J. R. Kim, Y. W. Jung, T. W. Park. "A cytokine study of pediatric Tourette's disorder without obsessive compulsive disorder." *Psychiatry Research* 247 (2017): 90–96.

Yong, Cheryl Qian Ying, Suresh Valiyaveettil, and Bor Luen Tang. "Toxicity of microplastics and nanoplastics in mammalian systems." *International Journal of Environmental Research and Public Health* 17, no. 05 (2020): 1509.

Young, Carly B., Vamsi Reddy, and James Sonne. "Neuroanatomy, basal ganglia." In *StatPearls [Internet].* StatPearls Publishing, 2021.'

3 Applications of Computational Models in OCD

Varun Varma Sangaraju[a,†] *and Senthilkumar Rajagopal*[b]
[a] Department of Information Technology, University of the Cumberlands, Williamsburg, KY, USA
[b] Department of Biotechnology, School of Applied Sciences, REVA University, Bangalore, KA, India

3.1 INTRODUCTION

Computational modeling is a technique that uses computers, mathematics, and computer science to simulate and research complex systems. Computational modeling helps to recreate real-time problems through datasets and find solutions through statistical models. It is highly impossible to collect numerous amounts of body samples from obsessive-compulsive disorder patients for research purposes. Computational modeling provides the ability to experiment with the variables of the sample test results. Scientists and researchers can simulate hundreds and thousands of inputs as a sample by changing variables and obtain results through computational modeling methods. The results from computational modeling can be used for more in-depth data analysis and drawing inferences.

Many infectious diseases are tracked through computational modeling. Real-time data is collected, and models predict the rate of diseases in particular geographic locations. For example, during the COVID-19 pandemic, countries like India tracked patients with the virus using mobile applications. Computational models would collect the data to alert nearby active members and predict the rate at which the disease increases or decreases. One of the pros of computational modeling is to provide clinical decision support to doctors or healthcare providers for their patients. All the visits of the patients, test results, and health history, which are recorded, can be used to predict the health trend of the patient, and doctors can make appropriate medical decisions. These suggestions from a doctor will allow the patient to avoid deterioration of health or identify symptoms of certain conditions early. Computational modeling also helps researchers to identify the side effects of prescription drugs. For example, companies like CVS Health in the US use models to create drug utilization reviews. Computation modeling is the best way to create safe prescription drugs by minimizing the combination of drugs that cause more side effects. There are various computational modeling approaches for solving the problems in OCD (obsessive-compulsive disorder).

3.2 DYNAMICAL SYSTEM APPROACHES

This section discusses the dynamical system approaches used in computational models of obsessive-compulsive disorder. The human brain consumes streams of information continuously and categorizes the experiences into a combination of discrete patterns or events (Rabinovich and Varona 2018). Dynamical system computational approaches provide mechanistic insights into pathological neural dynamics in OCD using custom nonlinear differential equation systems (Szalisznyó and Silverstein 2021).

The computational model using a stochastic neurodynamical framework helps to understand the relationship between the stability of attractor networks and glutamatergic activity in the brain

[†]This is the first author.

DOI: 10.1201/9781003387893-4

(Rolls 2012). Rolls (2012) stated that obsessive-compulsive disorder cases have over the stability of attractor neuronal networks in the cortical. The attractor framework used for this experiment is based on the dynamical systems approach. A network of interconnected neurons forms a memory pattern (memories, thoughts, or perceptual representations), which can be stored and later recalled using external inputs (Rolls 2012). There are two phases of the energy landscape in the neurons, one is a shallow unstable attractor, and the other is a deep stable attractor. Rolls (2012) identified that the cortical and associated attractor networks become very stable in obsessive-compulsive disorders, and the networks tend to remain in the deep stable attractor state for long periods. When the depth of basins in the attractors increases, the stability of the attractor networks increases. Various regions of the brain can experience the increased stability of attractor networks and higher neuronal firing rates, which leads to different types of symptoms (Rolls 2012). Glutamate acts as a transmitter between the neurons and has the capacity to influence the stability of attractor networks. The simulation evidence obtained from the computational model suggested that the increase in glutamatergic synaptic efficacy also increased the stability of attractor networks and proved the cause of some of the symptoms of OCD (Rolls 2012). The evidence also suggested that the gene-based transmission of OCD is possible. The SLC1A1 glutamate transporter gene limits the function of glutamate as an excitatory neurotransmitter and maintains healthy glutamate levels (Rolls 2012). When the levels of this gene are reduced, then glutamate levels increase and cause problems in worsening the symptoms of OCD. From this computation modeling methodology, Rolls (2012) indicated that antiglutamatergic activity agents such as riluzole can help reduce the levels of glutamate transmitter release. The model suggested that the use of combinations of drugs in low doses can be beneficial in combating OCD. One example of such a combination is hyperglutamatergia drug riluzole and antianxiety drug benzodiazepine (Rolls 2012).

STN-DBS (subthalamic nucleus-deep brain stimulation) technique is considered for the treatment-refractory OCD as per level I clinical evidence (Karamintziou et al. 2015). A stochastic dynamical model is used for computing the data obtained from microelectrode recording datasets, which were acquired from eight different surgical events for OCD (Karamintziou et al. 2015). The stochastic phase model used the Langevin equation with the application of simulation patterns (Karamintziou et al. 2015). There was a stronger desynchronizing effect on the results with varied frequencies (high and low) when compared to standard simulation (Karamintziou et al. 2015). Karamintziou et al. (2015) provided model-based indications for treatment-refractory OCD using optimal alternative patterns of STN-DBS.

The human brain's cognitive and emotional responses to positive and negative moments are governed by the cortex, the basal ganglia, the limbic system, the thalamus, and the dopamine system (Rădulescu and Marra 2017). Rădulescu and Marra (2017) created a computational model of brain circuitry that governs motion and rewards using continuous time dynamical systems and ordinary differential equations. All non-dopaminergic connections were considered linear terms, and the dopaminergic connections were termed nonlinear to represent mathematically. This computational model predicted that enhanced OFC (orbitofrontal cortex) projections to the NAc (nucleus accumbens) and decreased amygdala connectivity with the NAc correlate with typical OCD symptoms (Rădulescu and Marra 2017). The model also predicted that decreased VTA (ventral tegmental area) dopamine modulation of NAc and decreased inhibition of the amygdala correlate with OCD symptoms and increased anxiety, respectively (Rădulescu and Marra 2017). Rădulescu and Marra (2017) warned that the computational model is meaningful when all the components of the study are considered when conducting research, but the results might not be conducive when using part of the model. Rădulescu and Marra (2017) studied the OCD phenomenon and illustrated it with a mathematical model of connectivity in a brain network (Rădulescu and Marra 2017).

Cognitive brain activity is transformed into a low-dimensional dynamical process with discrete sequential information coding (Rabinovich and Varona 2018). The low-dimensional model is an

TABLE 3.1
Summary of Dynamic System Approach Models

Computational Modeling Methods	Contribution to OCD Studies	Reference
Stochastic neurodynamical framework	Finding the effects of glutamatergic activity and attractor networks	(Rolls 2012)
STN-DBS method with stochastic differential equations	For treatment-refractory OCD	(Karamintziou et al. 2015)
Differential equations and continuous time dynamical systems	Cognitive and emotional responses to positive and negative rewards correlate to OCD symptoms	(Rădulescu and Marra 2017)
Low dimensional computational model using differential equations	Cognitive activity, metastable states, and stable heteroclinic channels	(Rabinovich and Varona 2018)

appropriate tool to describe the underlying aspects of a dynamical theory of consciousness and cognition (Rabinovich and Varona 2018). Rabinovich and Varona (2018) built a model by considering these aspects.

i. Using low-frequency variables for information generation and information processing in the human brain.
ii. Sequential switching between metastable states in the cognitive processes.
iii. Information signals sensibility.

The cognitive information flows that are based on sequential dynamics are described by a mathematical image, such as a stable heteroclinic channel (Rabinovich and Varona 2018). Discrete sequential coding is a robust tool to explore brain functions and cognitive activity, which helps to analyze OCD symptoms. However, Rabinovich and Varona (2018) believed that predictability would only come from high-dimensional computational models.

3.3 BAYESIAN APPROACHES

This section belongs to the Bayesian approaches used in computational models contributing to obsessive-compulsive disorder. The human brain implements the Bayesian inferences of the current state by using the knowledge of prior states and evidence and weighting every state by relative uncertainty (Szalisznyó and Silverstein 2021).

Antidepressant medications are used by young patients (age<18) with OCD and anxiety to improve functional outcomes and symptoms (Mills and Strawn 2020). The most commonly used medications in these antidepressants are SSRIs (selective serotonin reuptake inhibitors) and SNRIs (serotonin norepinephrine reuptake inhibitors) (Mills and Strawn 2020). The Bayesian hierarchical modeling is used to estimate the probability of adverse events at each hierarchy, and the observed data is combined with sub-hierarchical models to handle uncertainty (Mills and Strawn 2020). The aggregation of the adverse events' data from various trials of SSRIs and SNRIs in pediatric OCD and anxiety led to evaluation with better statistical power than using the data from individual trials alone (Mills and Strawn 2020). Mills and Strawn (2020) conducted a meta-analysis of adverse events caused due to antidepressant medications using the Bayesian hierarchical model. This meta-analysis of randomized and placebo-controlled trials of SNRIs and SSRIs is one of the first experiments to address the adverse events in young patients using Bayesian hierarchical modeling (Mills and Strawn 2020). The study identified that young patients have more adverse effects with SSRIs compared to SNRIs and placebo (Mills and Strawn 2020). Due to the hierarchical nature of

Bayesian hierarchical modeling, the comparison of data events at each of the sub-hierarchical levels is preserved.

One of the characteristics of OCD is the difficulty in believing that past events will predict the OCD patient's future behavior (Fradkin et al. 2020). Fradkin et al. (2020) proposed a computational model which was developed on top of a standard Bayesian brain framework, and the impairment is mentioned as excessive uncertainty in state transitions. Though the state of the brain is not directly observable through this model, the assumption is that the brain will probabilistically gather various sources of information, such as predictions about state evolution and sensory inputs (Fradkin et al. 2020). The Bayesian brain framework also assumes that the relative weights in the inferential process are due to the perceived uncertainty of various information sources (Fradkin et al. 2020). The model suggested excessive state transition uncertainty, which led to different behavioral consequences and impaired goal-directed control (Fradkin et al. 2020). Fradkin et al. (2020) identified that computational models could trace back the obsessive-compulsive pathology to excessive state transitions' uncertainty.

A disorder like OCD has diverse symptoms and a large variety of heterogeneous behaviors, feelings, and cognitions (Schulze et al. 2018). The previous research used Y-BOCS (Yale-Brown Obsessive Compulsive Scale) to understand the heterogenous symptoms of OCD and identified four underlying dimensions (Schulze et al. 2018). However, Schulze et al. (2018) deduced that the variation in past research results could be due to aggregated scores, the exclusion of some of the symptoms, and the use of exploratory factor analyses. Schulze et al. (2018) tested the original Y-BOCS model's homogeneity with exploratory factor analysis and changed the categories. The data from this research was analyzed using BSEM (Bayesian structural equation modeling) to understand the multidimensional models and heterogeneity (Schulze et al. 2018). According to Schulze et al. (2018), BSEM inherits some of the characteristics of exploratory factor analysis but differentiates main and cross-loadings to provide means of testing model fit. The two-level factor model used by Schulze et al. (2018) identified ten homogenous first-order factors, which were reduced to four second-order factors, such as incompleteness, responsibility, taboo thoughts, and contamination. The results suggested that the two-factor model using BSEM outperformed other factor structure studies (Schulze et al. 2018).

The human brain is an integral part of decision-making, and people rethink or reevaluate their decisions frequently. Usually, people make a decision from two options but try to revisit the decision episode again the other day (Solway et al. 2021). Though this happens less in normal people, OCD patients are frequently troubled with decisions made even in small of the smallest things, such as

TABLE 3.2
Summary of Bayesian Approach Models

Computational Modeling Methods	Contribution to OCD Studies	Reference
Bayesian hierarchical models	Understanding the adverse drug effects in young patients	(Mills and Strawn 2020)
Bayesian brain framework with state transition capability	Uncertainty in state transitions of the human brain	(Fradkin et al. 2020)
Bayesian structural equation models and factor analysis	Finding four broader symptom dimensions in OCD	(Schulze et al. 2018)
Bayesian multilevel version of drift diffusion model	The behavior of OCD patients in repeated decision making	(Solway et al. 2021)
Quasi-optimal Bayesian learning model	Disassociation between action and confidence due to the ego-dystonic nature of OCD	(Vaghi et al. 2017)

locking the door or switching off the stove (Solway et al. 2021). Many studies are present on OCD patients' single decision episodes, but no attention on repeated decision-making in computational modeling work (Solway et al. 2021). Computational models on such data use integration models because the evidence for and against each option will be integrated into each step until it reaches a threshold (Solway et al. 20201). The option with the highest evidence will be selected by the computational model (Solway et al. 2021). Solway et al. (2021) used the dot motion perceptual model as a base for their computational model, where baseline decisions are taken based on the direction (right or left) of an array of dots shown to participants. The decisions made by participants were tagged with unique words, and the participants were asked to make a decision a second time to see if they selected the same word (Solway et al. 2021). Solway et al. (2021) used a Bayesian multilevel version of the drift diffusion model, which was implemented using a statistical programming language. Solway et al. (2021) tested various hypotheses related to information transfer and whether the information transfer and decision-making differed in OCD patients using the drift-diffusion framework.

Confidence and actions are two tightly coupled events of OCD (Vaghi et al. 2017). As compulsive actions are regarded as disproportionate in the ego-dystonic nature of OCD, Vaghi et al. (2017) hypothesized that confidence and action could be independent events while learning. Vaghi et al. (2017) implemented a computational model from the reduced quasi-optimal Bayesian observer with changes to input their circular data. Previous Bayesian models of belief systems did not include the computational complexity of searching the state space (Vaghi et al. 2017). The proposed model uses a dynamic learning rate to adapt to the inputs instead of reinforcement models using fixed learning rate (Vaghi et al. 2017). The dynamic learning rate helps the model to understand environmental changes such as change of location and disregarding outliers when beliefs mean estimates are established (Vaghi et al. 2017). Vaghi et al. (2017) utilized the quasi-optimal Bayesian learner, which is the computational, algorithmic version of a normal optimal Bayesian learner. Vaghi et al. (2017) compared the beliefs of the quasi-optimal Bayesian learner with human participants to understand the evolution of participants' beliefs. A predictive inference task was designed for participants to understand their behavior toward confidence and action in the ever-changing environment (Vaghi et al. 2017). The results of the study showcased a disassociation between action and confidence (Vaghi et al. 2017).

3.4 REINFORCEMENT LEARNING APPROACHES

This section belongs to the reinforcement learning approaches used in computational models contributing to obsessive-compulsive disorder. Reinforcement learning approaches are used on the data of OCD patients with impairment to simulate goal-directed behavior (Szalisznyó and Silverstein 2021).

OCD patients are known to have learning and decision-making deficits (Murray et al. 2019). Many experiments on animals showcased the evidence that dopamine neurons exhibited an increased rate of firing to unexpected or better rewards (positive prediction error) and decreased rate of firing when omission of expected rewards (negative prediction error) (Murray et al. 2019). OCD patients are bound to exhibit exaggerated error responses and prediction error learning signals in various EEG (Electroencephalogram) tests and fMRI (functional Magnetic Resonance Imaging) tasks (Murray et al. 2019). Murray et al. (2019) found considerable evidence that dopaminergic function is the link to prediction error processing. Murray et al. (2019) studied group differences in anterior cingulate and nucleus accumbens regions of interest using a Q-learning computational on the fMRI prediction error responses. Murray et al. (2019) considered each trial and estimated a value for both negative and positive reward prediction errors. The results of computational modeling showcased a marginally significant difference in learning rates between patients and controls, whilst OCD patients showcased strong cingulate signaling of prediction errors when omission of an expected reward (Murray et al. 2019).

People with a deficit in goal-directed control may be vulnerable to forming more rigid habits than normal individuals (Gillan et al. 2016). Goal-directed control means resisting certain temptations of habits to achieve a specific goal by an individual. For example, not sleeping when tired during the day to achieve a full night's sleep (Gillan et al. 2016). Evidence suggests that the reduction of goal-directed control in an individual can lead to compulsive disorders such as OCD and drug addiction (Gillan et al. 2016). Gillan et al. (2016) hypothesized that a dimensional approach to calculating the efficiencies of online data collected from healthy individuals could be implemented in patients associated with goal-directed deficiencies. Goal-directed control is extracted computationally from model-based learning, which is a form of reinforcement learning (Gillan et al. 2016). This model predicts an individual's likeliness of forming habits using difference measures (Gillan et al. 2016). Gillan et al. (2016) conducted the first experiment with a self-reported questionnaire measuring OCD symptoms' severity using model-based learning and found that obsessive-compulsive disorder itself was the first among nine identified symptoms (OCD, alcohol addiction, depression, trait anxiety, eating disorder, social anxiety, impulsivity, schizotypy, and apathy). Gillan et al. (2016) conducted a second experiment where the association with self-reported symptoms was replicated and identified that eating disorders, impulsivity, and OCD are the topmost out of nine recorded symptoms. Gillan et al. (2016) also conducted a factor analysis and categorized the symptoms into three latent symptom dimensions such as anxious depression, compulsive behavior and intrusive thought, and social withdrawal.

Decision-making in people is characterized by two different systems such as habitual and goal-directed (Voon et al. 2015; Deserno et al. 2015). People make some choices repeatedly, such as eating unhealthy food, even though they have strong negative consequences (Voon et al. 2015). Habitual decisions are taken based on previously rewarded actions, whereas goal-directed control decisions are made based on likely affective outcomes (Voon et al. 2015; Deserno et al. 2015). The computational models proposed that habitual and goal-directed learning mechanisms arise from model-free and model-based reinforcement learning, respectively (Voon et al. 2015). According to Voon et al. (2015), model-free reinforcement learning algorithms use outcomes and transitions to calculate the reward prediction error, which leads to executed actions and turns into habitual behavior (Voon et al. 2015). A decision choice would be repeatedly selected in this model when there is a reward in the habitual system regardless of the transition state (Voon et al. 2015). Voon et al. (2015) analyzed the data using the computational model, where the choices are weighted and fitted into a model-free learning algorithm (Voon et al. 2015).

PRL (probabilistic reversal learning) paradigm is used to capture the deficits in behavior flexibility (Kanen et al. 2019). Changing actions flexibly when there is a change in the environment and avoiding some rare actions when the environment is stable is known as Adaptive behavior (Kanen et al. 2019). The models using PRL usually challenges the participants with two choices and ask

TABLE 3.3
Summary of Reinforcement Learning Approach Models

Computational Modeling Methods	Contribution to OCD Studies	Reference
Q-learning computational model	Prediction error learning signals through EEG and fMRI	(Murray et al. 2019)
Two-step model-based reinforcement learning task, including factor analysis	Assessment of goal-directed learning in OCD or mental health patients	(Gillan et al. 2016)
Model-free reinforcement learning	Impact of disorders in habit formation	(Voon et al. 2015)
Probabilistic reversal learning using seven RL models	Reward and punishment-driven learning in OCD	(Kanen et al. 2019)
Hybrid reinforcement learning model (model-based and model-free)	Effect of dopamine in model-based and model-free learning	(Deserno et al. 2015)

them to learn by selecting the correct option through the trial-and-error method (Kanen et al. 2019). Kanen et al. (2019) used a hierarchical Bayesian approach to fit the seven reversal learning models to the PRL's behavioral data. PRL tasks consist of trials, and each trial goes through the computational model where information such as the subject's identity, group, response location, drug condition, stimuli, and whether rewarded/unrewarded was recorded (Kanen et al. 2019). From this computational model, Kanen et al. (2019) characterized reinforcement learning processes, which were patterns of PRL behavior in OCD and showcased the effects of dopamine D2/3 receptor behavior in pramipexole and amisulpride.

Behavior control is based on deliberate model-based and reflexive model-free systems (Deserno et al. 2015). Habitual and goal-directed decision systems are based on model-free (influence of past rewards) and model-based (actions leading to potential outcomes) systems, respectively (Deserno et al. 2015; Voon et al. 2015). When there is a change in model input, goal-directed systems are easily adaptable (Deserno et al. 2015). However, the habitual systems need more time to relearn the decisions based on the experiences collected from their previous memory (Deserno et al. 2015). Deserno et al. (2015) collected evidence stating that model-based learning signals are encoded in prefrontal-parietal cortices and model-free signals in the ventral striatum. The important task is to understand the factors affecting the individual to fall into one of the two computational models. Dopamine is one of the most linked neuromodulators that affects a person's choice; however, there are other notable neuromodulators also (Deserno et al. 2015). Deserno et al. (2015) demonstrated that the ventral striatum consisting of high presynaptic dopamine would see enhanced coding of model-based signatures in the lateral prefrontal cortex and deteriorated coding of model-free learning signals.

3.5 SUPERVISED AND UNSUPERVISED MACHINE LEARNING APPROACHES

This section belongs to the supervised and unsupervised machine learning approaches used in computational models contributing to obsessive-compulsive disorder. Machine learning techniques are mainly used to investigate neuropathophysiology and phenomenology in OCD patient data (Szalisznyó and Silverstein 2021).

It is absolutely necessary to identify or predict the symptoms of OCD using computational models and the latest technologies. Askland et al. (2015) conducted a study with the objective of identifying predictors to diagnose OCD or symptoms of OCD using machine learning methodologies. Machine learning algorithms such as random forest are widely used in biomedical datasets but very limited applications in psychiatric research (Askland et al. 2015). Askland et al. (2015) aimed to use machine learning methodology to analyze existing OCD data and examine predictors of remission. The other goal of this study was to apply a model on random forest analyses to multidimensional and psychiatric datasets (Askland et al. 2015). The basic Random Forest algorithm is an ensemble learning method that is non-parametric, and its analyses only depend on good predictors and not the noise variables of a dataset (Askland et al. 2015). Askland et al. (2015) found 24 features above the threshold in all their model runs which were highly consistent outcome predictors.

Frontostriatal hyperactivity is considered to be influencing deficits in decision-making, goal-directed planning, working memory, response inhibition, and task switching in neuropsychological models (Stamatis et al. 2021). However, the findings on OCD are divergent from this theory, and the deficits are found in non-verbal memory, sustained attention, and processing speed (Stamatis et al. 2021). The heterogeneity across neuropsychological models makes drawing inferences extremely difficult (Stamatis et al. 2021). The algorithms in machine learning train existing data to develop and use the models on test data for making predictions (Stamatis et al. 2021). The data set is usually divided into two sets, such as training (70–80%) and test (20–30%), using random selection. Supervised machine learning algorithms are beneficial to OCD's neuropsychological features due to their training models (Stamatis et al. 2021). Stamatis et al. (2021) extrapolated information on supervised machine learning algorithms.

i. Random forests and SVM (support vector machines) as the number of input variables are larger than the sample size.
ii. These algorithms rely on fewer sample sizes and eliminate noise from the neuropsychological datasets.
iii. Cross-validation procedures help to avoid overfitting problems.

Stamatis et al. (2021) used a series of supervised machine learning algorithms such as random forest, SVM, KNN (K-Nearest Neighbor), and logistic regression models to characterize the OCD's neuropsychological processes and account for sensory phenomena.

TMS (transcranial magnetic stimulation) is one of the noninvasive brain stimulation techniques which uses magnetic pulses to stimulate the brain cortex externally (Metin et al. 2020). TMS is commonly called rTMS due to the repetitive stimuli delivery to the brain (Metin et al. 2020). According to Metin et al. (2020), low-frequency rTMS stimulations suppress cortical excitability, while high-frequency stimuli increase excitability. The use of ANNs (artificial neural networks) in psychiatric research is extensive. ANNs are composed of complex interconnected neurons (nodes), and each node consists of three elements such as inputs to the node (weight or strength), the summing function (weighted sum of inputs), and the activation function (limits amplitude range of output) (Metin et al. 2020). The training process of ANN algorithms is to input vectors into the network and keep iterating the process by adjusting synaptic weights for model learning (Metin et al. 2020). According to Metin et al. (2020), BP (backpropagation) algorithm is a commonly used learning algorithm in ANNs, which uses the gradient descent rule to minimize the prediction error in ANNs and nonlinear search spaces. Metin et al. (2020) studied the right frontal rTMS in OCD patients and predicted treatment response using qEEG (quantitative EEG). The study's results showed that patients' rTMS response was not significant in their gender or age (Metin et al. 2020).

FC (functional connectivity) matrix is computed by correlations between each region of the brain and all regions of the brain, which is also known as connectome (Xing et al. 2020). Conventional classification algorithms consider connectome as a vector and load into machine learning models

TABLE 3.4
Summary of Supervised and Unsupervised Machine Learning Approach Models

Computational Modeling Methods	Contribution to OCD Studies	Reference
Random Forest Classification	24 high-priority features were identified to predict outcomes of OCD symptoms	(Askland et al. 2015)
SVM, Random Forest, KNN, Logistic Regression	Characterizing the OCD's neuropsychological processes and accounting for sensory phenomena	(Stamatis et al. 2021)
ANNs with Backpropagation Algorithm	Effects of right frontal rTMS in OCD patients using quantitative EEG	(Metin et al. 2020)
Reimann Manifold PCA and XGBoost Classifier	Finding correlations between brain regions and drawing meaningful OCD predictions	(Xing et al. 2020)

such as decision trees and SVM (Xing et al. 2020). However, these methods may disturb the FC matrix's topological structure and can lose useful connectome information (Xing et al. 2020). Xing et al. (2020) hypothesized that OCD diagnosis and fMRI result analysis could be achieved using artificial intelligence algorithms. Xing et al. (2020) proposed a connectome decomposition algorithm such as Riemann manifold PCA (principal component analysis) for feature extraction of the FC matrix. As the features were extracted from Reimann manifold PCA, the XGBoost classifier was used to retrieve meaning predictions on OCD (Xing et al. 2020).

3.6 CONCLUSION

This chapter has discussed the possible applications of computational models in OCD. Though efficient doctors are good at the diagnosis of OCD, computational models play a vital role in analyzing and predicting the symptoms of OCD in the early stages. Proper use of technology and data analysis in the medical field can do a lot of favors. Various computational model approaches, such as the dynamical systems approach, Bayesian approach, reinforcement learning approach, and supervised and unsupervised machine learning approach, were discussed in this chapter. These approaches are used to tackle various OCD problems, which is groundbreaking in the field of neurology.

REFERENCES

Askland, Kathleen D., Sarah Garnaat, Nicholas J. Sibrava, Christina L. Boisseau, David Strong, Maria Mancebo, Benjamin Greenberg, Steve Rasmussen, and Jane Eisen. "Prediction of remission in obsessive compulsive disorder using a novel machine learning strategy." *International Journal of Methods in Psychiatric Research* 24, no. 2 (2015): 156–169. https://doi.org/10.1002/mpr.1463

Deserno, Lorenz, Quentin JM Huys, Rebecca Boehme, Ralph Buchert, Hans-Jochen Heinze, Anthony A. Grace, Raymond J. Dolan, Andreas Heinz, and Florian Schlagenhauf. "Ventral striatal dopamine reflects behavioral and neural signatures of model-based control during sequential decision making." *Proceedings of the National Academy of Sciences* 112, no. 5 (2015): 1595–1600. https://doi.org/10.1073/pnas.1417219112

Fradkin, Isaac, Rick A. Adams, Thomas Parr, Jonathan P. Roiser, and Jonathan D. Huppert. "Searching for an anchor in an unpredictable world: A computational model of obsessive compulsive disorder." *Psychological Review* 127, no. 5 (2020): 672. https://doi.org/10.1037/rev0000188

Gillan, Claire M., Michal Kosinski, Robert Whelan, Elizabeth A. Phelps, and Nathaniel D. Daw. "Characterizing a psychiatric symptom dimension related to deficits in goal-directed control." *Elife* 5 (2016): e11305. http://dx.doi.org/10.7554/eLife.11305

Kanen, Jonathan W., Karen D. Ersche, Naomi A. Fineberg, Trevor W. Robbins, and Rudolf N. Cardinal. "Computational modelling reveals contrasting effects on reinforcement learning and cognitive flexibility in stimulant use disorder and obsessive-compulsive disorder: Remediating effects of dopaminergic D2/3 receptor agents." *Psychopharmacology* 236, no. 8 (2019): 2337–2358. https://doi.org/10.1007/s00213-019-05325-w

Karamintziou, Sofia D., Nick G. Deligiannis, Brigitte Piallat, Mircea Polosan, Stephan Chabardes, Olivier David, Pantelis G. Stathis et al. "Dominant efficiency of nonregular patterns of subthalamic nucleus deep brain stimulation for Parkinson's disease and obsessive-compulsive disorder in a data-driven computational model." *Journal of Neural Engineering* 13, no. 1 (2015): 016013. https://doi.org/10.1088/1741-2560/13/1/016013

Metin, Sinem Zeynep, Tugçe Balli Altuglu, Baris Metin, Turker Tekin Erguzel, Selin Yigit, Mehmet Kemal Arıkan, and Kasif Nevzat Tarhan. "Use of EEG for predicting treatment response to transcranial magnetic stimulation in obsessive compulsive disorder." *Clinical EEG and Neuroscience* 51, no. 3 (2020): 139–145 . https://doi.org/10.1177/1550059419879569

Mills, Jeffrey A., and Jeffrey R. Strawn. "Antidepressant tolerability in pediatric anxiety and obsessive-compulsive disorders: A Bayesian hierarchical modeling meta-analysis." *Journal of the American Academy of Child & Adolescent Psychiatry* 59, no. 11 (2020): 1240–1251 . https://doi.org/10.1016/j.jaac.2019.10.013

Murray, Graham K., Franziska Knolle, Karen D. Ersche, Kevin J. Craig, Sanja Abbott, Shaila S. Shabbir, Naomi A. Fineberg et al. "Dopaminergic drug treatment remediates exaggerated cingulate prediction error responses in obsessive-compulsive disorder." *Psychopharmacology* 236, no. 8 (2019): 2325–2336.

Rabinovich, Mikhail I., and Pablo Varona. "Discrete sequential information coding: Heteroclinic cognitive dynamics." *Frontiers in Computational Neuroscience* (2018): 73. https://doi.org/10.3389/fncom.2018.00073

Rădulescu, Anca, and Rachel Marra. "A mathematical model of reward and executive circuitry in obsessive compulsive disorder." *Journal of Theoretical Biology* 414 (2017): 165–175 . https://doi.org/10.1016/j.jtbi.2016.11.025

Rolls, Edmund T. "Glutamate, obsessive–compulsive disorder, schizophrenia, and the stability of cortical attractor neuronal networks." *Pharmacology Biochemistry and Behavior* 100, no. 4 (2012): 736–751. https://doi.org/10.1016/j.pbb.2011.06.017

Schulze, Daniel, Norbert Kathmann, and Benedikt Reuter. "Getting it just right: A reevaluation of OCD symptom dimensions integrating traditional and Bayesian approaches." *Journal of Anxiety Disorders* 56 (2018): 63–73. https://doi.org/10.1016/j.janxdis.2018.04.003

Solway, Alec, Zhen Lin, and Ekansh Vinaik. "Transfer of information across repeated decisions in general and in obsessive–compulsive disorder." *Proceedings of the National Academy of Sciences* 118, no. 1 (2021): e2014271118. https://doi.org/10.1073/pnas.2014271117

Stamatis, Caitlin A., Marcelo C. Batistuzzo, Tais Tanamatis, Euripedes C. Miguel, Marcelo Q. Hoexter, and Kiara R. Timpano. "Using supervised machine learning on neuropsychological data to distinguish OCD patients with and without sensory phenomena from healthy controls." *British Journal of Clinical Psychology* 60, no. 1 (2021): 77–98. https://doi.org/10.1111/bjc.1227

Szalisznyó, Krisztina, and David N. Silverstein. "Computational Predictions for OCD Pathophysiology and Treatment: A Review." *Frontiers in Psychiatry* (2021): 1052 . https://doi.org/10.3389/fpsyt.2021.687062

Vaghi, Matilde M., Fabrice Luyckx, Akeem Sule, Naomi A. Fineberg, Trevor W. Robbins, and Benedetto De Martino. "Compulsivity reveals a novel dissociation between action and confidence." *Neuron* 96, no. 2 (2017): 348–354 . https://doi.org/10.1016/j.neuron.2017.09.006

Voon, Valerie, Katherine Derbyshire, Chistian Rück, Michael A. Irvine, Yulia Worbe, Jesper Enander, Liana R. N. Schreiber et al. "Disorders of compulsivity: A common bias towards learning habits." *Molecular Psychiatry* 20, no. 3 (2015): 345–352 . https://doi.org/10.1038/mp.2014.44

Xing, Xiaodan, Lili Jin, Qingfeng Li, Qiong Yang, Hongying Han, Chuanyong Xu, Zhen Wei et al. "Modeling essential connections in obsessive–compulsive disorder patients using functional MRI." *Brain and Behavior* 10, no. 2 (2020): e01499 . https://doi.org/10.1002/brb3.1499

4 Link Between Obsessive-Compulsive Disorder and Glycemic Control

Kartheek. R. Balapala[a] *and Balakrishnan Subramanian*[b]
[a] Physiological Sciences Unit, Department of Basic Medical Sciences, MCS School of Medicine, Copperbelt University, Zambia
[b] Physiological Sciences (Biochemistry) Unit, Department of Basic Medical Sciences, MCS School of Medicine, Copperbelt University, Zambia

4.1 BACKGROUND

Over the past two decades, the prevalence of diabetes has nearly doubled. This disorder is accompanied by physical and mental complications associated with raised morbidity and mortality rates of afflicted patients (1–3). During recent years, the concomitant existence of psychiatric disorders and chronic diseases has gained critical attention among biomedical researchers. Diabetes and its comorbid mental disorders have gained primal significance because of the need for daily and accurate care and control of health (4). Concomitant psychiatric disorders could further deteriorate the clinical condition of patients by upsetting the quality of self-care regimens and treatment compliance (5).

4.2 GLYCEMIC CONTROL AND OCD

A meta-analysis indicated that patients with diabetes are afflicted with depression at the rate of twice than healthy individuals (6). Uncontrolled diabetes mellitus could increase depression by directly affecting brain function and increasing its complications, leading to further impairment of function indirectly or reduced quality of life (7). Past research has shown that depression is accompanied by hyperglycemia among patients having type I and II diabetes (8), while other studies have not confirmed this finding (9). Other studies have shown that adaptive treatment of psychiatric disorders associated with diabetes could lead to better glycemic control (10, 11).

In contrast to depression, some studies have been done on the relationship between diabetes and anxiety (12). This is while anxiety is highly prevalent in the normal population (13, 14). On the other hand, somewhat controversial results were found with regard to glycemic control and anxiety. A positive relationship between anxiety and worsening of blood glucose levels (15–17), while some studies shown that anxiety is associated with better glycemic control. Nevertheless, results did not provide adequate evidence and clarity on relationship between diabetes and anxiety due to differences in protocols of anxiety assessment, diabetes assessment, and confounder control (12). To the best of the author's knowledge, there is a single study on the relationship between obsessive-compulsive symptoms and glycemic control in patients having type II diabetes, showing a significant relationship between these two variables (16). In a longitudinal study, the risk of type II diabetes increased by two folds in subjects with high levels of anxiety (18) and post-traumatic stress disorder (19, 20).

Glycemic control is associated with increased oxytocin levels, and increased oxytocin levels reduces obsessive-compulsive symptoms (17). A study done by Kontaoangelos et al. (21) found a direct relationship between homocysteine and the psychopathology of subjects with type II diabetes.

DOI: 10.1201/9781003387893-5

Those who controlled their blood glucose significantly had lower levels of serum homocysteine. Atmaca et al. (22) research indicated that homocysteine levels were positively correlated with severity of obsessive-compulsive symptoms in patients with OCD. A few patients with diabetes become obsessed and preoccupied with controlling their glucose levels after diagnosis. Obsessive behaviors like this do not lead to better glycemic control and rather worsen their metabolic profile (16).

4.3 INSULIN SIGNALING AND OCS

Research analysis found that compulsive symptoms in the general population overlap with the genetic liability for clinical OCD (23), and incorporating compulsive symptom; genome-wide association analyses [GWAS] data in a meta-analysis with OCD GWAS data revealed new findings about gene-based and gene enrichment analyses (23). Hence, genetic studies of OCS in the general population could aid in the identification of susceptibility loci for clinical OCD, giving insight in specific symptom domains affected by individual genetic risk factors.

In a study aiming to identify molecular mechanisms underlying OCD, researchers performed integration of the top-ranked results from the existing GWASs with genes implicated in OCD through other evidence. Then resulted in a "molecular landscape" suggesting that the involvement of genes regulating postsynaptic dendritic spine formation and function through central nervous system (CNS) insulin-dependent signaling (24). Support for a role of dysregulated insulin signaling in OCD and OCS disclosed from studies showing increased OCS in men with type 1 diabetes (25) and from similar study indicating that OCD patients have a higher risk of developing type 2 diabetes (T2D) (26). In addition, OCS were found to be positively correlated with blood levels of glycosylated hemoglobin (HbA1c), a diagnostic measure of T2D (27), and OCD patients had markedly higher levels of fasting glucose (a characteristic of T2D) (28).

Past study demonstrated that bilateral deep brain stimulation (DBS), a safe and effective treatment option for pharmaco-resistant OCD, not only reduced OCD symptoms, also reduced fasting insulin levels in the blood of both OCD patients with T2D and nondiabetic OCD patients (29). Insulin in the CNS—either entering from the periphery by crossing the blood brain barrier (30) or synthesized in the CNS (31)—has critical non-metabolic functions, including modulating synaptic plasticity (32) and learning and memory (33). Although it appears elusive what the relative contributions are of dysregulated peripheral and CNS insulin signaling to OCD and OCS, study demonstrated that compulsivity observed in Tallyho (TH) mice, a rodent model of T2D, is potentially linked to disturbances in insulin signaling. TH mice displayed compulsive behavior and increased glucose levels in their dorsomedial striatum, which could be due to alleviated action of peripheral and/or CNS insulin and the glucose levels correlated with compulsivity (34, 35). Altered insulin signaling is quite related to somatic disorders but is also involved in the etiology of psychiatric disorders and related symptoms in the population, with OCD and OCS. Further research is needed to disentangle the contributions of peripheral and CNS insulin production and signaling to these disorders and symptoms. Finding that insulin-related signaling contributes to the pathogenesis of OCD could lead to the development of novel treatments for OCD patients aimed at modulating (CNS) insulin levels or insulin-related signaling.

4.4 BRAIN GLUCOSE METABOLISM AND OCD

Functional brain imaging studies of patients with obsessive-compulsive disorder (OCD) have repeatedly found elevated cerebral glucose metabolism and blood flow in the orbitofrontal cortex (OFC), anterior cingulate cortex (ACC), basal ganglia and thalamus (36–42) that reduced in response to treatment with serotonin reuptake inhibitor (SRI) medications (40–50) or by cognitive-behavioral therapy (CBT) (44, 51, 52). These studies have led to the theory that the symptomatic expression of OCD is mediated by hyperactivity along frontal-subcortical circuits connecting the OFC, caudate nucleus, globus pallidus and the medial dorsal nucleus of the thalamus in the human brain (53, 54).

Clinical response of OCD symptoms needs up to 12 weeks of treatment with SRI medications and standard, weekly outpatient CBT (55). The response of OCD symptoms to SRI medications depends on the downregulation of terminal serotonin 1dβ receptors and subsequent increase in serotonin release in the OFC, needing nearly eight weeks of SRI administration. Not much is known about the brain mediation of response to CBT in OCD. Baxter and colleagues (44, 51), by doing positron emission tomography (PET) scans on a total of 18 OCD patients before and after 8–12 weeks of weekly CBT, found that the 12 patients who responded to treatment showed explicit pre- to posttreatment decreases in normalized right caudate glucose metabolism. A study of 22 treatment-refractory OCD patients with Xenon-enhanced computerized tomography before and after seven to eight months of CBT has revealed significant reductions in right caudate blood flow and smaller decreases in left frontal cortex and thalamus as well (52).

However, it is not explicit whether the functional changes in OFC, basal ganglia, and thalamus associated with response to prolonged treatment in OCD can be produced or accelerated by an intensive, short-term treatment. If those changes could be demonstrated after only a few weeks of starting the treatment, it would be a significant advance in our understanding of the cerebral mechanisms and time course of treatment response in OCD that gives clinical implications. The effectiveness of brief, intensive, daily CBT using exposure and response prevention (ERP) for OCD is well established, thus making as one of the standard, frontline treatments for OCD (56). Intensive CBT gives improvement in 60–80% of OCD patients in as little as four weeks, with symptom improvement ranging from 50 to 80% (57–59).

Rapid response of OCD to brief intensive CBT, mediated by a distinct pattern of functional neuroanatomical changes: decreases in thalamic activity accompanied by an increase in dorsal anterior cingulate cortex (dACC) activity, correlated with the degree of symptomatic improvement. Decreasing thalamic activity represents a common pathway to response of OCD symptoms to a variety of treatment modalities, while activation of the dACC, a mechanism of action required for response to CBT across mood and anxiety disorders. The patients with compulsive hoarding syndrome had a different pattern of cerebral glucose metabolism than both the nonhoarding OCD patients and the normal comparison subjects. Although the nonhoarding OCD patients showed expected elevations in regional metabolism in the caudate and thalamus in relation to the comparison subjects, the compulsive hoarders did not. Instead, the hoarders showed a unique pattern of lower metabolism in the posterior cingulate cortex and the occipital cortex (cuneus) in relation to the comparison subjects. The hoarders and the nonhoarding OCD patients also differed from each other, with the hoarders having significantly lower metabolism in the dorsal anterior cingulate gyrus and the thalamus than the nonhoarding OCD patients. The findings suggest that compulsive hoarding syndrome may be a neurobiologically distinct variant of OCD. The symptoms of compulsive hoarding syndrome may be mediated by diminished activity in several parts of the cingulate cortex. Across all OCD patients studied, hoarding severity was significantly correlated with lower activity in the dorsal part of the anterior cingulate gyrus, its cognitive subdivision (60). Functions of the anterior cingulate cortex include motivation, executive control, focused attention, assigning emotional valence to stimuli, monitoring response to conflict, emotional self-control, problem solving, detecting errors, and selecting responses (60, 61). The anterior cingulate also plays a key role in decision-making, especially in choosing between multiple conflicting options (62, 63).

Compulsive hoarders had significantly lower activity in the right posterior cingulate cortex than the comparison subjects. The posterior cingulate cortex is involved in the monitoring of visual events, spatial orientation, episodic memory (64), and the processing of emotional stimuli (65). It modulates activity in several brain regions involved in the putative functional neurocircuitry of OCD, including the orbitofrontal cortex, the anterior cingulate cortex, and the striatum (66). Thus, lower activity in both the anterior and posterior parts of the cingulate gyrus may mediate the remarkable difficulty in making decisions, attentional problems, and other cognitive deficits seen in compulsive hoarders. Compulsive hoarders were found to have lower glucose metabolism in brain regions specifically associated with response to treatment. Lower pretreatment activity of the anterior cingulate gyrus has been strongly associated with poor response to antidepressant treatment (67–70), while lower pretreatment activity in the posterior cingulate gyrus has been found to correlate with worse response to fluvoxamine (71) and cingulotomy (66) in patients with OCD. Thus, the finding of low cingulate gyrus activity in patients with the compulsive hoarding syndrome is quite consistent with its poor response to standard treatments for OCD. Lower activity in the anterior and posterior cingulate cortex may not only mediate the symptoms of the compulsive hoarding syndrome but also its poor response to treatment. The findings of low metabolism in the occipital cortex in compulsive hoarders was consistent with the results of prior imaging studies of OCD patients that found decreased metabolism (72, 73) and gray matter density (74) in this area. Abnormalities in parietal-occipital regions may be related to deficits in visuospatial processing and visual memory seen in some OCD patients (75, 76). Study done by Sanjaya Saxena et al. (77) suggests that these abnormalities may be preferentially found in OCD patients with compulsive hoarding syndrome. OCD patients with the compulsive hoarding syndrome were found to have a different pattern of cerebral glucose metabolism than nonhoarding OCD patients and comparison subjects. Obsessive-compulsive hoarding is a neurobiologically distinct subgroup or variant of OCD whose symptoms and low response to anti-obsessional treatment are mediated by lower activity in the cingulate cortex (77).

4.5 GENETIC CORRELATIONS OF PSYCHIATRIC TRAITS AND GLYCEMIC TRAITS

Psychiatric disorders are complex traits with a myriad of genetic variants that act in concert with environmental factors (78, 79). Genome-wide association studies (GWASs) of psychiatric disorders have identified over 300 independent genomic loci (80, 81), informed biological follow-up studies (82) and may deliver promising targets for drug discovery and repurposing (83–85). Genome-wide summary statistics generated by GWASs aid in multiple diverse ways (86), including estimating single-nucleotide polymorphism-based heritability (h2 SNP), which is the phenotypic variance elucidated by common genomic variants. Values of h2 SNP range from 10 to 30% for psychiatric disorders and typically capture nearly a third of the heritability estimated by twin studies (87). Additionally, genetic correlations can be calculated using GWAS summary statistics via bivariate linkage disequilibrium score regression (LDSC), which estimates the genetic overlap (i.e., the shared genetic effects) between two traits (88, 89). Such GWAS-based genetic correlation analyses have revealed substantial genetic overlap among psychiatric disorders (90), providing evidence for an underlying "p factor" representing general liability for psychiatric illness (91, 92). For instance, genomic structural equation modelling of GWAS summary statistics for schizophrenia, bipolar disorder, major depressive disorder, posttraumatic stress disorder and anxiety indicated that they load onto one shared latent factor with loading estimates between 0.29 and 0.86. However, marked differences in the clinical presentation of psychiatric disorders that still exist for psychotic experiences or dysfunctional reward systems suggest the influence of additional disorder-specific genetic effects (90, 91, 93). Clinically, most psychiatric disorders are accompanied by disturbances in appetite regulation, eating behavior with altered physical activity. Such disturbances can alter body composition, resulting in comorbid overweight or underweight (94), mostly seen in eating disorders, such as binge-eating disorder and anorexia nervosa (95). Such intense weight dysregulation typically reduces patients' quality of life, associated with excess morbidity and mortality. Body composition traits, including body fat % (BF%) and fat-free mass (FFM), are quite complicated, with substantial twin heritabilities of ~70% (96, 97). Body mass index (BMI) is the most commonly studied body composition phenotype and its associated genetic variants found to be significantly overrepresented in genes and genomic regions active in brain cell types, indicating it may be a partially behavioral trait. Past research indicated negative genetic correlations of BMI with anorexia nervosa and schizophrenia (89, 98–100) and positive genetic correlations of BMI with attention-deficit/hyperactivity disorder (ADHD), including major depressive disorder (101, 102). Such observations suggest the critical need for an in-depth investigation of the shared genomics between psychiatric and body composition traits. Extreme overweight and extreme underweight both showed an explicit sex difference: females are not only disproportionately affected by anorexia nervosa (with ratios up to 15:1) but also by obesity (≥30 kg/m^2) (103–105). Sex differences are not limited to body composition alone. Major depressive disorder (106) and anxiety (107) are more common in females, whereas ADHD (108) and autism spectrum disorder (109) occur frequently in males. Sex differences in body composition, psychiatry, and their interplay need to be more understood. Hormones and sex chromosomes have explicitly revealed to play a critical role (110) but are insufficient to fully explain the sex differences (112).

4.6 INSULIN SIGNALING IN BRAIN AND OCD

Type 2 diabetes mellitus (T2DM) is a widely prevalent and disabling condition worldwide. Diabetes is increasingly recognized as a frequent comorbid condition across psychiatric disorders like mood and psychotic disorders (113, 114). The relationship between the two being complex and bidirectional. Longitudinal population-based studies indicated double risk of cardiovascular diseases and type 2 diabetes in psychiatric patients with a diagnosis of schizophrenia, bipolar disorder, or depression relative to the general population, even after controlling done for the use of psychotropic medications (115). Opposingly, large-scale multicenter trials revealed moderate-severe depressive symptoms are present in up to 17% of diabetic patients, and 10% of diabetic patients found to have a diagnosis of major depression (116). Recent meta-analyses indicated bipolar patients have a double risk of T2DM relative to non-psychiatric populations, and around 20% of bipolar patients have a diagnosis of diabetes, while 30% recorded a pre-diabetes condition (117). Several hypotheses have been proposed to elucidate the link between mood disorders and diabetes highlighting a putative role of inflammation, hypothalamic-pituitary-adrenal (HPA)-axis dysregulation, gut microbiota alterations, dysregulated circadian rhythms, genetic and lifestyle factors (118). These factors have influence on glucose metabolism and insulin signaling. The role of insulin and insulin-signaling in the brain gained a lot of attention in the last years. Insulin has relevant non-metabolic functions in the brain, like neural survival, synaptic and dendritic plasticity, learning and memory, neural circuitries formation and dopaminergic transmission modulation in the striatum and brain stem areas (119–127). In order to exert such functions, insulin reaches the brain from the periphery through the blood-brain barrier binding to specific insulin receptors in both neurons and glial cells (8) and can directly alter white matter microstructures in both type 1 and 2 diabetic patients and adults with hyperglycemia (35). Needful literature on diabetes and mood disorders has grown in the last decades. Still, little is known about the link between diabetes, insulin signaling, and obsessive-compulsive disorder (OCD). Factors mediating mood disorders in diabetes, such as inflammation, HPA-axis dysregulation, and gut microbiota alterations, have been found to be associated to OCD (128–134). OCD has relation to dysfunctional dopaminergic transmission in the striatum and insulin modulated dopaminergic mesolimbic pathways and vice versa (119, 135).

4.6.1 Clinical Evidence of Link Between OCD and Insulin Signaling

A large international cross-sectional survey on DSM-IV mental disorders and a series of physical chronic diseases, indicated an association between only depression, eating disorders, intermittent explosive disorder, and type 2 diabetes (136). Nevertheless, with regards to the Swedish study, this survey was not longitudinal, the OCD sample was almost one quarter of the Swedish one, and the diabetes diagnosis was only retrospectively self-reported. A German survey on the general population did not find any correlation between OCD symptoms and diabetes but found that subjects self-reporting both OCD symptoms and physical diseases (including diabetes) reported a higher number of days of disability because of physical or psychological problems relative to subjects with only OCD symptoms (137). Indeed, sparse studies focused on OCD in diabetic patients. An early study in 1990 revealed a higher prevalence of obsessive-compulsive symptoms in insulin-dependent diabetic patients relative to healthy controls (41% vs. 21.3%) (138). A study done in pregnant women indicated higher rates of OC symptoms (a mean difference up to 40%) during pregnancy among type 1 diabetic women compared to women with gestational diabetes without a previous history of diabetes (139). A sample study comparing patients with a history of rheumatic fever (a population suggested to have higher rates of OCD by a subsequent larger study) and type 1 diabetic patients revealed lower but similar rates of OCD and OC symptoms across the two groups (140). A Norwegian epidemiological study on mental health in adolescents with diabetes did not find notable differences between OCD symptoms and other psychiatric disorders between subjects with or without type 1 diabetes (141). Nevertheless, this study failed to include a validated tool for the assessment of OC symptoms. Evidence of link between OCD and diabetes in the literature exists for type 2 diabetes. Study done on more than 9,000 Chinese adults found a higher prevalence of obsessions in newly diagnosed diabetic and pre-diabetic subjects relative to non-diabetic controls (odd ratios of 1.20–1.29) (142). A Brazilian study on insulin-dependent type 2 diabetic patients indicated a greater prevalence of anxiety disorders and OCD than healthy controls (with an odd ratio of 2.47 for OCD) (143). Two independent studies showed higher prevalence of OC symptoms (OCS) in uncontrolled with respect to controlled diabetic patients and a correlation between OCS and blood levels of HbA1c (a medium-term glycemic index with higher levels indicating an uncontrolled diabetes) (27, 144).

An Italian cross-sectional study revealed a 21.2% prevalence of metabolic syndrome (including the presence of obesity, hypertension, diabetes, and dyslipidemia) in a sample of 104 OCD patients (145). This prevalence rate was higher than those reported in the Italian general population, even when the confidence interval encompassed the general population estimate report. In this study, the presence of metabolic syndrome was linked with lifetime exposure to antidopaminergic drugs (145). A subsequent Italian study reported a similar rate of metabolic disorders (19.8% of metabolic syndrome and 6.2% of diabetes) in a sample of 162 OCD patients and a correlation between the presence of any medical condition and older age and the absence of any physical activity (146). A large longitudinal population-based study in Sweden on around 25,000 OCD patients followed for 40 years indicated that OCD patients have a higher risk (up to 45% higher) of obesity, type 2 diabetes mellitus, and circulatory system diseases compared to their relatives and the general population (odds ratio of 1.22 for T2DM). Similar results were noted when different groups of psychiatric comorbidities were excluded and the increased risk of metabolic and circulatory disease for OCD patients was evident from the beginning of the follow-up. Selective serotonin reuptake inhibitors (SSRIs) (with or without antidopaminergic augmentations) decreased the observed risk in a dose and time-dependent manner, in this study (147).

4.6.2 Biological Evidence for a Link Between Diabetes, Insulin Signaling, and OCD

4.6.2.1 Genetic Studies

In 2016, van de Vondervoort and colleagues aimed to identify the molecular mechanisms underlying OCD by integrating the top-ranked results from the existing GWAS association studies with genes implicated in OCD through other evidence (e.g., a gene associated with OCD in at least a human and an animal genetic study) (24). The researchers selected the top-ranked GWAS finding (89 genes presenting the most significant p values) and other 26 candidate genes (selected from other genetic studies) and therefore conducted a literature search in order to define the putative function and neuronal expression of these genes. This study results showed more than 40% of the proteins encoded by the selected genes (implicated in OCD pathophysiology) functionally interacted within a distinct "hypothetical" molecular landscape. This molecular OCD landscape contains a number of signaling cascades involved in the endogenous synthesis, secretion, and extracellular signaling of insulin in and around postsynaptic dendritic spines, and that regulated the formation of these spines, which in turn affects and regulates synaptic plasticity. In this molecular landscape, two additional signaling cascades center on receptors for the neurotransmitters glutamate and serotonin (both linked to OCD pathophysiology), and both cascades regulate the secretion of insulin having either autocrine or paracrine effects on (neighboring) neurons, leading to the modulation of dendritic spine formation. Hence this model suggests that OCD is associated to genes regulating postsynaptic dendritic spine formation and function through insulin-dependent serotonergic and glutamatergic signaling (24).

In line with this, a recent study by the same group identified a shared genetic risk between OCD/OCS and insulin-related traits (148). In this study, the scientists compared the polygenic risk scores of 51 CNS insulin-linked genes (extracted from the molecular landscape mentioned previously), five peripheral insulin-signaling related traits (presence of T2D, increased Hb1Ac levels, raised fasting insulin levels, increased fasting glucose levels, and increased level after 2 h of a glucose challenge) and OCD-related genes (extracted from previous large GWAS association studies) in a population-based sample of 650 children and adolescents and in a second replication sample of over 5,000 adolescents (aged 7–17 years) (148). The results revealed a correlation between both CNS and peripheral insulin-linked genes and OC symptoms. A shared genetic risk was noticed between two of the four T2D blood markers (increased blood levels of fasting insulin and increase 2 h glucose levels) and OCD (148).

4.6.2.2 Preclinical Studies

TALLYHO/JngJ (TH) mice are a reliable murine model for diabetes as these mice develop hyperglycemia, hyperinsulinemia, and enlargement of the islets of Langerhans in pancreas (149). In a recent study, TH mice showed heightened compulsive-like and anxiety behaviors (expressed by reduced spontaneous alternation behavior and decreased time spent in the open arm respectively) relative to control mice (35). These compulsive-like behaviors in TH mice were correlated to blood glucose levels. TH mice exhibited increased levels of glucose in the dorsal-medial striatum (DMS) as measured with magnetic resonance spectroscopy (MRS) and levels correlated with compulsive-like behaviors like the higher glucose levels correlated with reduced spontaneous alternation behavior. Diffusion tensor imaging (DTI) showed altered structural connectivity (reduced fractional anisotropy) in the DMS (and other brain regions) and correlated to behavioral flexibility, and hence suggesting the hypothesis that increased glucose levels in DMS may increase compulsive-like behaviors through altered structural connectivity (despite the studies did not report any correlation between MRS and DTI results). TH mice had higher levels of glutathione, production of which is stimulated by insulin, in the anterior cingulate cortex (a region previously implicated in OCD). Last but not least, the researchers found decreased levels of insulin grow factor 1 (IGF-1) in the mice's cerebellum and increased IGF-1 blood levels (35). This latter result is specifically intriguing in the light of two recent studies showing increased blood levels of IGF-1 in adult OCD patients relative to controls and a positive correlation between baseline IGF-1 levels and SSRIs response in drug-naïve OCD patients (150, 151).

Another research study indicated streptozotocin (STZ)-induced diabetic mice show higher compulsive behaviors (expressed as increased marble burying behaviors) compared to non-diabetic mice (35). Treatment with combined metformin and genistein (an isoflavonoid phytoestrogen known to have neuroprotective, anti-inflammatory, and antidiabetic effects) significantly reduced compulsive behaviors to a comparable degree, as did fluoxetine (152).

4.6.2.3 Deep Brain Stimulation Studies

In 2018, a study in OCD patients undergoing deep brain stimulation (DBS) of the ventral anterior limb of the internal capsule (vALIC) emphasized a putative role of striatal dopamine transmission in glucose metabolism and insulin sensitivity (153). This study started from the observation of increased peripheral insulin sensitivity in an OCD patient with insulin-dependent diabetes after vALIC-DBS. Later, the researchers found that vALIC-DBS increased hepatic and peripheral insulin sensitivity in a sample of 14 non-diabetic OCD patients. These effects could be potentially related to an increase of dopamine transmission in the striatum. Indeed, it was revealed that vALIC-DBS increased striatal dopamine transmission as confirmed by reduced dopamine D2/3 receptor availability in the striatum after acute and chronic DBS (154). Insulin sensitivity was assessed in a group of healthy controls after the induction of an acute dopamine depletion, through a pharmacological inhibition of tyrosine hydroxylase, the rate-limiting enzyme of endogenous dopamine synthesis. Acute dopamine depletion reduced peripheral insulin sensitivity in this sample of healthy subjects supporting the role of central dopamine in insulin sensitivity. Such findings were further corroborated by a preclinical investigation in which optogenetic activation of dopamine D1 receptor–expressing neurons in the nucleus accumbens (NAcc) increased glucose tolerance and insulin sensitivity in mice (132). Altogether, findings pinpoint toward a central role of striatal dopamine transmission in glucose metabolism and insulin sensitivity. A potential mechanism for NAcc modulation of glucose metabolism seems to be NAcc-hypothalamus interaction. DBS of the shell of the NAcc increases peripheral glucagon and glucose levels in mice that is coupled with an activation in lateral hypothalamic area (155).

4.6.3 Hypotheses Linking Diabetes, Insulin Signaling, and OCD

Several factors displayed a potential link between diabetes, insulin-signaling, and OCD.

Lifestyle factors associated with OCD potentially played a role in the emergence of metabolic disorders. For example, reduced physical activity, disruption of circadian rhythms, and sleep disorders are associated to OCD with a negative impact on glucose metabolism and insulin sensitivity (156, 157). In line with this, research studies showed a correlation between absence of physical activity and the presence of a general medical condition (including metabolic disorders) (146). Mood and anxiety comorbidities were common in OCD and have been associated to diabetes and impaired insulin-signaling, both directly and indirectly through changes in lifestyle habits (112, 117, 158). Unfortunately, not all of the reviewed studies controlled for these factors, suggesting that these hypotheses should be addressed in future studies.

Other several biological factors could play a role in mediating insulin-signaling alterations in OCD. First, according to human and preclinical studies, insulin is known to modulate dopaminergic transmission in the mesolimbic circuitry and vice versa (119). OCD is associated to impaired dopaminergic transmission that impaired insulin sensitivity similar to what has been found in healthy controls (135, 153). In line with this hypothesis, non-overweight OCD patients treated with vALIC-DBS, known to increase dopaminergic transmission in the striatum, revealed increased insulin sensitivity after treatment (153). Additionally, dopamine blockade induced by antidopaminergic medications for OCD could also contribute to dysregulated insulin-signaling (145). Secondly, the presence of inflammation, HPA-axis dysregulation, and gut microbiota alteration, factors that all have been related to OCD, have also been related to insulin-signaling dysregulation with impaired peripheral insulin sensitivity through different mechanisms (128, 133, 134, 157, 159–163). So stress-related activation of the HPA-axis altered the insulin-signaling (through the release of cortisol) and to induce inflammatory response, as well as inflammation, has been linked to both HPA-axis dysregulation and insulin resistance (157, 162). Alteration of gut microbiota can also play a critical role in this triangle by increasing systemic inflammation and altering insulin-signaling and brain functions (163). Although the role of inflammation, HPA-axis dysregulation, and gut microbiota alteration on diabetes and metabolic disorders' pathophysiology has been largely studied (164), their potential role in OCD pathophysiology and neuroinflammation remains speculative and unclear. However, it could have several potential treatment implications. For example, it's quite captivating to comprehend if anti-inflammatory agents (e.g., non-steroidal anti-inflammatory drugs, NSAIDs, that show preliminary positive effects in OCD trials) may induce a marked improvement on OCD symptoms in the context of a metabolic illness (159). It will be fascinating to investigate the potential anti-obsessive effects of anti-diabetic drugs. In this updated perspective, one animal study indicated an anti-compulsive effect of metformin, and at least one case report in the literature showed improvement of OC-like symptoms in an autistic patient after treatment with liraglutide, a glucagon-like peptide-1 analog (165). Alternatively, future medical research might explore the effects of cognitive-behavioral therapy (CBT) on metabolic features in OCD patients, as data from samples with chronic disorders revealed that CBT outcomes may be associated with an improvement in the function of the immune system (166, 167) and brain-gut-microbiome axis (168).

4.6.4 Clinical Implications and Open Issues

As a whole, the epidemiological, clinical, and animal studies converge toward a link between insulin signaling, glucose metabolism, and OCD. This preliminary evidence makes a relevant point that has been partly neglected in the literature as follows:

The relevance of metabolic assessments in OCD patients: Clinicians should assess familial risk of metabolic disorders and monitor metabolic parameters like body mass index, blood glucose levels, dietary habits, physical activity, and sleep patterns during the course of treatment, specifically when prescribing antidopaminergic agents, in order to understand and prevent their impact on OCD severity and patient trajectory. Literature again confirms the importance of providing OCD patients with proper treatment. While the metabolic impact of SSRIs is still debated, with multi-studies showing a negative metabolic impact (169), OCD patients treated with medium-high doses of SSRIs showed a reduced metabolic risk over time, suggesting at least an indirect positive metabolic effect of a proper pharmacological treatment of the disorder. As for example, inducing lifestyle changes that follow a sustained symptom reduction. vALIC-DBS, an established treatment for refractory patients, showed a positive effect on insulin sensitivity and glucose metabolism. Some studies have demonstrated that physical activity (mainly moderate aerobic activity) significantly reduced OCD symptoms at least in the short term (170, 171). However, the available literature also showed multiple limitations and open issues. First of all, the studies did not control for comorbidities like mood and anxiety disorders which are linked to diabetes. For instance, one large epidemiological study still found significant increased *odds ratios* of diabetic risk even after removing most other disorders (including mood disorders) but not when removing anxiety disorders, that are highly comorbid with OCD. Data on the prevalence of food addiction as a proposed behavioral addiction in OCD are largely lacking in the current literature, and none of the presented studies assessed behavioral addictions as moderator factors. Moreover, except for one study showing a correlation between the absence of physical activity and the presence of medical conditions, not many studies analyzed

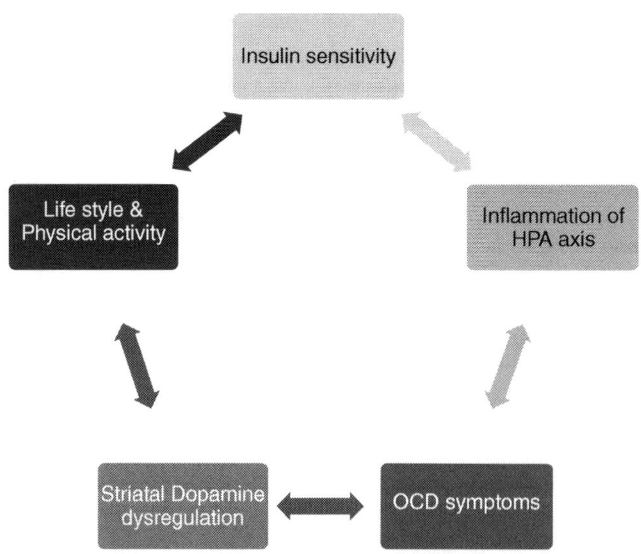

FIGURE 4.1 Factors affecting insulin sensitivity.

the impact of lifestyle habits, such as the level of daily physical activity, sleep patterns, and diet, on metabolic profiles of patients with OCD. An issue regarding the temporal relationship between OCD and diabetes as research study indicated higher OCD symptoms in uncontrolled respect to controlled diabetes, but no prospective study investigated a potential correlation between improved diabetic control and improved OCD symptoms over time (143, 144). Finally, little is known about how other clinical characteristics, for example, illness duration, age at onset, duration of untreated illness, symptom dimensions, moderate the link with diabetes and how the presence of diabetes could affect OCD contents, for example, inducing compulsive checking of glycemic parameters (144). Future research based on prospective designs with long-term follow-up assessments are needed to clarify the role of comorbidities, lifestyle habits, biological and genetic factors of inflammatory status and familial risk factors, and medications like antidopaminergic agents in the emergence of insulin-signaling dysregulation in OCD patients.

4.7 BRAIN CIRCUIT INVOLVED IN SUGAR ADDICTION

Although obesity and type 2 diabetes are major problems in human society, most treatments do not tackle the primary cause: unhealthy eating habits, says Kay Tye of the Massachusetts Institute of Technology. His research findings are exciting since they raise the possibility for developing a treatment option that selectively curbs compulsive overeating without altering healthy eating behavior. Compulsive overeating, a type of reward-seeking behavior, akin to drug addiction. A major difference between the two behaviors is that eating is required for survival, underscoring the need to tease apart brain circuits involved in compulsive overeating versus normal feeding to create safe and effective therapies. Tye and her team speculated a neural pathway from the lateral hypothalamus to the ventral tegmental area played an important role in compulsive overeating because these brain regions have been implicated in reward related behaviors such as eating, sexual activity, and drug addiction. Tye and her team used a technique called optogenetics, involving genetically modifying specific populations of neurons to express light-sensitive proteins that control neural excitability, and then delivering either blue or yellow light through an optic fiber to activate or inhibit those cells, respectively. Activation of the pathway from the lateral hypothalamus to the ventral tegmental area caused well-fed mice to spend greater time feeding and increased the number of times mice poked their nose into a port that gave sugar reward, even when they had to cross a platform that delivered foot shocks to get to the reward. In opposition, inhibition of the same pathway reduced this compulsive sugar-seeking behavior with no decrease in food consumption in hungry mice, suggesting that different neural circuits control feeding in hungry animals (172).

Garret Stuber from the University of North Carolina School of Medicine and his team also used an optogenetic approach in mice to identify neurons in the lateral hypothalamus that control both feeding and reward-seeking behavior. By imaging the activity of hundreds of individual lateral hypothalamus neurons as the mice freely explored an area with food or worked to obtain a sweet reward, they further discovered a distinct subset of neurons that either mediated food-seeking behavior or responded to reward consumption (173). Brain circuits evolved to support binging on scarce, sugary foods whenever these valuable sources of energy become transiently available during some seasons. It might be adaptive for separate neural circuits to drive hungry animals to eat whatever type of food is available but to consume less overall to ration out limited resources, in the winter months. In our modern-day society, there is no scarcity of palatable foods, and high-sugar or high-fat foods are much more available than fresh produce or proteins. Humans have not yet adapted to a world where there is an overabundance of sugar. These circuits that drive human brain to stuff ourselves with sweets are now undermining the birth of novel health issues. The discovery of a specific neural circuit underlying compulsive sugar consumption could pave the way for the development of targeted drug therapies to effectively treat this major issue. Compulsive overeating and sugar addiction are critical threats to human health. Potential treatments face the risk of impairing normal feeding behaviors that are crucial for human survival. A new research revealed a reward-related neural circuit that specifically controls compulsive sugar consumption in mice without preventing feeding necessary for survival, providing a novel target for the safe and effective treatment of compulsive overeating in humans (174).

REFERENCES

1. Zimmet, P.Z., Alberti, K.G.M.M. 2016. Epidemiology of diabetes—status of a pandemic and issues around metabolic surgery. *Diabetes Care* 39(6):878–83. doi: 10.2337/dc16-0273.
2. Mundet, X., Pou, A., Piquer, N., Sanmartin, M.I.F., Tarruella, M., Gimbert, R., et al. 2008. Prevalence and incidence of chronic complications and mortality in a cohort of type 2 diabetic patients in Spain. *Prim. Care Diabetes* 2(3):135–40. doi: 10.1016/j.pcd.2008.05.001.
3. Song, S.H., Hardisty, C.A. 2008. Type 2 diabetes mellitus: A high-risk condition for cardiovascular disease irrespective of the different degrees of obesity. *QJM* 101(11):875–9. doi: 10.1093/qjmed/hcn109.
4. Zimmet, P.Z., Magliano, D.J., Herman, W.H., Shaw, J.E. 2014. Diabetes: A 21st century challenge. *Lancet Diabetes Endocrinol* 2(1):56–64. doi: 10.1016/S2213-8587(13)70112-8. [PubMed: 24622669].
5. Tuomi, T., Santoro, N., Caprio, S., Cai, M., Weng, J., Groop, L. 2014. The many faces of diabetes: A disease with increasing heterogeneity. *Lancet* 383(9922):1084–94. doi: 10.1016/s0140-6736(13)62219-9.
6. Anderson, R.J., Freedland, K.E., Clouse, R.E., Lustman, P.J. 2006. The prevalence of comorbid depression in adults with diabetes: A meta-analysis. *Diabetes Care* 24(6):1069–78. doi: 10.1111/j.1464-5491.2006.01943.x.
7. Chen, L., Magliano, D.J., Zimmet, P.Z. 2011. The worldwide epidemiology of type 2 diabetes mellitus—present and future perspectives. *Nat Rev Endocrinol.* 8(4):228–36. doi: 10.1038/nrendo.2011.183.
8. Lustman, P.J., Anderson, R.J., Freedland, K.E., de Groot, M., Carney, R.M., Clouse, R.E. 2000. Depression and poor glycemic control: A meta-analytic review of the literature. *Diabetes Care* 23(7):934–42. [PubMed: 10895843].
9. Trief, P.M., Grant, W., Elbert, K., Weinstock, R.S. 1998. Family environment, glycemic control, and the psychosocial adaptation of adults with diabetes. *Diabetes Care* 21(2):241–5. [PubMed: 9539989].
10. Lustman, P.J., Griffith, L.S., Clouse, R.E., Freedland, K.E., Eisen, S.A., Rubin, E.H., et al. 1997. Effects of nortriptyline on depression and glycemic control in diabetes: Results of a double-blind, placebo-controlled trial. *Psychosom Med* 59(3):241–50. [PubMed: 9178335].
11. Lustman, P.J., Griffith, L.S., Freedland, K.E., Kissel, S.S., Clouse, R.E. 1998. Cognitive behavior therapy for depression in type 2 diabetes mellitus. A randomized, controlled trial. *Ann. Intern. Med* 129(8):613–21. [PubMed: 9786808].
12. Smith, K.J., Beland, M., Clyde, M., Gariepy, G., Page, V., Badawi, G., et al. 2013. Association of diabetes with anxiety: A systematic review and meta-analysis. *J. Psychosom Res* 74(2):89–99. doi: 10.1016/j.jpsychores.2012.11.013.
13. Kessler, R.C., Berglund, P., Demler, O., Jin, R., Merikangas, K.R., Walters, E.E. 2005. Lifetime prevalence and age-of-onset distributions of DSM-IV disorders in the national comorbidity survey replication. *Arch. Gen. Psychiatry* 62(6):593. doi: 10.1001/archpsyc.62.6.593.
14. Leray, E., Camara, A., Drapier, D., Riou, F., Bougeant, N., Pelissolo, A., et al. 2011. Prevalence, characteristics and comorbidities of anxiety disorders in France: Results from the "mental health in general population" Survey (MHGP). *Eur. Psychiatry* 26(6):339–45. doi: 10.1016/j.eurpsy.2009.12.001.
15. Balhara, Y.P.S., Sagar, R. 2011. Correlates of anxiety and depression among patients with type 2 diabetes mellitus. *Indian J. Endocrinol. Metab* 15(5):50. doi: 10.4103/2230-8210.83057.
16. Kontoangelos, K., Raptis, A.E., Papageorgiou, C.C., Papadimitriou, G.N., Rabavilas, A.D., Dimitriadis, G., et al. 2012. The association of the metabolic profile in diabetes mellitus type 2 patients with obsessive-compulsive symptomatology and depressive symptomatology: New insights. *Int. J. Psychiatry Clin. Pract* 17(1):48–55. doi: 10.3109/13651501.2012.697563.
17. Kontoangelos, K., Raptis, A.E., Papageorgiou, C.C., Tsiotra, P.C., Papadimitriou, G.N., Rabavilas, A.D., et al. 2012. Oxytocin and psychological factors affecting type 2 diabetes mellitus. *Exp. Diabetes Res* 2012:1–7. doi: 10.1155/2012/560864.
18. Engum, A. 2007. The role of depression and anxiety in onset of diabetes in a large population-based study. *J. Psychosom Res* 62(1):31–8. doi: 10.1016/j.jpsychores.2006.07.009.
19. Agyemang, C., Goosen, S., Anujuo, K., Ogedegbe, G. 2011. Relationship between post-traumatic stress disorder and diabetes among 105 180 asylum seekers in the Netherlands. *Eur. J. Public Health* 22(5):658–62. doi: 10.1093/eurpub/ckr138.

20. Boyko, E.J., Jacobson, I.G., Smith, B., Ryan, M.A.K., Hooper, T.I., Amoroso, P.J., et al. 2010. Risk of diabetes in U.S. Military service members in relation to combat deployment and mental health. *Diabetes Care* 33(8):1771–7. doi: 10.2337/dc10-0296.
21. Kontoangelos, K., Papageorgiou, C.C., Raptis, A.E., Tsiotra, P., Lambadiari, V., Papadimitriou, G.N., et al. 2015. Homocysteine, cortisol, diabetes mellitus, and psychopathology. *J. Diabetes Res* 2015:1–10. doi: 10.1155/2015/354923.
22. Atmaca, M., Tezcan, E., Kuloglu, M., Kirtas, O., Ustundag, B. 2005. Serum folate and homocysteine levels in patients with obsessive-compulsive disorder. *Psychiatry Clin. Neurosci* 59(5):616–20. doi: 10.1111/j.1440-1819.2005.01425.x. [PubMed: 16194269]
23. Smit, D.J.A., et al. 2019. Genetic meta-analysis of obsessive-compulsive disorder and self-report compulsive symptoms. *Am. J. Med. Genet. B Neuropsychiatr. Genet.* https://doi.org/10.1002/ajmg.b.32777. [Epub ahead of print].
24. van de Vondervoort, I., et al. 2016. An integrated molecular landscape implicates the regulation of dendritic spine formation through insulin-related signalling in obsessive-compulsive disorder. *J. Psychiatry Neurosci* 41:280–5.
25. Winocour, P.H., Main, C.J., Medlicott, G., Anderson, D.C. 1990. A psychometric evaluation of adult patients with type 1 (insulin-dependent) diabetes mellitus: Prevalence of psychological dysfunction and relationship to demographic variables, metabolic control and complications. *Diabetes Res* 14:171–6.
26. Isomura, K., et al. 2018. Metabolic and cardiovascular complications in obsessivecompulsive disorder: A total population, sibling comparison study with longterm follow-up. *Biol. Psychiatry* 84:324–31.
27. Kontoangelos, K., et al. 2013. The association of the metabolic profile in diabetes mellitus type 2 patients with obsessive-compulsive symptomatology and depressive symptomatology: New insights. *Int. J. Psychiatry Clin. Pract* 17:48–55.
28. Albert, U., Aguglia, A., Chiarle, A., Bogetto, F., Maina, G. 2013. Metabolic syndrome and obsessive-compulsive disorder: A naturalistic Italian study. *Gen. Hospital Psychiatry* 35:154–9.
29. Ter Horst, K.W., et al. 2018. Striatal dopamine regulates systemic glucose metabolism in humans and mice. *Sci. Transl. Med* 10, eaar3752.
30. Margolis, R.U., Altszuler, N. 1967. Insulin in the cerebrospinal fluid. *Nature* 215:1375–6.
31. Clarke, D.W., Mudd, L., Boyd, F.T. Jr., Fields, M., Raizada, M.K. 1986. Insulin is released from rat brain neuronal cells in culture. *J. Neurochem* 47:831–6.
32. Chiu, S.L., Chen, C.M., Cline, H.T. 2008. Insulin receptor signaling regulates synapse number, dendritic plasticity, and circuit function in vivo. *Neuron* 58:708–19.
33. Dou, J.T., Chen, M., Dufour, F., Alkon, D.L., Zhao, W.Q. 2005. Insulin receptor signaling in long-term memory consolidation following spatial learning. *Learn Mem* 12:646–55.
34. Zhao, W., et al. 1999. Brain insulin receptors and spatial memory. Correlated changes in gene expression, tyrosine phosphorylation, and signaling molecules in the hippocampus of water maze trained rats. *J. Biol. Chem* 274:34893–902.
35. van de Vondervoort, I., et al. 2019. Converging evidence points towards a role of insulin signaling in regulating compulsive behavior. *Transl. Psychiatry* 9, 225.
36. Baxter, L.R., Phelps, M.E., Mazziotta, J.C., Guze, B.H., Schwartz, J.M., Selin, C.E. 1987. Local cerebral glucose metabolic rates in obsessive-compulsive disorder—a comparison with rates in unipolar depression and in normal controls. *Arch. Gen. Psychiatry* 44:211–18. [PubMed: 3493749]
37. Nordahl, T.E., Benkelfat, C., Semple, W.E., Gross, M., King, A.C., Cohen, R.M. 1989. Cerebral glucose metabolic rates in obsessive-compulsive disorder. *Neuropsychopharmacology* 2:23–28. [PubMed: 2803479]
38. Swedo, S., Schapiro, M.G., Grady, C.L., Cheslow, D.L., Leonard, H.L., Kumar, A., et al. 1989. Cerebral glucose metabolism in childhood onset obsessive-compulsive disorder. *Arch. Gen. Psychiatry* 46:518–23. [PubMed: 2786402]
39. Sawle, G.V., Hymas, N.F., Lees, A.J., Frackowiak, R.S. 1991. Obsessional slowness: Functional studies with positron emission tomography. *Brain* 114:2191–202. [PubMed: 1933241]
40. Perani, D., Colombo, C., Bressi, S., Bonfanti, A., Grassi, F., Scarone, S., et al. 1995. [18F]FDG PET study in obsessive-compulsive disorder: A clinical/metabolic correlation study after treatment. *Br. J. Psychiatry* 166:244–50. [PubMed: 7728370]
41. Saxena, S., Brody, A.L., Ho, M.L., Alborzian, S., Ho, M.K., Maidment, K.M., et al. 2001. Cerebral metabolism in major depression and obsessive-compulsive disorder occurring separately and concurrently. *Biol. Psychiatry* 50:159–70. [PubMed: 11513814]
42. Kwon, J.S., Kim, J.J., Lee, D.W., Lee, J.S., Lee, D.S., Kim, M.S., et al. 2003. Neural correlates of clinical symptoms and cognitive dysfunctions in obsessive-compulsive disorder. *Psychiatry Res: Neuroimaging* 122:37–47.

43. Benkelfat, C., Nordahl, T.E., Semple, W.E., King, A.C., Murphy, D.L., Cohen, R.M. 1990. Local cerebral glucose metabolic rates in obsessive-compulsive disorder: Patients treated with clomipramine. *Arch. Gen. Psychiatry* 47:840–8. [PubMed: 2393342]
44. Baxter, L.R., Schwartz, J.M., Bergman, K.S., Szuba, M.P., Guze, B.H., Mazziota, J.C., et al. 1992. Caudate glucose metabolic rate changes with both drug and behavior therapy for obsessive-compulsive disorder. *Arch. Gen. Psychiatry* 49:681–9. [PubMed: 1514872]
45. Swedo, S.E., Pietrini, P., Leonard, H.L., Schapiro, M.B., Rettew, D.C., Goldberger, E.L., et al. 1992. Cerebral glucose metabolism in childhood-onset obsessive-compulsive disorder: Revisualization during pharmacotherapy. *Arch. Gen. Psychiatry* 49:690–4. [PubMed: 1514873]
46. Saxena, S., Brody, A.L., Ho, M.L., Alborzian, S., Maidment, K.M., Zohrabi, N., et al. 2002. Differential cerebral metabolic changes with paroxetine treatment of obsessive-compulsive disorder versus major depression. *Arch. Gen. Psychiatry* 59:250–61. [PubMed: 11879163]
47. Hansen, E.S., Hasselbach, S., Law, I., Bolwig, T.G. 2001. The caudate nucleus in obsessive-compulsive disorder. Reduced metabolism following treatment with paroxetine: A PET study. *Int J Neuropsychopharmacol* 5:1–10. [PubMed: 12057027] Saxena et al. Page 9 Mol Psychiatry. Author manuscript; available in PMC 2010 June 29.
48. Kang, D.H., Kwon, J.S., Kim, J.J., Youn, T., Park, H.J., Kim, M.S., et al. 2003. Brain glucose metabolic changes associated with neuropsychological improvements after 4 months of treatment in patients with obsessive-compulsive disorder. *Acta Psychiatr Scand* 107:291–7. [PubMed: 12662252]
49. Diler, R.S., Kibar, M., Ayse, A. 2004. Pharmacotherapy and regional cerebral blood flow in children with obsessive-compulsive disorder. *Yonsei. Med. J* 45:90–9. [PubMed: 15004874]
50. Ho Pian, K.L., van Megan, H.J.G.M., Ramsy, N.F., Mandl, R., van Rijk, P.P., Wynee, H.J., et al. 2005. Decreased thalamic blood flow in obsessive-compulsive disorder patients responding to fluvoxamine. *Psychiatry Res: Neuroimaging* 138:89–97.
51. Schwartz, J.M., Stoessel, P.W., Baxter, L.R., Martin, K.M., Phelps, M.E. 1996. Systematic changes in cerebral glucose metabolic rate after successful behavior modification treatment of obsessive-compulsive disorder. *Arch. Gen. Psychiatry* 53:109–13. [PubMed: 8629886]
52. Nakatani, E., Nakgawa, A., Ohara, Y., Goto, S., Uozumi, N., Iwakiri, M., et al. 2003. Effects of behavior therapy on regional cerebral blood flow in obsessive-compulsive disorder. *Psychiatry Res: Neuroimaging* 124:113–20.
53. Alexander, GE., DeLong, M.R., Strick, P.L. 1986. Parallel organization of functionally segregated circuits linking basal ganglia and cortex. *Ann. Rev. Neurosci* 9:357–81. [PubMed: 3085570]
54. Saxena, S., Brody, A.L., Schwartz, J.M., Baxter, L.R. 1998. Neuroimaging and frontal-subcortical circuitry in obsessive-compulsive disorder. *Br. J. Psychiatry* 173(Suppl 35):26–38.
55. Greist, J.H., Jefferson, J.W., Kobak, K.A., Katzelnick, D.J., Serlin, R.C. 1995. Efficacy and tolerability of serotonin transport inhibitors in obsessive-compulsive disorder. A meta-analysis. *Arch. Gen. Psychiatry* 52:53–60.
56. Marks, I. 1997. Behaviour therapy for obsessive-compulsive disorder: A decade of progress. *Can. J. Psychiatry* 42:1021–7. [PubMed: 9469234]
57. Foa, E.B., Goldstein, A. 1978. Continuous exposure and complete response prevention in the treatment of obsessive-compulsive neurosis. *Behav. Ther* 9:821–29.
58. Lindsay, M., Crino, R., Andrews, G. 1996. Controlled trial of exposure and response prevention in obsessivecompulsive disorder. *Br. J. Psychiatry* 171:135–9. [PubMed: 9337948]
59. Foa, E.B., Liebowitz, M.R., Kozak, M.J., Davies, S., Campeas, R., Franklin, M.E., et al. 2005. Randomized, placebocontrolled trial of exposure and ritual prevention, clomipramine, and their combination in the treatment of obsessive-compulsive disorder. *Am. J. Psychiatry* 162:151–61. [PubMed: 15625214]
60. Devinsky, O., Morrell, M.J., Vogt, B.A. 1995. Contributions of anterior cingulate cortex to behavior. *Brain* 118:279–306.
61. Awh, E., Gehring, W.J. 1999. The anterior cingulate lends a hand in response selection. *Nature Neurosci* 2:853–4.
62. Carter, C.S., Braver, T.S., Barch, D.M., Botvinik, M.M., Noll, D., Cohen, J. 1998. Anterior cingulate cortex, error detection, and online monitoring of performance. *Science* 280:747–9.
63. Krawczyk, D.C. 2002. Contributions of the prefrontal cortex to the neural basis of decision making. *Neurosci Biobehav Rev* 26:631–64.
64. Vogt, B.A., Finch, D.M., Olson, C.R. 1992. Functional heterogeneity in the cingulate cortex: The anterior executive and posterior evaluative regions. *Cereb Cortex* 2:435–43.
65. Maddock, R.J. 1999. The retrosplenial cortex and emotion: New insight from functional neuroimaging of the human brain. *Trends Neurosci* 22:310–16.

66. Rauch, S.L., Dougherty, D.D., Cosgrove, G.R., Cassem, E.H., Alpert, N.M., Price, B.H., Nierenberg, A.A., Mayberg, H.S., Baer, L., Jenike, M.A., Fischman, A.J. 2001. Cerebral metabolic correlates as potential predictors of response to anterior cingulotomy for obsessive-compulsive disorder. *Biol. Psychiatry* 50:659–67.
67. Mayberg, H.S., Brannan, S.K., Mahurin, R.K., Jerabek, P.A., Brickman, J.S., Tekell, J.L., Silva, J.A., McGinnis, S., Glass, T.G., Martin, C.C., Fox, P.T. 1997. Cingulate function in depression: A potential predictor of treatment response. *Neuroreport* 8:1057–61.
68. Wu, J., Buchsbaum, M.S., Gillin, J.C., Tang, C., Cadwell, S., Wiegand, M., Najafi, A., Klein, E., Hazen, K., Bunney, W.E. Jr., Fallon, J.H., Keator, D. 1999. Prediction of antidepressant effects of sleep deprivation by metabolic rates in the ventral anterior cingulate and medial prefrontal cortex. *Am. J. Psychiatry* 156:1149–58; correction, 1999; 156:1666.
69. Pizzagalli, D., Pascual-Marqui, R.D., Nitschke, J.B., Oakes, T.R., Larson, C.L., Abercrombie, H.C., Schaefer, S.M., Koger, J.V., Benca, R.M., Davidson, R.J. 2001. Anterior cingulate activity as a predictor of degree of treatment response in major depression: Evidence from brain electrical tomography analysis. *Am. J. Psychiatry* 158: 405–15.
70. Saxena, S., Brody, A.L., Ho, M.L., Zohrabi, N., Maidment, K.M., Baxter, L.R. Jr. 2003. Differential brain metabolic predictors of response to paroxetine in obsessive-compulsive disorder versus major depression. *Am. J. Psychiatry* 160:522–32.
71. Mataix-Cols, D., Cullen, S., Lange, K., Zelaya, F., Andrew, C., Amaro, E., Brammer, M.J., Williams, S.C.R., Speckens, A., Phillips, M.L. 2003. Neural correlates of anxiety associated with obsessive-compulsive symptom dimensions in normal volunteers. *Biol. Psychiatry* 53:482–93.
72. Nordahl, T.E., Benkelfat, C., Semple, W.E., Gross, M., King, A.C., Cohen, R.M. 1989. Cerebral glucose metabolic rates in obsessive-compulsive disorder. *Neuropsychopharm* 2:23–28.
73. Kwon, J.S., Kim, J-J., Lee, D.W., Lee, J.S., Lee, D.S., Kim, M-S., Lyoo, I.K., Cho, M.J., Lee, M.C. 2003. Neural correlates of clinical symptoms and cognitive dysfunctions in obsessive-compulsive disorder. *Psychiatry Res Neuroimaging* 122:37–47.
74. Kim, J-J., Lee, M.C., Kim, J., Kim, I.Y., Kim, S.I., Han, M.H., Chang, K-H., Kwon, J.S. 2001. Grey matter abnormalities in obsessive-compulsive disorder: SPM of segmented magnetic resonance images. *Br. J. Psychiatry* 179:330–4.
75. Purcell, R., Maruff, P., Kyrios, M., Pantelis, C. 1998. Cognitive deficits in obsessive-compulsive disorder in tests of Fronto-striatal function. *Biol. Psychiatry* 43:348–57.
76. Savage, C.R., Baer, L., Keuthen, N.J., Brown, H.D., Rauch, S.L., Jenike, M.A. 1999. Organizational strategies mediate nonverbal memory impairment in obsessive-compulsive disorder. *Biol. Psychiatry* 45:905–16.
77. Sanjaya, Saxena, Brody, Maidment, et al. 2004. Cerebral glucose metabolism in obsessive-compulsive hoarding. *Am. J. Psychiatry* 161:1038–48.
78. Geschwind, D.H. 2018. Evolving views of human genetic variation and its relationship to neurologic and psychiatric disease. *Handb. Clin. Neurol* 147:37–42.
79. Polderman, T.J.C., et al. 2015. Meta-analysis of the heritability of human traits based on fifty years of twin studies. *Nat. Genet* 47:702–9.
80. Schizophrenia Working Group of the Psychiatric Genomics Consortium. et al. 2014. Biological insights from 108 schizophrenia-associated genetic loci. *Nature* 511:421–27.
81. Howard, D.M., et al. 2019. Genome-wide meta-analysis of depression identifies 102 independent variants and highlights the importance of the prefrontal brain regions. *Nat. Neurosci.* https://doi.org/10.1038/s41593-018-0326-7.
82. Visscher, P.M., et al. 2017.10 Years of GWAS discovery: Biology, function, and translation. *Am. J. Hum. Genet* 101:5–22.
83. Breen, G., et al. 2016. Translating genome-wide association findings into new therapeutics for psychiatry. *Nat. Neurosci* 19:1392–6.
84. Gaspar, H.A., Breen, G. 2017. Drug enrichment and discovery from schizophrenia genome-wide association results: An analysis and visualisation approach. *Sci. Rep* 7:12460.
85. Gaspar, H.A., et al. 2019. Using genetic drug-target networks to develop new drug hypotheses for major depressive disorder. *Transl. Psychiatry* 9:117.
86. Maier, R.M., Visscher, P.M., Robinson, M.R., Wray, N.R. 2017. Embracing polygenicity: A review of methods and tools for psychiatric genetics research. *Psychol. Med* 48:1–19.
87. Sullivan, P.F., et al. 2018. Psychiatric genomics: An update and an agenda. *Am. J. Psychiatry* 175:15–27.
88. Bulik-Sullivan, B.K., et al. 2015. LD Score regression distinguishes confounding from polygenicity in genome-wide association studies. *Nat. Genet* 47:291–5.

89. Bulik-Sullivan, B.K., et al. 2015. An atlas of genetic correlations across human diseases and traits. *Nat. Genet* 47:1236–41.
90. Brainstorm Consortium et al. 2018. Analysis of shared heritability in common disorders of the brain. *Science* 360. https://doi.org/10.1126/science.aap8757.
91. Selzam, S., Coleman, J.R.I., Caspi, A., Moffitt, T.E., Plomin, R. 2018. A polygenic p factor for major psychiatric disorders. *Transl. Psychiatry* 8:205.
92. Pettersson, E., Larsson, H., Lichtenstein, P. 2016. Common psychiatric disorders share the same genetic origin: A multivariate sibling study of the Swedish population. *Mol. Psychiatry* 21:717–21.
93. Grotzinger, A.D., et al. 2019. Genomic structural equation modelling provides insights into the multivariate genetic architecture of complex traits. *Nat. Hum. Behav* 1:513–25.
94. Kahl, K.G., Deuschle, M., Stubbs, B., Schweiger, U. 2018. Visceral adipose tissue in patients with severe mental illness. *Horm. Mol. Biol. Clin. Investig* 33:1–7.
95. Schaumberg, K., et al. 2017. The science behind the academy for eating disorders' nine truths about eating disorders. *Eur. Eat. Disord. Rev* 25:432–50.
96. Tarnoki, A.D., et al. 2014. Bioimpedance analysis of body composition in an international twin cohort. *Obes. Res. Clin. Pract* 8:e201–98.
97. Schousboe, K., et al. 2004. Twin study of genetic and environmental influences on adult body size, shape, and composition. *Int. J. Obes. Relat. Metab. Disord* 28:39–48.
98. Ikeda, M., et al. 2018. Re-evaluating classical body type theories: Genetic correlation between psychiatric disorders and body mass index. *Psychol. Med* 48:1745–8.
99. Duncan, L., et al. 2017. Significant locus and metabolic genetic correlations revealed in genome-wide association study of anorexia nervosa. *Am. J. Psychiatry* 174:850–8.
100. Watson, H.J., et al. 2019. Anorexia nervosa genome-wide association study identifies eight loci and implicates Metabo-psychiatric origins. *Nat. Genet* 51:1207–14.
101. Demontis, D., et al. 2019. Discovery of the first genome-wide significant risk loci for attention deficit/ hyperactivity disorder. *Nat. Genet* 51:63–75.
102. Wray, N.R., et al. 2018. Genome-wide association analyses identify 44 risk variants and refine the genetic architecture of major depression. *Nat. Genet* 50:668–1.
103. Kelly, T., Yang, W., Chen, C.-S., Reynolds, K., He, J. 2008. Global burden of obesity in 2005 and projections to 2030. *Int. J. Obes* 32:1431–7.
104. Mauvais-Jarvis, F. 2015. Sex differences in metabolic homeostasis, diabetes, and obesity. *Biol. Sex. Differ* 6:14.
105. Yao, S., et al. 2016. Familial liability for eating disorders and suicide attempts: Evidence from a population registry in Sweden. *JAMA Psychiatry* 73:284–91.
106. Fernandez-Pujals, A.M., et al. 2015. Epidemiology and heritability of major depressive disorder, stratified by age of onset, sex, and illness course in Generation Scotland: Scottish Family Health Study (GS:SFHS). *PLoS One* 10:e0142197.
107. Bandelow, B., Michaelis, S. 2015. Epidemiology of anxiety disorders in the [21]st century. *Dial. Clin. Neurosci* 17:327–35.
108. Fayyad, J., et al. 2017. The descriptive epidemiology of DSM-IV Adult ADHD in the world health organization world mental health surveys. *Atten. Defic. Hyperact. Disord* 9:47–65.
109. Loomes, R., Hull, L., Mandy, W.P.L. 2017. What is the male-to-female ratio in autism spectrum disorder? A systematic review and meta-analysis. *J. Am. Acad. Child Adolesc. Psychiatry* 56:466–74.
110. Papathanasiou, A., Nolen-Doerr, E., Farr, O., Mantzoros, C.S. 2018. Geoffrey harris prize 2018: Novel pathways regulating neuroendocrine function, energy homeostasis and metabolism in humans. *Eur. J. Endocrinol.* https://doi.org/10.1530/EJE-18–0847.
111. McCarthy, M.M., Nugent, B.M., Lenz, K.M. 2017. Neuroimmunology and neuroepigenetics in the establishment of sex differences in the brain. *Nat. Rev. Neurosci.* https://doi.org/10.1038/nrn.2017.61.
112. Mukherjee, N., Chaturvedi, S.K. 2019. Depressive symptoms and disorders in type 2 diabetes mellitus. *Curr. Opin. Psychiatry* 32(5):416–21. https://doi.org/10.1097/YCO.0000000000000528 [PMID: 31135489].
113. Chung, J., Miller, B.J. 2020. Meta-analysis of comorbid diabetes and family history of diabetes in non-affective psychosis. *Schizophr. Res* 216:41–7. https://doi.org/10.1016/j.schres.2019.10.062.
114. Khaledi, M., Haghighatdoost, F., Feizi, A., Aminorroaya, A. 2019. The prevalence of comorbid depression in patients with type 2 diabetes: An updated systematic review and meta-analysis on huge number of observational studies. *Acta Diabetol.* 56(6): 631–50. https://doi.org/10.1007/s00592-019-01295-9.
115. Bent-Ennakhil, N., C´ecile P´erier, M., Sobocki, P., Gothefors, D., Johansson, G., Milea, D., et al. 2018. Incidence of cardiovascular diseases and type-2-diabetes mellitus in patients with psychiatric disorders. *Nord J. Psychiatry* 72(7):455–61. https://doi.org/10.1080/08039488.2018.1463392 [Epub 2018 Dec 4. PMID: 30513230].

116. Lloyd, C.E., Nouwen, A., Sartorius, N., et al. 2018. Prevalence and correlates of depressive disorders in people with type 2 diabetes: Results from the international prevalence and treatment of diabetes and depression (INTERPRET-DD) study, a collaborative study carried out in 14 countries. *Diabet. Med.* 35:760–9.
117. Hajek, T., McIntyre, R., Alda, M. 2016. Bipolar disorders, type 2 diabetes mellitus, and the brain. *Curr. Opin. Psychiatry* 29(1):1–6. https://doi.org/10.1097/YCO.0000000000000215 [PMID: 26575297].
118. Moulton, C.D., Pickup, J.C., Ismail, K. 2015. The link between depression and diabetes: The search for shared mechanisms. *Lancet Diabetes Endocrinol* 3(6):461–71. https://doi.org/10.1016/S2213-8587(15)00134-5 [Epub 2015 May 17. PMID: 25995124].
119. Kullmann, S., Kleinridders, A., Small, D.M., Fritsche, A., Haring, H.U., Preissl, H., et al. 2020. Central nervous pathways of insulin action in the control of metabolism and food intake. *Lancet Diabetes Endocrinol.* 8(6):524–34. https://doi.org/10.1016/S2213-8587(20)30113-3 [PMID: 32445739]
120. Beattie, E.C., Carroll, R.C., Yu, X., Morishita, W., Yasuda, H., von Zastrow, M., et al. 2000. Regulation of AMPA receptor endocytosis by a signaling mechanism shared with LTD. *Nat. Neurosci* 3(12):1291–300. https://doi.org/10.1038/81823[PMID: 11100150].
121. Chiu, S.L., Chen, C.M., Cline, H.T. 2008. Insulin receptor signaling regulates synapse number, dendritic plasticity, and circuit function in vivo. *Neuron* 58(5):708–19. https://doi.org/10.1016/j.neuron.2008.04.014 [PMID: 18549783; PMCID: PMC3057650].
122. Chiu, S.L., Cline, H.T. 2010. Insulin receptor signaling in the development of neuronal structure and function. *Neural Dev* 5:7. https://doi.org/10.1186/1749-8104-5-7 [PMID: 20230616; PMCID: PMC2843688].
123. Dou, J.T., Chen, M., Dufour, F., Alkon, D.L., Zhao, W.Q. 2005. Insulin receptor signaling in longterm memory consolidation following spatial learning. *Learn. Mem* 12(6):646–55. https://doi.org/10.1101/lm.88005 [Epub 2005 Nov 14. PMID: 16287721; PMCID: PMC1356184].
124. Valenciano, A.I., Corrochano, S., de Pablo, F., de la Villa, P., de la Rosa, E.J. 2006. Proinsulin/insulin is synthesized locally and prevents caspase- and cathepsin-mediated cell death in the embryonic mouse retina. *J. Neurochem* 99(2):524–36. https://doi.org/10.1111/j.1471-4159.2006.04043.x [PMID: 17029604].
125. Wan, Q., Xiong, Z.G., Man, H.Y., Ackerley, C.A., Braunton, J., Lu, W.Y., et al. 1997. Recruitment of functional GABA(a) receptors to postsynaptic domains by insulin. *Nature* 388(6643):686–90. https://doi.org/10.1038/41792 [PMID: 9262404].
126. Zhao, W., Chen, H., Xu, H., Moore, E., Meiri, N., Quon, M.J., et al. 1999. Brain insulin receptors and spatial memory. Correlated changes in gene expression, tyrosine phosphorylation, and signaling molecules in the hippocampus of water maze trained rats. *J. Biol. Chem* 274(49):34893–902. https://doi.org/10.1074/jbc.274.49.34893 [PMID: 10574963].
127. Lee, C.C., Huang, C.C., Hsu, K.S. 2011. Insulin promotes dendritic spine and synapse formation by the PI3K/Akt/mTOR and Rac1 signaling pathways. *Neuropharmacology* 61(4):867–79. https://doi.org/10.1016/j.neuropharm.2011.06.003 [Epub2011 Jun 12. PMID: 21683721].
128. Attwells, S., Setiawan, E., Wilson, A.A., Rusjan, P.M., Mizrahi, R., Miler, L., et al. 2017. Inflammation in the Neurocircuitry of obsessive-compulsive disorder. *JAMA Psychiat* 74(8):833–40. https://doi.org/10.1001/jamapsychiatry.2017.1567. PMID: 28636705; PMCID: PMC5710556.
129. Grassi, G., Cecchelli, C., Mazzocato, G., Vignozzi, L. 2021. Early onset obsessive-compulsive disorder: The biological and clinical phenotype. *CNS Spectr* Feb 1:1–7. https://doi.org/10.1017/S1092852921000122. Epub ahead of print, 33517936.
130. Singh, R.K., Chang, H.W., Yan, D., Lee, K.M., Ucmak, D., Wong, K., et al. 2017. Influence of diet on the gut microbiome and implications for human health. *J. Transl. Med* 15(1):73. https://doi.org/10.1186/s12967-017-1175-y. PMID: 28388917; PMCID:PMC5385025.
131. Labad, J., Melia, C.S., Segalas, 'C., Alonso, P., Salvat-Pujol, N., Real, E., et al. 2020. Sex differences in the association between obsessive-compulsive symptom dimensions and diurnal cortisol patterns. *J. Psychiatr. Res* 133:191–6. https://doi.org/10.1016/j.jpsychires.2020.12.017. Epub 2020 Dec 14. PMID: 33352399.
132. Mataix-Cols, D., Frans, E., P´erez-Vigil, A., et al. 2018. A total-population multigenerational family clustering study of autoimmune diseases in obsessive-compulsive disorder and Tourette's/chronic tic disorders. *Mol. Psychiatry* 23(7):1652–8. https://doi.org/10.1038/mp.2017.215. Epub 2017 Nov 14. PMID: 29133949; PMCID:PMC5951741.
133. Faravelli, C., Lo Sauro, C., Godini, L., Lelli, L., Benni, L., Pietrini, F., et al. 2012. Childhood stressful events, HPA axis and anxiety disorders. *World J. Psychiatry* 2 (1):13–25. https://doi.org/10.5498/wjp.v2.i1.13. PMID: 24175164; PMCID:PMC3782172.
134. Turna, J., Grosman Kaplan, K., Anglin, R., et al. 2020. The gut microbiome and inflammation in obsessive-compulsive disorder patients compared to age- and sex-matched controls: A pilot study. *Acta Psychiatr. Scand* https://doi.org/10.1111/acps.13175. Epub ahead of print, 32307692.

135. Figee, M., Vink, M., de Geus, F., Vulink, N., Veltman, D.J., Westenberg, H., et al. 2011. Dysfunctional reward circuitry in obsessive-compulsive disorder. *Biol. Psychiatry* 69(9):867–74. https://doi.org/10.1016/j.biopsych.2010.12.003. Epub 2011 Jan 26. PMID: 21272861.
136. Ashfaq-U-Rahaman., Janardhan Reddy, Y.C., Prabhavathi Pal, P.K. 2007. Obsessive compulsive disorder in adults with rheumatic heart disease. *Acta Neuropsychiatr* 19(2):118–21. https://doi.org/10.1111/j.1601-5215.2006.00169.x [PMID: 26952823].
137. Witthauer, C.T., Gloster, A., Meyer, A.H., Lieb, R. 2014. Physical diseases among persons with obsessive compulsive symptoms and disorder: A general population study. *Soc. Psychiatry Psychiatr. Epidemiol* 49(12):2013–22. https://doi.org/10.1007/s00127-014-0895-z. Epub 2014 Jun 8. PMID: 24907897; PMCID:PMC4228109.
138. Winocour, P.H., Main, C.J., Medlicott, G., Anderson, D.C. 1990. A psychometric evaluation of adult patients with type 1 (insulin-dependent) diabetes mellitus: Prevalence of psychological dysfunction and relationship to demographic variables, metabolic control and complications. *Diabet. Res* 14(4):171–6 [PMID: 2132190].
139. Ilias, I., Papageorgiou, C., Katsadoros, K., Zapanti, E., Anastasiou, E. 2005. Preliminary report: Psychological assessment of Greek women with diabetes during pregnancy. *Percept. Mot. Skills* 101(2):628–30. https://doi.org/10.2466/pms.101.2.628-630 [PMID: 16383101]
140. Asbahr, F.R., Ramos, R.T., Costa, A.N., Sassi, R.B. 2005. Obsessive-compulsive symptoms in adults with history of rheumatic fever, Sydenham's chorea and type I diabetes mellitus: Preliminary results. *Acta Psychiatr. Scand* 111(2):159–61. https://doi.org/10.1111/j.1600-0447.2004.00455.x [PMID: 15667436].
141. Sivertsen, B., Petrie, K.J., Wilhelmsen-Langeland, A., Hysing, M. 2014. Mental health in adolescents with type 1 diabetes: Results from a large population-based study. *BMC Endocr. Disord* (14):83. https://doi.org/10.1186/1472-6823-14-83. PMID: 25303963; PMCID: PMC4197324.
142. Tsai, C.H., Wu, J.S., Chang, Y.F., Lu, F.H., Yang, Y.C., Chang, C.J. 2012. The relationship between psychiatric symptoms and glycemic status in a Chinese population. *J. Psychiatr. Res* 46(7):927–32. https://doi.org/10.1016/j.jpsychires.2012.04.003. Epub 2012 May 19. PMID: 22608774.
143. dos Santos, M.A., Ceretta, L.B., R´eus, G.Z., Abelaira, H.M., Jornada, L.K., Schwalm, M.T., et al. 2014. Anxiety disorders are associated with quality of life impairment in patients with insulin-dependent type 2 diabetes: A case-control study. *Braz. J. Psychiatry* 36(4):298–304. https://doi.org/10.1590/1516-4446-2013-1230. Epub 2014 Jun 24. Erratum in: Rev Bras Psiquiatr 2015 Jan-Mar;37(1):88–9. PMID:25003555.
144. Shams Alizadeh, N., Maroufi, A., Rostami, S., Ghaderi, E. 2018. The relationship between obsessive-compulsive symptoms and glycemic control in patients with diabetes. *Iran. J. Psychiatry Behav. Sci* 12(3):e11611. https://doi.org/10.5812/ijpbs.11611.
145. Albert, U., Aguglia, A., Chiarle, A., Bogetto, F., Maina, G. 2013. Metabolic syndrome and obsessive-compulsive disorder: A naturalistic Italian study. *Gen. Hosp. Psychiatry* 35(2):154–9. https://doi.org/10.1016/j.genhosppsych.2012.10.004. Epub 2012 Nov 13, 23158675.
146. Aguglia, A., Signorelli, M.S., Albert, U., Maina, G. 2018. The impact of general medical conditions in obsessive-compulsive disorder. *Psychiatry Investig* 15(3):246–53. https://doi.org/10.30773/pi.2017.06.17.2. Epub 2018 Feb 22. PMID: 29475243; PMCID: PMC5900370.
147. Isomura, K., Brander, G., Chang, Z., Kuja-Halkola, R., Rück, C., Hellner, C., et al. 2018. Metabolic and cardiovascular complications in obsessive-compulsive disorder: A total population, sibling comparison study with long-term follow-up. *Biol. Psychiatry* 84(5):324–31. https://doi.org/10.1016/j.biopsych.2017.12.003. Epub 2017 Dec 21, 29395042.
148. Bralten, J., Widomska, J., Witte, W., Yu, D., Mathews, C.A., Scharf, J.M., et al. 2020. Shared genetic etiology between obsessive-compulsive disorder, obsessive-compulsive symptoms in the population, and insulin signaling. *Transl. Psychiatry* 10(1):121. https://doi.org/10.1038/s41398-020-0793-y. PMID: 32341337; PMCID: PMC7186226.
149. Kim, J.H., Stewart, T.P., Soltani-Bejnood, M., Wang, L., Fortuna, J.M., Mostafa, O.A., et al. 2006. Phenotypic characterization of polygenic type 2 diabetes in TALLYHO/JngJ mice. *J. Endocrinol* 191(2):437–46. https://doi.org/10.1677/joe.1.06647 [PMID: 17088413].
150. Rosso, G., Zanardini, R., Chiodelli, D.F., Ferrari, C., Gennarelli, M., Bocchio-Chiavetto, L. 2016. Serum levels of insulin-like growth Factor-1 and obsessive-compulsive disorder: A case-control study. *Neuropsychobiology* 74(1):15–21. https://doi.org/10.1159/000446918. Epub 2016 Jul 27. PMID: 27459640.
151. Narayanaswamy, J.C., Jose, D., Shivakumar, V., Shrinivasa, B., Kaur, M., Kalmady, S.V., et al. 2017. Plasma insulin-like growth factor-1 levels and response to selective serotonin reuptake inhibitor treatment: A prospective study of medication-naïve OCD patients. *Asian J. Psychiatr* 28:65–6. https://doi.org/10.1016/j. ajp.2017.03.013. Epub 2017 Mar 18. PMID: 28784399.

152. Phadnis, P., Dey Sarkar, P., Rajput, M.S. 2018. Improved serotonergic neurotransmission by genistein pretreatment regulates symptoms of obsessive-compulsive disorder in streptozotocin-induced diabetic mice. *J. Basic Clin. Physiol. Pharmacol* 29(4):421–5. https://doi.org/10.1515/jbcpp-2017-0155 [PMID: 29561730].
153. Ter Horst, K.W., Lammers, N.M., Trinko, R., Opland, D.M., Figee, M., Ackermans, M.T., et al. 2018. Striatal dopamine regulates systemic glucose metabolism in humans and mice. *Sci. Transl. Med* May 23;10(442):eaar3752. https://doi.org/10.1126/scitranslmed.aar3752. PMID: 29794060.
154. Figee, M., de Koning, P., Klaassen, S., Vulink, N., Mantione, M., van den Munckhof, P., et al. 2014. Deep brain stimulation induces striatal dopamine release in obsessivecompulsive disorder. *Biol. Psychiatry* Apr 15;75(8):647–52. https://doi.org/10.1016/j.biopsych.2013.06.021. Epub 2013 Aug 12. PMID: 23938318.
155. Diepenbroek, C., van der Plasse, G., Eggels, L., Rijnsburger, M., Feenstra, M.G., Kalsbeek, A., et al. 2013. Alterations in blood glucose and plasma glucagon concentrations during deep brain stimulation in the shell region of the nucleus accumbens in rats. *Front. Neurosci* 7:226. https://doi.org/10.3389/fnins.2013.00226. PMID: 24339800; PMCID: PMC3857552.
156. Cox, R.C., Parmar, A.M., Olatunji, B.O. 2020. Sleep in obsessive-compulsive and related disorders: A selective review and synthesis. *Curr. Opin. Psychol* 34:23–6. https://doi.org/10.1016/j.copsyc.2019.08.018. Epub 2019 Aug 24. PMID: 31539831.
157. Nguyen, T.T.L., Chan, L.C., Borreginne, K., Kale, R.P., Hu, C., Tye, S.J. 2018. A review of brain insulin signaling in mood disorders: From biomarker to clinical target. *Neurosci. Biobehav. Rev* 92:7–15. https://doi.org/10.1016/j. neubiorev.2018.05.014. Epub 2018 May 26. PMID: 29758232.
158. Lin, E.H., Von Korff, M., Alonso, J., Angermeyer, M.C., Anthony, J., Bromet, E., et al. 2008. Mental disorders among persons with diabetes–results from the World Mental Health Surveys. *J. Psychosom. Res* 65(6):571–80. https://doi.org/10.1016/j. jpsychores.2008.06.007. Epub 2008 Oct 29. PMID: 19027447; PMCID: PMC3672403.
159. Grassi, G., Cecchelli, C., Vignozzi, L., Pacini, S. 2021. Investigational and experimental drugs to treat obsessive-compulsive disorder. *J. Exp. Pharmacol* 12:695–706. https://doi.org/10.2147/JEP.S255375. PMID: 33447096; PMCID: PMC7801912.
160. Quagliariello, A., Del Chierico, F., Russo, A., et al. 2018. Gut microbiota profiling and gutbrain crosstalk in children affected by pediatric acute-onset neuropsychiatric syndrome and pediatric autoimmune neuropsychiatric disorders associated with streptococcal infections. *Front. Microbiol* 9:675. https://doi.org/10.3389/fmicb.2018.00675. PMID: 29686658; PMCID: PMC5900790.
161. Bilgiç, A., Sertdemir, M., Kılınç, I., Akça, O.F. 2021. Increased serum brain-derived neurotrophic factor and adrenocorticotropic hormone levels are associated with obsessive compulsive disorder in medication-free children. *Eur. Child Adolesc. Psychiatry* Jan 2. https://doi.org/10.1007/s00787-020-01690-6. Epub ahead of print, 33389158.
162. Zand, H., Morshedzadeh, N., Naghashian, F. 2017. Signaling pathways linking inflammation to insulin resistance. *Diabetes Metab. Syndr* 11(Suppl. 1):S307–9. https://doi.org/10.1016/j.dsx.2017.03.006. Epub 2017 Mar 10. PMID: 28365222.
163. Cryan, J.F., O'Riordan, K.J., Cowan, C.S.M., Sandhu, K.V., Bastiaanssen, T.F.S., Boehme, M., et al. 2019. The microbiota-gut-brain Axis. *Physiol. Rev* 99(4):1877–2013. https://doi.org/10.1152/physrev.00018.2018 [PMID: 31460832].
164. Tsalamandris, S., Antonopoulos, A.S., Oikonomou, E., Papamikroulis, G.A., Vogiatzi, G., Papaioannou, S., et al. 2019. The role of inflammation in diabetes: Current concepts and future perspectives. *Eur. Cardiol* 14(1):50–9. https://doi.org/10.15420/ecr.2018.33.1. PMID: 31131037; PMCID: PMC6523054.
165. Jarvinen, A., Laine, M.K., Tikkanen, R., Castren, M.L. 2019. Beneficial effects of GLP-1 agonist in a male with compulsive food-related behavior associated with autism. *Front. Psychol* 10:97. https://doi.org/10.3389/fpsyt.2019.00097. PMID: 30881319; PMCID: PMC6405420.
166. Shields, G.S., Spahr, C.M., Slavich, G.M. 2020. Psychosocial interventions and immune system function: A systematic review and Meta-analysis of randomized clinical trials. *JAMA Psychiat* 77(10):1031–43. https://doi.org/10.1001/jamapsychiatry.2020.0431. PMID: 32492090; PMCID: PMC7272116.
167. Nemirovsky, A., Ilan, K., Lerner, L., Cohen-Lavi, L., Schwartz, D., Goren, G., et al. 2021. Brainimmune axis regulation is responsive to cognitive behavioral therapy and mindfulness intervention: Observations from a randomized controlled trial in patients with Crohn's disease. Brain Behav. *Immun. Health* 19: 100407. https://doi.org/10.1016/j.bbih.2021.100407. PMID: 35024638; PMCID:PMC8728050.
168. Jacobs, J.P., Gupta, A., Bhatt, R.R., Brawer, J., Gao, K., Tillisch, K., et al. 2021. Cognitive behavioral therapy for irritable bowel syndrome induces bidirectional alterations in the brain-gut-microbiome axis associated with gastrointestinal symptom improvement. *Microbiome* 9(1):236. https://doi.org/10.1186/s40168-021-01188-6. PMID: 34847963; PMCID: PMC8630837.

169. Salvi Virginio., et al. 2017. The risk of new-onset diabetes in antidepressant users—A systematic review and meta-analysis. *PLoS One* 12(7):e0182088. https://doi.org/10.1371/journal.pone.0182088.
170. Abrantes, A.M., Farris, S.G., Brown, R.A., Greenberg, B.D., Strong, D.R., McLaughlin, N.C., et al. 2019. Acute effects of aerobic exercise on negative affect and obsessions and compulsions in individuals with obsessive-compulsive disorder. *J. Affect. Disord* 245:991–7. https://doi.org/10.1016/j.jad.2018.11.074. Epub 2018 Nov 14. PMID: 30699885; PMCID: PMC7037579.
171. Freedman, D.E., Richter, M.A. 2021. A narrative review of exercise and obsessivecompulsive disorder. *Gen. Hosp. Psychiatry* 71:1–10. https://doi org/10.1016/j.genhosppsych.2021.03.014. Epub 2021 Apr 2, 33887525.
172. Edward, H. Nieh, Gillian, A. Matthews, Stephen, A. Allsop, Kara, N. Presbrey, Christopher, A. Leppla, Romy, Wichmann, Rachael, Neve, Craig, P. Wildes, Kay, M. Tye. 2015. Decoding neural circuits that control compulsive sucrose seeking. *Cell* 160 (3):528. DOI: 10.1016/j.cell.2015.01.003
173. Joshua, H. Jennings, Randall, L. Ung, Shanna, L. Resendez, Alice, M. Stamatakis, Johnathon, G. Taylor, Jonathan, Huang, Katie, Veleta, Pranish, A. Kantak, Megumi, Aita, Kelson, Shilling-Scrivo, Charu, Ramakrishnan, Karl, Deisseroth, Stephani, Otte, Garret, D. Stuber. 2015. Visualizing hypothalamic network dynamics for appetitive and consummatory behaviors. *Cell* 160 (3): 516. DOI: 10.1016/j.cell.2014.12.026
174. Cell Press. Brain circuit that controls compulsive overeating and sugar addiction discovered. *Science Daily*. www.sciencedaily.com/releases/2015/01/150129125459.htm (accessed November 8, 2022).

5 Gut Microbiota Associated with OCD and its Impact Mediated by Probiotics

Sumathi C Samiappan and Arunkumar Dhailappan

5.1 INTRODUCTION

Microbiota are communities of bacteria, fungi, algae, protozoa, and viruses. Such microbiota may be single or multicellular and exist as commensals, symbionts, or parasites in the human body. Microbiota are commonly found in skin, oral cavity, and nose and especially in gut as normal flora. They are known to have association with human health. Under diseased condition, the microbiota characteristics and composition are changed. Gut microbiome is defined as genes of intestinal microorganisms; confers several structural, metabolic, and protective functions; and regulates entire body health. Gut microflora is gaining an importance by functionally influencing the human body metabolism. The group of microorganisms communicate among themselves and human body by producing chemical signals. The functional role of gut microflora is a direct defense against bacterial pathogens and virus including influenza, stimulates host defense mechanism by inducing antibody production, and simply help for digestion and excretion of waste materials from the body. Gut microbiota make the human body fight against viruses through the immune system (Ichinohe et al., 2009; Wu and Wu, 2012). The bacterial community in the human intestine metabolize complex starch materials into simple structures (Macfarlane and Macfarlane, 2003). For example, xylan, a complex indigestible carbohydrate, is degraded by xylanases of microbial gut flora. The cloning and characterization of xylanase in various gut microbes such as *Bacteroidete ovatus* ATCC 8483, *Bacteroidete ovatus* V975, B *acteroidete cellulosilyticus* WH2, and *Bifidobacterium adolescentis* ATCC 15703 were studied (Zhang et al., 2022). The inhabitants activate the immune system and protect from diseases.

The gastrointestinal tract (GIT) serves as a habitat for most bacteria particularly strict anaerobes, including archae, virus, fungi, and protozoa (Gill et al., 2006). The collection of microorganisms present in the human gut is called gut microbiota. The gut of a fetus is devoid of microbiota. During and after childbirth, it is colonized by maternal microbiota. The development of intestinal mucosa is contributed by foods digested, nutrients, regulation of metabolism, and gut microbiota. Early colonization of gut by microorganisms depends on mode and time of birth, diet, host genetics, health status, and other environmental factors. Bifidobacterium colonizes the intestines of newborn babies after one to two months of birth. Other enteric bacteria comes after six months of birth. Roseburia and Faecalibacterium colonizes at the end of 12 months after birth. The human gut is normally colonized by four bacteria phyla: Firmicutes, Bacteroidetes, Actinobacteria, and Proteobacteria. The dominating bacterial genus are *Clostridium*, *Bacteroids*, *Faecalibacterium*, *Eubacterium*, *Peptostreptococcus*, *Ruminococcus*, *Peptococcus*, and *Bifidobacterium*. Thirty percent of the human gut is colonized by species of genus *Bacteroids*. The bacterial genus *Faeclibacterium*, *Roseburia*, and *Bacteroids* were identified in the human gut as butyric acid producers (Manichanh et al., 2012). The microbial diversity and load in the gut is influenced by various factors, such as age, type of food intake, malnutrition, individuals, and geography. The other factors known to

DOI: 10.1201/9781003387893-6

influence the bacterial population are acidity of stomach, intestinal transit, use of antibiotics, and bacterial communication between bacteria, shaping the neuronal development and intestinal epithelium (Ahern et al., 2014; Ashida et al., 2011; Diaz Heijtz et al., 2011). Fifty-five percent of the fecal solids contain bacteria, and fecal flora is considered as a marker of intestinal flora. Gut bacteria interacts with the host in several ways by regulating bowel movement, produce vitamins, and play a vital role in central nervous system development. The microorganisms in the gut of animals are characterized by qualitative and quantitative methods. The microbial cultivation methods are followed to study the colony and cultural characteristics (Shanahan, 2010). Molecular approaches are currently followed to study various aspects on the functions of gut microorganisms (Balakrishnan and Ramakrishna, 2007). Nucleotide sequencing methods such as fluorescent labeled oligonucleotide probes (FISH) and quantitative PCR, 16S rRNA gene sequencing of the healthy and diseased human feces that represents thousands of different bacterial species are loaded regularly.

Human guts typically contain about 540,000 microbial genes, which constitute the major bacteria in this ecosystem, according to metagenomic studies of the human intestinal microbiome. These genes, which are shared by at least 50% of people, make up about 55% of the core metagenome. Numerous other genes seem to be rare, only found in 20% of people, or both. The current book chapter focuses on interactions between the GI microbiota (especially in the colon), brain-gut axis, central nervous systems, and mental disorders and summarizes the findings on the effects of probiotics and prebiotics in improving mental health, under animal and human clinical trials.

5.1.1 Gut Dysbiosis

Gut dysbiosis can be defined as a condition that the bacteria in the stomach and intestine gets imbalanced. Microbial imbalance results in temporary or mild stomach upset. The symptoms arise from gas bloating, abdominal cramping, diarrhea, constipation, and mucus in the stool. Other common symptoms are also recorded, like halitosis, fatigue, chest pain, and rashes are also recorded. Change in the microbiota is associated with local infection, allergies, autoimmune disease, and cancer. This change also results in systemic diseases like obesity, autoimmunity, and other metabolic disorders. A change in the population diversity of the gut flora may cause or cure the diseases (Costello et al., 2009). Alteration in the composition and load of microbiota in gut results in stress. Those bacteria and other molecules from them enter into the bloodstream and make the immune system respond. Initially, pro-inflammatory molecules like interleukin-1 (IL-1) and interleukin-6 (IL-6) are synthesized in high levels (Dinan et al., 2013). Neuropsychiatric disorders like autism spectrum disorder, depression, and anxiety are linked to microbial dysbiosis, which is also linked to high levels of pro-inflammatory cytokines, elevated oxidative stress, modified gastrointestinal (GI) function, and decreased micronutrient and omega-3 fatty acid status (Dawson et al., 2016).

Gut dysbiotic state, that is, decrease in the population of *Bifidobacterium* and *Brevibacterium* facilitates the adhesion of pathogens associated with irritable bowel syndrome. *Lactobacillus* and *Bifidobacterium* are the two important genus of gram-positive bacteria forms microbiome in healthy individuals. Meta-analysis also showed significant high IL-6 concentrations in individuals with major depression than control (Dowlati et al., 2010). Gut dysbiosis occurs before the appearance of symptoms in OCD (Rees, 2014). The symptoms arise as a part of streptococcal infections particularly by beta haemolytic Streptococci. Even the use of antibiotics to treat streptococcal infections would change the gut microbiota. In human phase studies, dysbiotic state has led to rising of psychiatric symptoms like anxiety and depression. OCD is obsessive-compulsive disorder, in which anxiety is considered a primary feature. Several literature supports the relationship of anxiety and gut microflora with OCD (Jasmine, 2017). The imbalance in the neurotransmission molecules is a vital cause for OCD. The neurotransmission molecules are produced and altered by gut microflora. The recent research is focused on replacing the gut flora of a patient with a healthy flora by surgical methods. The treatment for dysbiotic condition is intake of pre or probiotic for balancing the

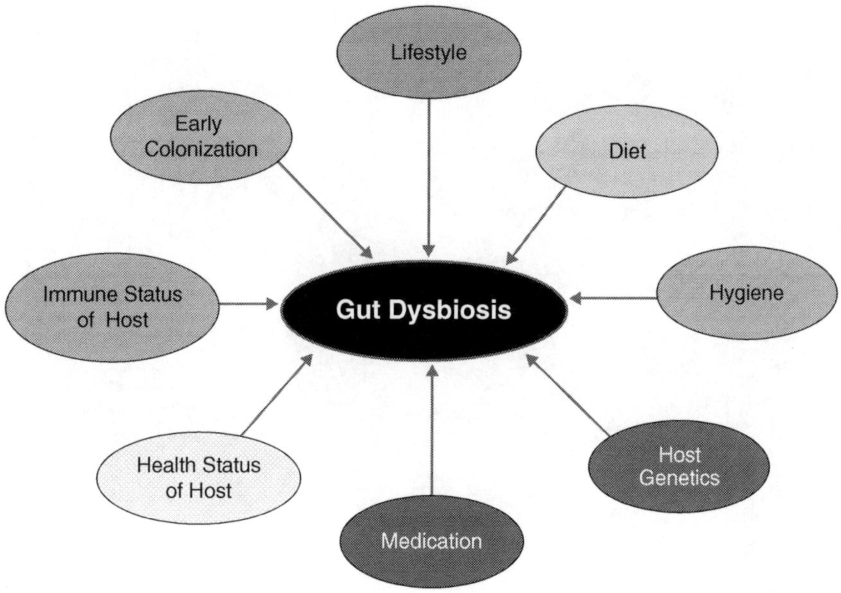

FIGURE 5.1 Factors influencing the gut dysbiosis.

intestinal microflora. Use of probiotics augment the positive effects on gut microflora (Moayyedi et al., 2010; Rees, 2014). The factors influencing the gut dysbiosis is represented in Figure 5.1.

The following probiotic strains belonging to the genus *Bifidobacterium*, *Bacillus*, *Enterococcus*, *Lactobacillus*, *Leuconostoc*, *Saccharomyces*, *Streptococcus*, *Pediococcus*, *Escherichia coli* are suggested to correct dysbiotic state (Fijan, 2014). Intake of certain antibiotics such as neomycin, bacitracin, and pimaricin also induce microbial imbalance in the gut. Gut dysbiosis is studied by next generation sequencing methods in different age groups, viz. children, teenagers, and mid aged.

5.1.2 Influence of Gut Microbiota in Mental Health

Researchers report that the gut microbiota is connected to brain by a gut-brain axis. To maintain body homeostasis, gut microbiota and brain are involved in bidirectional communication occurring between the central nervous system (CNS) and gastrointestinal tract (GIT) (Carabotti et al., 2015; Wang and Kasper, 2014). Under stress conditions, certain molecules are released, that is, catecholamines and norepinephrine into gastrointestinal tract causing dysregulation in the gut-brain axis leading to various gut diseases (Groot et al., 2000; Bonaz and Sabate, 2009). Stress also induces the symptoms of irritable bowel syndrome (Mayer et al., 2001). The gut microbiota is known to bring out the behavior changes known to influence the vagus nerve. In addition, good microbiota boosts up the immune power and neurotransmitters. This microbiota also helps in the development of the central nervous system (CNS) (Latalova et al., 2017). Animal studies were carried out in rodents, in which the response of animals changes if the gut microorganisms are modified (Figure 5.2). Those animals expressed anxiety behavior changes compared normal mice. Probiotics change the biomolecules like brain-derived neurotrophic factor (BDNF), gamma-aminobutyric acid (GABA), dopamine (DA), and serotonin (5 hydroxytryptamine—5 HT) of CNS directly, such as affecting mind and behavior. Probiotics also influence the vagus and enteric nerves. The indirect mechanism that improves CNS function is alteration of metabolites like tryptophan and short fatty acids (Wang et al., 2016). The probiotic treatment of animals observed to have reduced anxiety

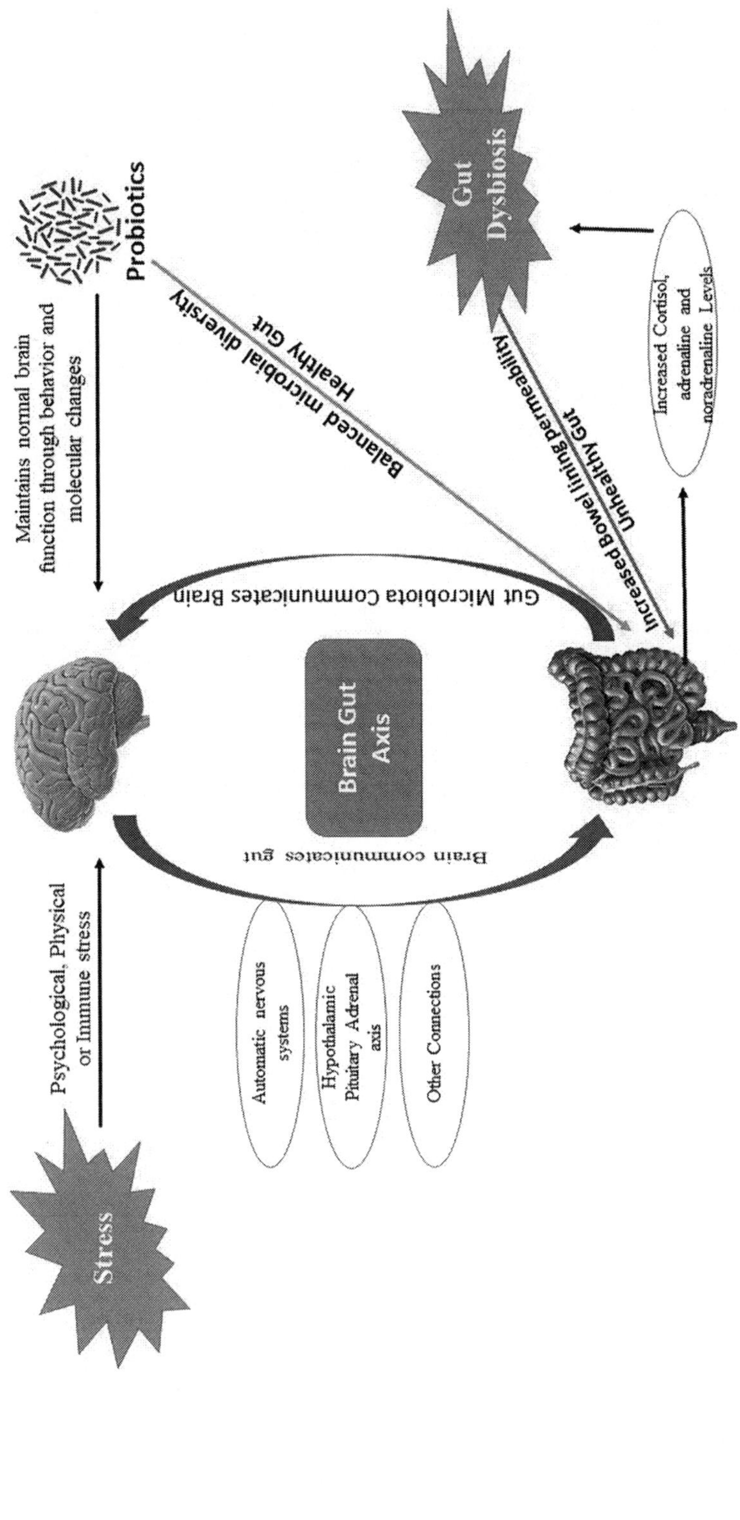

FIGURE 5.2 Impact of gut microbiota, probiotic bacteria on brain activity.

and depression-like symptoms (Bravo et al., 2011; Desbonnet et al., 2010). Similarly, a germ-free mice study revealed that the hypothalamus pituitary adrenal axis (HPA) was hyperactive towards stress. Such type of animals showed lower expression levels of reduced brain-derived neurotrophic factor (BDNA) in the cortex and hypothalamus (Sudo et al., 2004) leading to memory impairments (Gareau et al., 2011). The function of CNS is expressed as behaviors. The recent research interest is to study the involvement of microorganisms in neuropsychiatric conditions and chemical signaling of CNS (Grenham et al., 2011). Certain mechanisms of action proposed includes signals to brain passing through ascending neural pathways (i.e., vagus nerve) by gut hormones released by activating the enteric nervous system (Crumeyrolle-Arias et al., 2014). These bacteria produce signaling molecules such as serotonin, dopamine, gamma aminobutyric acid, norepinephrine, and acetylcholine enabling neurotransmission (Lyte, 2013). The signaling molecules produced by bacteria is represented in the Table 5.1 (Cryan and Dinan, 2012; Thakur et al., 2014).

Anxiety coexists with depression often. It is prevalent based on the time and type of the disorder (OCD). A neurotransmitter, gamma-aminobutyric acid (GABA) is functioning as brain inhibitory neurotransmitter as well as modulates the anxiety response. The other neurotransmitter molecules are endocannabinoids, neuropeptide Y, opioid peptides, oxytocin, and corticotropin releasing hormone (Nuss, 2015). The release of neurotransmitters and balancing of energy is influenced by gut microbiota, modulates several physiological and pathological conditions of mental health (Ajit Kumar et al., 2019). The metabolism of tryptophan is influenced by gut microbiota, which is a precursor of serotonin. The signals between GIT and brain is well regulated, known as brain-gut axis. The gut-brain axis includes gut microorganisms, central nervous system (CNS), the neuroendocrine and neuroimmune systems, autonomic and enteric nervous system (Evrensel and Ceylan, 2015; Thakur et al., 2014). The factors such as neural, hormonal, and immunological levels influence the signals of CNS. The seeding of microflora has critical influence on the growth, maturation, and development of neurons in the young brain. The gut bacteria control the maturation and function of microglia in CNS (Macfarlane et al., 2009; Erny et al., 2015). The symbiotic microflora in the gut plays a vital role in the functions of CNS directly or indirectly (Reider et al., 2017). Probiotics produce neuropeptides and neurotransmitters. Neuropeptides are released from the brain, stimulates various receptors allowing neuron-neuron communication. Neurotransmitters transfer the chemical signals from one neuron to another particular neuron, gland cell, or muscle cell (Slavin, 2013; Wang et al., 2016).

Intestinal microorganisms have direct influence with the brain functions, and the brain influences the microbial population in the intestine (Collins et al., 2012). To enrich the population of intestinal microbiota, some efforts are taken by oral administration of probiotics or prebiotics under

TABLE 5.1
Neurotransmitters Produced by Gut Microbiota

S. No.	Gut Bacteria	Neurotransmitter
1.	*Lactobacillus* sp. *Bifidobacterium* sp.	Gamma-aminobutyric acid (GABA)
2.	*Escherichia* sp., *Bacillus* sp. *Saccharomyces* sp.	Noradrenalin
3.	*Candida* sp., *Streptococcus* sp., *Escherichia* sp. *Enterococcus* sp.	Serotonin
4.	*Bacillus* sp.	Dopamine
5.	*Lactobacillus* sp.	Acetylcholine

controlled clinical trials. Probiotics are live formulations of microorganisms capable of promoting the human health by activating intestinal microbiota. They are usually administered orally and allowed to colonize in the colon (Cani et al., 2007).

Probiotic bacteria have a lot of health promoting activities and are considered as "Generally Regarded as Safe (GRAS)" (Walter, 2008). There are various species and strains of probiotic bacteria. Probiotics have several properties: they are resistant towards stomach and bile acids, adhere to the mucosal surface of the intestines, confers antipathogenic activity, and enhances the viability of probiotics (Sarkar, 2010; Seale and Miller, 2013; Anwar et al., 2014; Ganguly et al., 2011). They are fermented products suitable for consumption and serve as potential therapeutic molecules. According to European Food Safety Authority (EFSA), the health benefits of probiotics are more widespread and high quality than other food supplements like vitamins and minerals. There exists a large gap in understanding the relationship of probiotics, prebiotics, gut microbiota, and mental health. The long-term studies reveal that prebiotics are known to induce probiotic bacterial groups such as genus *Bifidobacterium* and genus *Lactobacillus* (Bedu-Ferrari et al., 2022). Animal studies and human trials on the use of probiotics found it reduces the symptoms of gastrointestinal and psychological problems (Saulnier et al., 2013). The colonization by probiotic bacteria in the gut prevents the entry of pathogens. Those pathogens try to adhere with the receptors and penetrate the membranes that are protected by gut beneficial microorganisms. Probiotics have a variety of application; in particular, they are used in treating diseases like necrotizing enterocolitis of premature infants, antibiotic associated diarrhea, maintaining urogenital hygiene, combating infectious diseases and allergic responses to improve respiratory health (Deshpande et al., 2011; Bernardo et al., 2013; Goldenberg et al., 2013; Macklaim et al., 2012).

The probiotic bacteria are able to break lactose sugar into lactic acid and lower the pH. As a result of acidic pH, growth of harmful bacteria is inhibited. They are considered as indigenous flora of human vagina and gastrointestinal tract (Makarova et al., 2006). Lactobacillus is the most interested probiotic organisms, having no significant difference on depression. It also has no effect on anxiety when given probiotic alone or in combination with other prebiotics. There are several drugs which regulate the imbalance of neurotransmitters used to treat depression. Meta-analysis studies confirms that antidepressants are more efficient to treat major depressive disorder. Antidepressants are also recommended for treatment of OCD (Bandelow et al., 2012; Cipriani et al., 2018). A promising animal study was conducted with *Lactobacillus salivarius* strain that produced bacteriocin against *Listeria monocytogenes* infection (Corr et al., 2007). A probiotic bacteria *Lactobacillus rhamnosus* used to treat OCD in mice models for a period of two to four weeks has reduced the symptoms of OCD. The combination of probiotic mixture containing *Lactobacillus helveticus* and *Bifidobacterium longum* was reported to reduce OCD symptoms on a daily dose intake for 30 days (Messaoudi et al., 2011). The colonization of microorganisms in the animal gut induces change in the neurochemical process of the brain and disturbs the neuronal pathways of motor control. Initiation of such signals make animals express anxiety. In germ-free mice studies under stressed conditions, abnormalities were found associated with hypothalamus-pituitary-adrenal (HPA) axis hyperactivity (Sudo et al., 2004). When the germ-free mice were restored with intestinal microbiota, anxiety was reduced. In mice, administration of *Lactobacillus rhamnosus* GG has been associated with decreased anxiety, obsessive-compulsive disorder (OCD), and depressive behaviors (Aljumaah et al., 2022).

The genus *Bifidobacterium* are gram-positive rod-shaped bacteria that are non-motile and anaerobic. They live symbiotically inside the intestines and vagina of humans. The important probiotic species of Bifidobacterium includes *B. animalis* subsp *animalis*, *B. animalis* subsp *lactis*, *B. adolescentis*, *B. breve*, *B. bifidum*, *Bifidobacterium infantis*, and *B. longum*. They are used in the treatment of various diseases. A species of bacteria *Bifidobacterium infantis* was used to treat such conditions and showed promising results than control group of animals. This confirms the significance of microbiota in reducing stress levels. The 30-day treatment of 66 healthy individuals were subjected to probiotics containing *Bifidobacterium longum* and *Lactobacillus helveticus* to study the mood behaviors. They showed antipsychotic effects or antidepression effects. Everyday consumption of *Lactobacillus casei*, a milk containing probiotics, for 20 days in 132 healthy individuals showed a happy response instead of depression (Benton et al., 2007). Many strains of *Lactobacillus* and *Bifidobacterium* showed promising beneficial effects in improving depression and mental health (Mayer et al., 2014). Similarly, a combination of three probiotic strains such as *Lactobacillus acidophilus, L. casei*, and *Bifidobacterium bifidum* were given for a group of 40 people and observed for reduction in depression symptoms. Another group that received *Lactobacillus casei* or placebo did not produce convincing results to reduce depression (Rao et al., 2009). Studies with probiotic preparation (containing *Lactobacillus helveticus* and *Bifidobacterium longum*) showed no evidence that they reduced the symptoms of depression, stress, and anxiety (Rao et al., 2009). *B. longum, B. breve, B. infantis, L. helveticus, L. rhamnosus*, and *L. casei* reduce the symptoms of CNS disorders in humans and animals (Wang et al., 2016).

Prebiotics are nondigestible substances and are fermented by probiotics or gastrointestinal microbiota. Many polymeric substances are used as prebiotics and include resistant starch, ß-glucan, fiber gums, inulin, raftilin, stachyose, and xylan. Prebiotics do not undergo any changes by digestive enzymes and secretions when they reach the colon. The administration of prebiotics is also capable of enriching the microflora in the intestines under controlled conditions. Several nondigestible oligosaccharides (inulin and oligofructose) administered are fermented by intestinal microorganisms. Such biochemical activities particularly enhance the bacteria belonging to the genus *Bifidobacterium* sp., and *Lactobacillus* sp., which exist as commensals in the intestine (Cani et al., 2007). Only carbohydrate diet alone was incapable of increasing the population of *Bifidobacterium* sp., and *Lactobacillus* sp., regulate brain and immune functions. The combination of carbohydrate diet and prebiotic administration was effective in improving the preserved tight junction integrity, intestinal barrier function, lowered endotoxinemia, systemic and hepatic cytokines, oxidative stress, and population of *Bifidobacterium* sp., and *Lactobacillus* sp. (Cani et al., 2008). In a group of healthy women, supplementation with probiotic bacteria *Lactobacillus rhamnosus* HN001 (HN001) showed reduced post-natal symptoms of depression and anxiety. A combined probiotic and weight reduction program showed reduced depression symptoms in obese women (Sanchez et al., 2017). Intake of foods with high vegetables, fruits, whole grains, meat, and fish significantly reduce depression (Lai et al., 2013). Gut microbiota can actively carry out catabolic and anabolic reactions of herbal drugs in the gastrointestinal tract supporting treatment of various ailments (Laparra and Sanz, 2010). According to the review studies of Fereshteh et al., probiotics and prebiotics can be used as innovative medicines to treat mental and psychological disorders, however, more clinical studies are required to prove the mechanisms and functions of probiotics. The individual or combined intake of probiotics and prebiotics has shown an important role in neuro-immune process (Savignac et al., 2016; Slavin, 2013).

5.2 CONCLUSION

The important probiotic bacteria species of Lactobacillus are *Lactobacillus acidophilus* and *Lactobacillus rhamnosus* and genus *Bifidobacterium* are *Bifidobacterium bifidum* and *Bifidobacterium infantis.* Functional foods contain beneficial microorganisms with the population above 10–^{7} CFU g^{-1} probiotic bacteria. Prebiotics are nondigestible food substances encouraging the growth and metabolism of colon bacteria. To combat host illness, probiotics and prebiotics function together and reduce the risk associated with mental disorders and other infections. Probiotics use prebiotics as a source of nutrients by fermentation. In the absence of prebiotics, probiotics lack nutrients for survival and cause complications like leaky gut in the host intestine, constipation, and compromised immune disorder. In the absence of probiotics, the effectiveness of prebiotics in the gut have little importance. Certainly probiotics and prebiotics affect the gastrointestinal microflora. The combination of probiotics and prebiotics can reduce the severity of OCD, a type of anxiety disorder. The medicines that are suggested to treat mental disorders like anxiety, depression, schizophrenia, Alzheimer, autism, and OCD are called "psychobiotics." These medicines are preferred for the reason that they exert very little side effects as antidepressants.

ABBREVIATIONS

5 HT—5-Hydroxytryptamine;
BDNF—Brain-Derived Neurotrophic Factor;
CFU—Colony Forming Units;
DA—Dopamine;
EFSA—European Food Safety Authority;
GABA—Gamma-Aminobutyric Acid;
GIT—Gastrointestinal Tract;
GRAS—Generally Regarded As Safe;
HPA—Hypothalamus Pituitary Adrenal Axis;
IL—Interleukin;
OCD—Obsessive-Compulsive Disorder

REFERENCES

Ahern P.P., Faith J.J., and Gordon J.I. (2014). Mining the human gut microbiota for effector strains that shape the immune system. Immunity 40, 815–823.

Ajit Kumar T., Sakshi T., and Nikhila S. (2019). Comorbid brain disorders associated with diabetes: Therapeutic potentials of prebiotics, probiotics and herbal drugs. Translational Medicine Communications 4, 12.

Aljumaah M.R., Urja B., Jeffery R., John G., and Andrea A.P. (2022). The gut microbiome, mild cognitive impairment, and probiotics: A randomized clinical trial in middle-aged and older adults. Clinical Nutrition. 41, 2565–2576.

Anwar A.A., Thikra A.A., and Saeed A.M. (2014). Adhesion, autoaggregation and hydrophobicity of six *Lacto bacillus* strains. British Microbiology Research Journal 4, 381–391.

Ashida H., Ogawa M., Kim M., Hitomi, and Chihiro S. (2011). Bacteria and host interactions in the gut epithelial barrier. Nature Chemical Biology 8, 36–45.

Balakrishnan S., and Ramakrishna, S. (2007). The normal bacterial flora of the human intestine and its regulation. Journal of Clinical Gastroenterology 41, S2–S6.

Bandelow B., Sher L., Bunevicius R., Hollander E., Kasper S., Zohar J., and Moller H.J., (2012). Care W.T.F.M.D.P. WFSBP Task Force on Mental Disorders in Primary Care; WFSBP Task Force on Anxiety Disorders, OCD and PTSD. Guidelines for the pharmacological treatment of anxiety disorders, obsessive-compulsive disorder and posttraumatic stress disorder in primary care. International Journal of Psychiatry and Clinical Practice 16, 77–84.

Bedu-Ferrari C., Biscarrat P., Langella P., Cherbuy C. (2022). Prebiotics and the human gut microbiota: From breakdown mechanisms to the impact on metabolic health. *Nutrients* 14, 2096.

Benton D., Williams C., and Brown A. (2007). Impact of consuming a milk drink containing a probiotic on mood and cognition. European Journal of Clinical Nutrition 61, 355–361.

Bernardo W.M., Aires F.T., Carneiro R.M., Fernando P., Vera Esteves V.R., and Dennis A. (2013). Effectiveness of probiotics in the prophylaxis of necrotizing enterocolitis in preterm neonates: A systematic review and meta-analysis. Journal of Pediatrics (Rio. J.) 89, 18–24.

Bonaz B., and Sabate J.M. (2009). Brain-gut axis dysfunction. Gastroenterology Clinical Biology 33, S48–S58.

Bravo J.A., Forsythe P., Chew M.V., Emily E., Helene M.S., Timothy G.D., John B., and John F.C. (2011). Ingestion of Lactobacillus strain regulates emotional behavior and central GABA receptor expression in a mouse via the vagus nerve. Proceedings of the National Academy of Sciences USA 108, 16050–16055.

Cani P.D., Bibiloni R., Knauf C., Aurélie W., Audrey M.N., Nathalie M.D. and Rémy B. (2008). Changes in gut microbiota control metabolic endotoxemia-induced inflammation in high-fat diet-induced obesity and diabetes in mice. Diabetes 57, 1470–81.

Carabotti M., Scirocco A., Maselli M.A., and Severi C. (2015). The gut-brain axis: Interactions between enteric microbiota, central and enteric nervous systems. Annals of Gastroenterology 28, 203–209.

Cipriani A., Furukawa T.A., Salanti G., Chaimani A., Atkinson L.Z., Ogawa Y., Leucht S., Ruhe H.G., Turner E.H., Higgins J.P.T., Egger M., Takeshima N., Hayasaka Y., Imai H., Shinohara K., Tajika A., Ioannidis J.P.A., and Geddes J.R. (2018). Comparative efficacy and acceptability of 21 antidepressant drugs for the acute treatment of adults with major depressive disorder: A systematic review and network meta-analysis. Lancet 391, 1357–1366.

Collins S.M., Surette M., and Bercik P. (2012). The interplay between the intestinal microbiota and the brain. Nature Reviews of Microbiology 10, 735–42.

Corr S.C., Li Y., Riedel, C.U., O'Toole P.W., Hill C., and Gahan C.G.M. (2007). Bacteriocin production as a mechanism for the antiinfective activity of *Lactobacillus salivarius* UCC118. Proceeding of the National Academy of Sciences. U.S.A 104, 7617–7621.

Costello E.K., Lauber C.L., Hamady M., Fierer N., Gordon J.I., and Knight R. (2009). Bacterial community variation in human body habitats across space and time. Science 326, 1694–1697.

Cryan J.F., and Dinan T.G. (2012). Mind-altering microorganisms: The impact of the gut microbiota on brain and behaviour. Nature Reviews of Neuroscience 13, 701–712.

Crumeyrolle-Arias M., Jaglin M., Bruneau A., Vancassel S., Cardona A., Daugé V., Naudon L., and Rabot S. (2014). Absence of the gut microbiota enhances anxiety-like behavior and neuroendocrine response to acute stress in rats. Psychoneuroendocrinology 42, 207–217.

Dawson S.L., Dash S.R., and Jacka F.N. (2016). The importance of diet and gut health to the treatment and prevention of mental disorders. International Reviews of Neurobiology 131, 325–346.

Desbonnet L., Garrett L., Clarke G., Kiely B., Cryan J.F., and Dinan T.G. (2010). Effects of the probiotic *Bifidobacterium infantis* in the maternal separation model of depression. Neuroscience 170, 1179–1188.

Deshpande G., Rao S., and Patole S. (2011). Progress in the field of probiotics. Current Opinion in Gastroenterology 27, 13–18.

Diaz Heijtz R., Wang S., Anuar F., Yu Q., Britta B., Annika S., Martin L.H., Hans F., and Sven P. (2011). Normal gut microbiota modulates brain development and behavior. Proceedings of the National Academy of Sciences, USA 108, 3047–3052.Dinan T.G., Stanton C., and Cryan J.F. (2013). Psychobiotics: A novel class of psychotropic. Biological Psychiatry 74, 720–726.

Dowlati Y., Herrmann N., Swardfager W., Liu H., Sham L., Reim E.K., and Lanctot K.L. (2010). A meta-analysis of cytokines in major depression. Biology Psychiatry 67, 446–457.

Erny D., Hrabe de Angelis A.L., Jaitin D., Wieghofer P., Staszewski O., David E., Keren-Shaul H., Mahlakoiv T., Jakobshagen K., Buch T., et al. (2015). Host microbiota constantly control maturation and function of microglia in the CNS. Nature Neuroscience 18, 965–977.

Evrensel A., and Ceylan M.E. (2015). The gut-brain Axis: The missing link in depression. Clinical Psychopharmacology Neuroscience 13, 239–244.

Fijan S. (2014). Microorganisms with claimed probiotic properties: An overview of recent literature. International Journal of Environment Research and Public Health 11, 4745–4767.

Ganguly N.K., Bhattacharya S.K., Sesikeran B., Nair G.B., Ramakrishna B.S., Sachdev H.P.S., Batish V.K., Kanagasabapathy A.S., Muthuswamy V. and Kathuria S.C., (2011). ICMR-DBT guidelines for evaluation of probiotics in food. Indian Journal of Medical Research 134, 22–25.

Gareau M.G., Wine E., Rodrigues D.M., Cho J.H., Whary M.T., Philpott D.J., MacQueen, G., and Sherman, P.M. (2011). Bacterial infection causes stress-induced memory dysfunction in mice. Gut 60, 307–317.

Gill S.R., Pop M., Deboy R.T., Eckburg P.B., Turnbaugh P.J., Samuel B.S., Gordon J.I., Relman D.A., Fraser-Liggett C.M., and Nelson K.E. (2006). Metagenomic analysis of the human distal gut microbiome. Science 312, 1355–1359.

Goldenberg J.Z., Ma S.S., Saxton J.D. Mark R.M., Per O.V., Kristian T., Gordon H.G., and Bradley C.J. (2013). Probiotics for the prevention of *Clostridium difficile*-associated diarrhea in adults and children. Cochrane Database Systems Review 5, CD006095.

Grenham S., Gerard C., John F.C., and Timothy G.D. (2011). Brain–gut–microbe communication in health and disease. Frontiers in Physiology 2, 1–15.

Groot J., Bijlsma P., Van Kalkeren A., Kiliaan A., Saunders P., and Perdue M. (2000). Stress-induced decrease of the intestinal barrier function. The role of muscarinic receptor activation. Annals of New York Academy of Sciences 915, 237–246.

Ichinohe T., Lee H.K., Ogura Y., Flavell R., and Iwasaki A. (2009). Inflammasome recognition of influenza virus is essential for adaptive immune responses. Journal of Experimental Medicine 206, 79–87.

Jasmine T. (2017). An update on the relationship between the gut microbiome and obsessive- compulsive disorder. Psychiatric Annals 47, 542–551.

Lai J.S., Hiles S., Bisquera A., Hure A.J., McEvoy M., and Attia J. (2013). A systematic review and meta-analysis of dietary patterns and depression in community-dwelling adults. American Journal of Clinical Nutrition 99, 181–197.

Laparra J.M., and Sanz Y. (2010). Interactions of gut microbiota with functional food components and nutraceuticals. Pharmacology Research 61, 219–25.

Latalova K., Miroslav H., and Jan P. (2017). Can gut microbes play a role in mental disorders and their treatment? Psychiatria Danubina 29, 28–30.

Lyte M. (2013). Microbial endocrinology in the microbiome-gutbrain axis: How bacterial production and utilization of neurochemicals influence behavior. PLoS Pathogens 9, e1003726.

Macfarlane G., Blackett K., Nakayama T., Steed H., and Macfarlane S. (2009). The gut microbiota in inflammatory bowel disease. Current Pharmaceutical Design 15, 1528–1536.

Macfarlane, S., and Macfarlane, G.T. (2003). Regulation of short-chain fattyacid production. Proceeding of Nutrition Society 62, 67–72.

Macklaim J.M., Cohen C.R., Donders G., Gregory B.G., Janet E.H., Groesbeck P.P., Jacques R., Gregory S., Janneke W., and Gregor R. (2012). Exploring a road map to counter misconceptions about the cervicovaginal microbiome and disease. Reproductive Sciences 19, 1154–1162.

Makarova K., Slesarev A., Wolf Y., Sorokin A., Mirkin B., Koonin E., Pavlov A., Pavlova N., Karamychev V., Polouchine N., et al. (2006). Comparative genomics of the lactic acid bacteria. Proceedings of National Academy of Sciences. USA 103, 15611–15616.

Manichanh C., Borruel N., Casellas F., and Guarner F. (2012). The gut microbiota in IBD. Nature Reviews Gastroenterology and Hepatology 9, 599–608.

Mayer E.A., Knight R., Mazmanian S.K., Cryan J.F., and Tillisch K. (2014). Gut microbes and the brain: Paradigm shift in neuroscience. The Journal of Neuroscience 34, 15490–15496.

Mayer E.A., Naliboff B.D., Chang L., and Coutinho S.V.V. (2001). Stress and irritable bowel syndrome. American Journal of Physiology Gastrointestinal Liver Physiology 280, G519–G524.

Messaoudi M., Violle N., Bisson J., Desor D., Javelot H., and Rougeot C. (2011). Beneficial psychological effects of a probiotic formulation (*Lactobacillus helveticus* R0052 and *Bifidobacterium longum* R0175) in healthy human volunteers. Gut Microbes 2, 256–261.

Moayyedi P., Ford A.C., Talley N.J., Cremonini F., Foxx-Orenstein A.E., Brandt L.J., and Quigley, E.M. (2010). The efficacy of probiotics in the treatment of irritable bowel syndrome: A systematic review. Gut 59, 325–32.

Nuss P. (2015). Anxiety disorders and GABA neurotransmission: A disturbance of modulation. Neuropsychiatric Disease and Treatment 11, 165–75.

Rao, A.V., Bested, A.C., Beaulne, T.M., Katzman, M.A., Iorio, C., and Berardi, J.M. (2009). A randomized, double-blind, placebo controlled pilot study of a probiotic in emotional symptoms of chronic fatigue syndrome. Gut Pathogens 1, 6.

Rees J.C. (2014). Obsessive-compulsive disorder and gut microbiota dysregulation. Med Hypotheses 82, 163–166.

Rieder R., Wisniewski P.J., Alderman B.L., and Campbell S.C. (2017). Microbes and mental health: A review. Brain Behaviour Immunology 66, 9–17.

Sanchez M., Darimont C., Panahi S., Drapeau V., Marette A., Taylor V.H., Doré J., and Tremblay A., (2017). Effects of a diet-based weight-reducing program with probiotic supplementation on satiety efficiency, eating behaviour traits, and psychosocial behaviours in obese individuals. Nutrients 9, 284.

Sarkar S. (2010). Approaches for enhancing the viability of probiotics: A review. British Food Journal 112, 329–349.

Saulnier D.M., Yehuda R., Melvin B.H., Jane A.F., Premysl B., Robert J.S., James V., Elena F.V., Ted G.D., Gail H., and Francisco G. (2013). The intestinal microbes in probiotics and prebiotics in neurogastroenterology. Gut Microbes 4, 17–27.

Savignac H.M., Couch Y., Stratford M., Bannerman D.M., Tzortzis G., Anthony D.C., Burnet P.W.J. (2016). Prebiotic administration normalizes lipopolysaccharide (LPS)-induced anxiety and cortical 5-HT2A receptor and IL1-β levels in male mice. Brain Behaviour Immunology 52, 120–131.

Seale J.V., Millar M. (2013). Probiotics: A new frontier for infection control. Journal of Hospital Infections 84, 1–4.

Shanahan F. (2010). Gut microbes: From bugs to drugs. American Journal of Gastroenterology 105, 275–279.

Slavin J. (2013). Fiber and prebiotics: Mechanisms and health benefits. Nutrients 5, 1417–1435.

Sudo N., Chida Y., Aiba Y., Sonoda J., Oyama N., Yu X.N., Kubo, C., and Koga, *Y*. (2004). Postnatal microbial colonization programs the hypothalamic-pituitary-adrenal system for stress response in mice. Journal of Physiology 558, 263–275.

Thakur A.K., Shakya A., Husain G.M., Emerald M., and Kumar V. (2014). Gut-microbiota and mental health: Current and future perspectives. Journal of Pharmacology Clinical Toxicology 2, 1–15.

Walter J. (2008). Ecological role of *Lactobacilli* in the gastrointestinal tract: Implications for fundamental and biomedical research. Applied Environmental Microbiology 74, 4985–4996.

Wang H., Lee I.S., Braun C., Enck P. (2016). Effect of probiotics on central nervous system functions in animals and humans: A systematic review. Journal of Neurogastroenterology Motility 22, 589–605.

Wang Y., and Kasper L.H. (2014). The role of microbiome in central nervous system disorders. Brain Behaviour Immunology 38, 1–12.

Wu HJ, Wu E. (2012). The role of gut microbiota in immune homeostasis and autoimmunity. Gut Microbes 1, 4–14.

Zhang B., Zhong Y., Dejun, D., Zhitian, Z., Hu J. (2022). Gut microbial utilization of xylan and its implication in gut homeostasis and metabolic response. Carbohydrate Polymers, 286, 119271.

Section 2

Diagnosis of OCD

6 Current Understanding of Biotechnology in Obsessive-Compulsive Disorder and its Treatment

Thabitha Aavula,[a] *Santhi Priya Sobha,*[b] *Saravanan Ramachandran,*[c,*] *and Kumar Ebenezar Kesavarao*[d]

[a] Native Medicine and Marine Pharmacology Lab, Faculty of Allied Health Sciences, Chettinad Hospital and Research Institute, Chettinad Academy of Research and Education, Chettinad Health City, Kelambakkam, Chennai-603103

[b] Natural Medicine and Molecular Physiology Lab, Faculty of Allied Health Sciences, Chettinad Hospital and Research Institute, Chettinad Academy of Research and Education, Chettinad Health City, Kelambakkam, Chennai-603103

[c] Associate Professor, Faculty of Allied Health Sciences, Chettinad Hospital and Research Institute, Chettinad Academy of Research and Education, Chettinad Health City, Kelambakkam, Chennai-603103

[d] Professor, Faculty of Allied Health Sciences, Chettinad Hospital and Research Institute, Chettinad Academy of Research and Education, Chettinad Health City, Kelambakkam, Chennai-603103

6.1 INTRODUCTION

Obsessive-compulsive disorder characterized by persistent thoughts/obsession combined with compulsion is linked to neurodevelopmental or psychiatric disorder (Zheng *et al.*, 2020). Obsessions are regarded as intrusive thoughts, while compulsions are repetitive behaviors which are clearly excessive and senseless (Goodman *et al.*, 2014). The *Diagnostic and Statistical Manual for Mental Disorders* (*DSM-5*) outlines the criteria for OCD diagnosis and defines that OCD is not attributed to any substance use or any other medical condition (American Psychiatric Association, Diagnostics and statistical manual of mental disorders, 2013). OCD was initially thought to be a rare condition, but several community surveys found OCD to be the principal cause of disability with a lifetime prevalence ranging from 1.9 to 3.0% and males having a higher chance of early onset of OCD (Goodman *et al.*, 2014). OCD can affect at any stage of life and is known to affect the patient's

*Corresponding author: Saravanan Ramachandran, Associate Professor, Faculty of Allied Health Sciences, Chettinad Hospital and Research Institute, Chettinad Academy of Research and Education, Chettinad Health City, Kelambakkam, Chennai-603103.

DOI: 10.1201/9781003387893-8

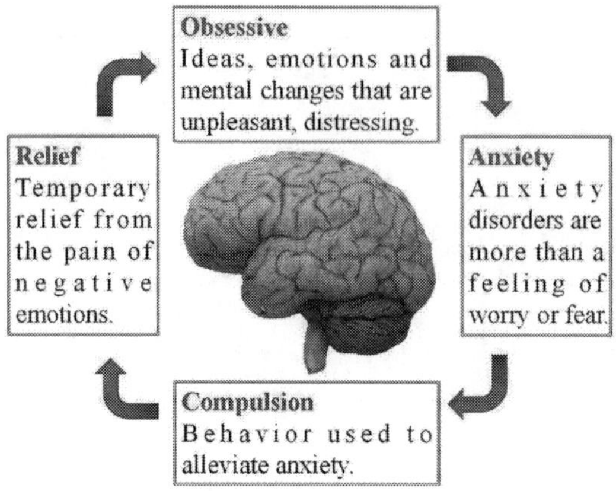

FIGURE 6.1 Major characteristics associated with obsessive-compulsive disorder.

normal work, social life, and family functioning (Wu *et al.*, 2016). OCD includes several themes ranging from fear of illness or contamination to aggressive thoughts, and individuals with OCD tend to neutralize the distress of their obsession by doing excessive labor such as cleaning, checking, etc. OCD is closely related to anxiety and was earlier considered as an anxiety disorder, (Goodman *et al.*, 2014). Figure 6.1 depicts that major characteristics associated with obsessive-compulsive disorder.

The symptoms of OCD are heterogeneous and can vary from individual to individual with non-overlapping symptom pattern. Bloch et al. have identified four major symptoms: (i) symmetry: symmetry obsessions and repeating, ordering, and counting compulsions; (ii) forbidden thoughts: aggressive, sexual, religious, and somatic obsessions and checking compulsions; (iii) cleaning: cleaning and contamination; and (iv) hoarding: hoarding obsessions and compulsions (Bloch *et al.*, 2008). The recent research recommends an additional symptom, namely, higher-order disturbing thoughts, which include harm, inflated responsibility, and checking. They found body focus, superstition, transformation, and loss/separation as unexplored symptom dimensions (Cervin *et al.*, 2021). OCD is also associated with certain comorbidities like personality disorder, social disability, work disability, experience of suicidal thoughts and behaviors, depression, anxiety, feelings of hopelessness, etc., as shown in Figure 6.2 (Mataix-Cols *et al.*, 2005; Angelakis *et al.*, 2015). Figure 6.2 shows the comorbid conditions associated with obsessive-compulsive disorder.

The prevalence of depression and bipolar disorder (BPD) among OCD patients were higher with a frequency of 19–90% and 21.5% respectively. OCD with comorbid depression showed several OCD symptomology with enhanced conflicts within family, while OCD with BPD showed increased hoarding, anxiety, social phobia, and poor functioning (Pallanti *et al.*, 2011). Among the comorbidities, the presence of Tourette syndrome, which is also known as tics, was also found to be evident. Individuals with severe spectrum of OCD showed tic disorder as the first major manifestation of the disease, and individuals with OCD + tic disorder showed enhanced mood swings, anxiety, substance abuse, and increased OCD spectrum (De Mathis *et al.*, 2013). Henceforth the presence of comorbidities among OCD individuals was higher and is considered to be the most important factor involved in chronic worsening of OCD (Mataix-Cols *et al.*, 2005; Angelakis *et al.*, 2015). The presence of comorbid condition can affect the treatment process; for instance, the treatment resistant group was found to have comorbid conditions like BPD and attention deficit hyperactivity disorder (ADHD) (Angelakis *et al.*, 2015). The major symptoms and their comorbid conditions are tabulated in Table 6.1.

The critical and accurate assessment of OCD is required for determining the diagnosis, treatment required for the patients as there are higher chances of under diagnosis. The presence and severity of OCD is widely assessed using the Yale-Brown Obsessive-Compulsive Scale (Y-BOCS), which was first developed in 1989 (Rodriguez-Salgado *et al.*, 2006). The Y-BOCS is a semi-structured interview which includes the symptom checklist with dichotomous questions based on their aggression,

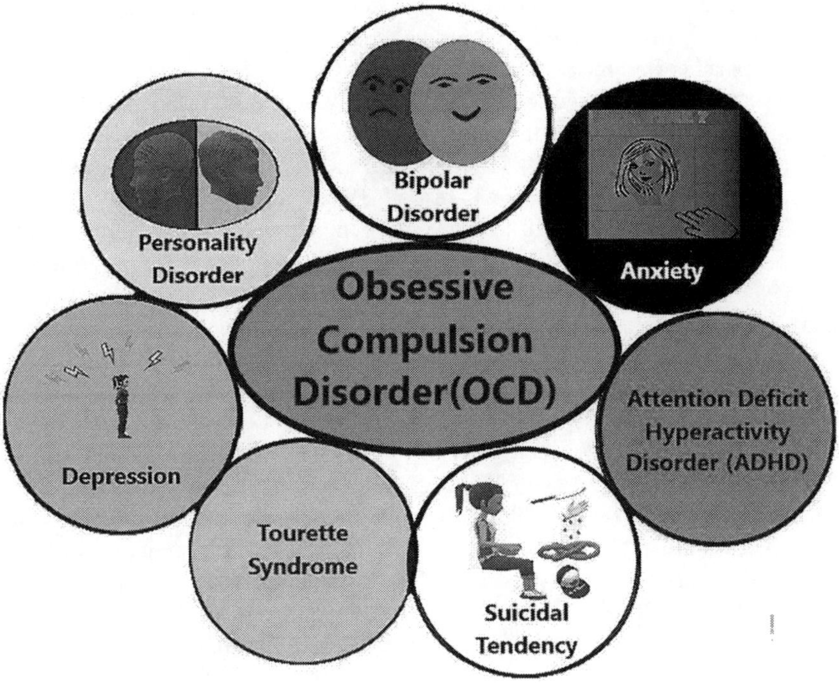

FIGURE 6.2 Obsessive-compulsive disorder and its associated comorbid conditions.

TABLE 6.1
The comorbid condition associated with major symptoms of OCD. These comorbid conditions are commonly found even among other types of obsession/compulsion.

Symptoms	Comorbidities
Symmetry/Hoarding	Tic disorder, work and social disability, pathological grooming condition (skin pricking, nail biting, trichotillomania), dementia, schizophrenia
Symmetry/Ordering	Tic disorder, reduced regional cerebral blood flow (rCBF)
Contamination/Cleaning	Depression, anxiety
Obsession/Checking	Tic disorder
Obsession/Washing	Increased rCBF
Pure Obsession	Depression, anxiety

contamination, sexual, saving/hoarding, religious, need for symmetry or exactness, somatic and miscellaneous obsession. The symptom checklist also contains rating for compulsions based on cleaning, checking, repeating ritual (Goodman *et al.*, 1989). Apart from the symptom checklist, there is a severity scale with 0–4 rate to evaluate the time, distractions, distress, suppression, and control related to the symptomatology. A total value of obsession, compulsion, and severity scale is calculated, and individuals with higher score indicate the presence of severe OCD. Earlier study conducted on 61 OCD patients found Y-BOCS-II is an extension and modified version of Y-BOCS to have good psychometric properties, internal consistency, inter-rater reliability, test-retest reliability, and convergent/divergent validity (Wu *et al.*, 2016). The Y-BOCS symptom checklist questionnaire is shown in Annexure I; however the Y-BOCS require further improvements. The development of OCD is attributed to several underlying etiological factors such as biological, behavioral, neurological, genetic, and immunological across different age groups, as shown in Table 6.2.

OCD has been associated with deviant cortico-striato-thalamo-cortical (CSTC) circuits, which governs flexible behavior and is involved in the normal cognitive, motivational process. Several brain imaging techniques supported the presence of abnormal CSTC axis among individuals with OCD (Van den Heuvel *et al.*, 2016; Nazeer *et al.*, 2020). A lack of communication, imbalance between the frontal area and deeper parts of the brain which uses neurotransmitters for communication was observed (Gershkovich *et al.*, 2018). Psychological theories suggest that obsession is a learned response against avoidance and compulsion in response to anxiety. Parenting was also an important etiology in OCD development, especially among children, and parents of children with OCD were found to be authoritative, overprotective, rejecting, and less emotionally warm (Brakoulias *et al.*, 2018). OCD individuals also showed increased pro-inflammatory markers, and there is a proposed link between OCD development and PANDAS (pediatric autoimmune neuropsychiatric disorders associated with streptococcal infections) which require further confirmation. Genetics also plays a vital role in the development of OCD, and the risk is increased with first-degree OCD relative, and several genes are studied for its association with OCD development (Nazeer *et al.*, 2020).

Several factors like gender, age, occupation are known to affect OCD. Female are more prone to OCD development compared to male; however males are having an increased chance of early onset by 10 years of age. OCD can develop at any stage of life, but the individuals among the age period of 18–29 years are at the highest risk of developing OCD compared to other age groups (Ruscio *et al.*, 2008). In contrast, a population-based Canadian study found that there were no significant changes in OCD prevalence between males and females but was dominant among younger age group (Osland *et al.*, 2018). The recent COVID-19 pandemic also had a worsening impact on OCD, and a recent study conducted during the epidemic period found that prevalence of OCD increased to 17.93%. Other comorbid conditions like occupation, marital status, family history, sleep latency, and presence of other mental disorders were also found to be associated with OCD (Zheng *et al.*, 2020). Occupation was a relevant factor of OCD, and it was reported that the prevalence of OCD among healthcare workers (HCW) was higher compared to non-HCW. Marital status was another predictor with unmarried individuals having a 1.83 higher OCD risk compared to unmarried individuals (Goodman *et al.*, 2014). The current medications of OCD include psychotherapy and pharmacological interventions. The first-line medication of OCD include cognitive-behavioral therapy (CBT), and pharmacological therapy includes the use of clomipramine, fluvoxamine, sertraline, paroxetine, selective serotonin reuptake inhibitor (SSRIs) etc. (Nazeer *et al.*, 2020). However, these drugs need to cross the blood brain barrier (BBB), and small molecules are restricted to enter the BBB. Hence, there is an urgent need to develop efficient delivery

TABLE 6.2
The Major Etiology and its Associated Symptoms

Etiology	Symptoms
Biological	Behavioral inflexibility, aberrant CSTC anatomy and physiology, set shifting, increased glutamate in CSF, frontoparietal-limbic dysfunction, orbitofrontal dysfunction, brain volume reduction, basal ganglia dysfunction, error-related brain activity, altered glutamatergic system, altered amygdala, larger right orbitofrontal complex
Structural	Thick gray matter, cortical surface area, cortical folding, cortical surface and folding measurements, decreased volume of the dorsomedial prefrontal cortex, and an increased volume of the bilateral lenticular nuclei
Psychological	Socioeconomic disadvantages, anxiety, obsession concern
Parenting	Hoarding, contamination/cleaning symptoms
Immunological	Damaged monocytes, TNF-α, IL-12
Genetics	SLC64A, HTR2A, HTR1B, HTR2C, 5-HTTLPR, HTR2A rs6311(rs6313), SLC1A1

system to cross BBB for which nanoparticles may act as a solution. Nanoparticles are small molecules which are in the size of 10–^{9}, and due to its smaller size, it can cross the BBB. Therefore, this chapter details the current nanotechnology developed and used in the diagnosis and treatment of OCD and its comorbidities.

6.2 NANOTECHNOLOGY AND OCD

Nanotechnology is an emerging field and has several applications in the diagnosis and treatment of diseases. Nanotechnology also modifies the pathway of drug delivery and target specific areas of the body by increasing the permeability factor for the release of the drug. Classical drugs cannot pass through the BBB; however nanoparticles, due to its small size, can easily penetrate BBB and has an added advantage by reducing the side effect of the drug injected through BBB. Studies in rat model have shown that nano-encapsulation technology allows nano-capsules to penetrate BBB through ultrasonic wave technology and plasma protein surfactants (Kumar and Alukal, 2020). However, the use of nano-encapsulation for treatment of OCD is still vague.

Different techniques like deep brain stimulations (DBS), neuromodulations are studied as a treatment option for individuals with severe OCD symptoms and are unresponsive to traditional OCD treatments like drugs and behavioral therapy. A systematic review of selected literature found that DBS can act as a promising add-on treatment for patients with OCD (Rapinesi *et al.*, 2019). Nanoscale piezoelectric materials, which are wireless, battery-free implantable, were found to modulate neural activity stimulating neuromodulation. Noninvasive acoustic neuromodulation using piezoelectric barium titanate nanoparticles were found to activate neuron cells which require further confirmation in animal models. ZnS nanoparticles light emitters can act as an optogenetic neuromodulation, which activates neuronal cells using light which passes through the skull. These light emitters can circulate within the systemic circulation and can be recharged using excitation light (Yang *et al.*, 2021). A nanoparticle brain implant of 22nm size coated with polymer was found to increase the compatibility of DBS. These implants were also found to be more specific to its target and have a long life, which was confirmed in studies conducted in mice model. Thus, these nanoparticle implants can increase the efficacy of DBS (Kumar and Alukal, 2020). Nanotechnology is also used in the development of models to study human structure, such as the biomimetic circuit, which demonstrates both typical brain circuits and atypical conditions of OCD. Carbon nano tubes (CNT) transistors was used to study both excitatory and inhibitory synapses involved in OCD as CNT were found to be biocompatible and have lower weight. CNT in combination with CMOS as control unit enabled complex neural networks to be created (Barzegarjalali *et al.*, 2016). For the better understanding of neuronal networks, nano-gel which target specific part of the brain and convert the underdeveloped area to mature cells using stem cell regeneration was found to be a better alternative. The brain cell then surrounds the nano-gel, forming wirelike structures called nanowires which transmits electrical signaling required for the development of neurons (Kumar and Alukal, 2020). Certain recreationally used drugs like cannabinoids are also used to treat OCD and its comorbidities, like ADHD, BPD, tics, etc. The dry powder of cannabinoids is spray dried to form an oil-in-water nano-emulsion which may be administered to the patients through pulmonary inhalation. The pulmonary inhalation was reported to have rapid absorption and have high ability to overcome the first pass effect, thereby increasing the concentration of available drugs (Cookson, 2017). The regulation of neuropeptidergic system using neuropeptide was found to be having safety range, relative potency, and selectivity. Activation of corticotrophin releasing factor (CRF2) receptor was associated with fear, anxiety, and increased serotonin activity, which are primarily seen in patients with OCD. The CRF2 antagonist Antisauvagine-30 (ASV-30) when combined with iron oxide nanoparticle (Fe_2O_3) reduced anxiety like behavior in rats undergoing amphetamine withdrawal with less probability of iron accumulation and thereby neurotoxicity. Direct treatment of ASV-30 in mice also reduced anxiety behavior and did not show any side effect or affect any locomotor function among control mice. Hence the APTES-coated Fe_2O_3 nanoparticle is a promising candidate against anxiety-like disorder (Vinzant *et al.*, 2017). Buspirone hydrochloride (BUH) is used to treat general anxiety disorder by acting as an agonist and partial agonist of presynaptic serotonergic receptors in brain hippocampal and postsynaptic serotonergic receptors in raphe nucleus respectively. Formulated polymeric BUH nanoparticles when administered through nasal capacity using chitosan nanoparticle showed higher concentration in brain compared to oral administration of the same (Bari *et al.*, 2015).

Nanotechnology can be used as an alternative for targeted gene delivery, and PEG and polyethylenimine nano-gels were found to efficiently transport oligonucleotides across the BBB. In rats induced with Parkinson's disease (PD), PEG liposomes with tyrosine hydroxylase plasmid conjugated with anti-transferrin receptor were found to restore the tyrosine hydroxylase activity deficiency and also reversed the motor impairments. Poly L-Lactic acid nano-scaffolds support neurite outgrowth and can be applied for neuron regenerative, which is commonly observed in neurodegenerative diseases including OCD. Studies have been conducted to study the efficacy of nano-carriers and genetic nano-carriers which helps to deliver drugs to the brain and were found to significantly increase the intracerebral concentration of drugs. They were found to efficiently deliver drugs to tumor cells within the brain. However, the delivery of drugs specific to OCD has not yet been studied (Srikanth and Kessler, 2012). Another noninvasive technology developed for the better delivery of protein and peptide drugs to the brain is the nose-to-brain delivery using polymer and lipid based nano-carriers. Nevertheless, further conclusive studies are required to confirm the use of nose to brain (Samaridou and Alonso, 2018). Liposomes encapsulating fluoxetine, which is commonly used as a treatment for OCD, was developed as a co-crystal showed better solubility and aqueous solubility of co-crystal was increased. The liposome encapsulating fluoxetine was found to have improved anti-cancer activity but its efficacy was in treatment for OCD. Trans-dermally administered paroxetine liposomes was found to reduce side effects compared to those of orally administered drugs (Dening *et al.*, 2016). However, the role of various types of nanoparticles conjugated with drugs used to treat OCD is still unknown.

6.3 CONCLUSION

Obsessive-compulsive disorder (OCD) is a neurodevelopmental disorder with a prevalence rate of 1.9 to 3.0%. OCD is also associated with several comorbid conditions like depression, attention deficit hyperactive disorder (ADHD), Tourette syndrome, etc. Cognitive-behavioral therapy (CBT) and pharmacological therapy are currently used as a first-line medication against OCD, but these medications were found to have reduced bioavailability after crossing blood brain barrier (BBB). Nanotechnology can help to overcome these barriers due to its small sizes, and certain technologies like nano-encapsulation can help drugs cross BBB. Nanoscale piezoelectric materials, brain implants, nano-gel, nano-emulsion, PEG liposomes were found to have improved neuronal activity. However, certain limitations associated with nanoparticles is that they remain in systemic circulation for a longer period, lower renal clearance, nano-toxicities, induce oxidative stress, genotoxicity, and modulation of immune system. Further research on the pharmacokinetics and pharmacodynamics of nano-medicines in OCD target can help to better understand the brain targeting efficiency of theses nano-system and their mechanism. Nanotechnology can help in clinical improvement of brain and its diseases. However, the use of these nanomedicines in treatment of OCD needs to be studied in both in vivo and in vitro models to study their physiological and psychological impact.

A

Annexure I

Instructions: Generate a Target Symptoms List from the attached Y-BOCS Symptom Checklist by asking the patient about specific obsessions and compulsions. Check all that apply. Distinguish between current and past symptoms. Mark principal symptoms with a "p." These will form the basis of the Target Symptoms List. Items marked "*" may or may not be OCD.

AGGRESSIVE OBSESSIONS

Current Past

_______ Fear might harm self
_______ Fear might harm others
_______ Violent or horrific images
_______ Fear of blurting out obscenities or insults
_______ Fear of doing something else embarrassing*
_______ Fear will act on unwanted impulses (e.g., to stab friend)
_______ Fear will steal things
_______ Fear will harm others because not careful enough (e.g., hit/run motor vehicle accident)
_______ Fear will be responsible for something else terrible happening (e.g., fire, burglary)
_______ Other_______________________

CONTAMINATION OBSESSIONS

_______ Concerns or disgust with bodily waste or secretions (e.g., urine, feces, saliva)*
_______ Concern with dirt or germs
_______ Excessive concern with environmental contaminants (e.g., asbestos, radiation, toxic waste)
_______ Excessive concern with household items (e.g., cleansers, solvents)
_______ Excessive concern with animals (e.g., insects)
_______ Bothered by sticky substances or residues
_______ Concerned will get ill because of contaminant
_______ Concerned will get others ill by spreading contaminant (Aggressive)
_______ No concern with consequences of contamination other than how it might feel
_______ Other_______________________

Current Past

_______ Fear of losing things
_______ Intrusive (nonviolent) images
_______ Intrusive nonsense sounds, words, or music
_______ Bothered by certain sounds/noises*
_______ Lucky/unlucky numbers
_______ Colors with special significance
_______ Superstitious fears
_______ Other_______________________

SOMATIC OBSESSIONS

_______ Concern with illness or disease*
_______ Excessive concern with body part or aspect of appearance (e.g., dysmorphophobia)*
_______ Other_______________________

CLEANING/WASHING COMPULSIONS

_______ Excessive or ritualized handwashing
_______ Excessive or ritualized showering, bathing, toothbrushing, grooming, or toilet routine
_______ Involves cleaning of household items or other inanimate objects
_______ Other measures to prevent or remove contact with contaminants
_______ Other_______________________

CHECKING COMPULSIONS

________ Checking locks, stove, appliances, etc.

________ Checking that did not/will not harm others

________ Checking that did not/will not harm self

________ Checking that nothing terrible did/will not happen

________ Checking that did not make mistake

________ Checking tied to somatic obsessions

________ Other________________________ ________

SEXUAL OBSESSIONS

________ Forbidden or perverse sexual thoughts, images, or impulses

________ Content involves children or incest

________ Content involves homosexuality*

________ Sexual behavior towards others (Aggressive)*

________ Other________________________ ________

HOARDING/SAVING OBSESSIONS

(distinguish from hobbies and concern with objects of monetary or sentimental value)

____ ________

RELIGIOUS OBSESSIONS (Scrupulosity)

________ Concerned with sacrilege and blasphemy

________ Excess concern with right/wrong, morality

________ Other________________________ ________

OBSESSION WITH NEED FOR SYMMETRY OR EXACTNESS

________ Accompanied by magical thinking (e.g., concerned that another will have an accident unless things are in the right place)

________ Not accompanied by magical thinking

MISCELLANEOUS OBSESSIONS

________ Need to know or remember

________ Fear of saying certain things

________ Fear of not saying just the right thing

REPEATING RITUALS

________ Rereading or rewriting

________ Need to repeat routine activities (e.g., in/out door, up/down from chair)

________ Other________________________ ________

COUNTING COMPULSIONS

________ ________ ________________________ __________

ORDERING/ARRANGING COMPULSIONS

________ ________ ________________________ __________

HOARDING/COLLECTING COMPULSIONS

(distinguish from hobbies and concern with objects of monetary or sentimental value, e.g., carefully reads junk mail, piles up old newspapers, sorts through garbage, collects useless objects)

________ ________ ________________________ __________

MISCELLANEOUS COMPULSIONS

________Mental rituals (other than checking/counting)

________ Excessive list making

________ Need to tell, ask, or confess

________ Need to touch, tap, or rub*

________ Rituals involving blinking or staring*

________ Measures (not checking) to prevent harm to self_____; harm to others_____; terrible consequences_____

________ Ritualized eating behaviors*

________ Superstitious behaviors

________ Trichotillomania*

________ Other self-damaging or self-mutilating behaviors*

________ Other________________________ ______________

REFERENCES

American Psychiatric Association. 2013. *Diagnostic and Statistical Manual of Mental Disorders*, 5th Edition. American Psychiatric Publishing, Washington, DC, pp. 237.

Angelakis, Ioannis, Patricia Gooding, Nichoas Tarrier, and Maria Panagioti. 2015. Suicidality in Obsessive Compulsive Disorder (OCD): A Systematic Review and Meta-Analysis. *Clinical Psychology Review* 39: 1–15. doi:10.1016/j.cpr.2015.03.002.

Bari, Naimat Kalim, Mohammad Fazil, Md Quamrul Hassan, *et al.* 2015. Brain Delivery of Buspirone Hydrochloride Chitosan Nanoparticles for the Treatment of General Anxiety Disorder. *International Journal of Biological Macromolecules* 81: 49–59. doi:10.1016/j.ijbiomac.2015.07.041.

Barzegarjalali, Saeid, Kun Yue, and Alice C. Parker. 2016. Noisy Neuromorphic Circuit Modeling Obsessive Compulsive Disorder. In *2016 29th IEEE International System-on-Chip Conference (SOCC)*. IEEE.

Bloch, Michael H., Angeli Landeros-Weisenberger, Maria C. Rosario, Christopher Pittenger, and James F. Leckman. 2008. Meta-Analysis of the Symptom Structure of Obsessive-Compulsive Disorder. *The American Journal of Psychiatry* 165 (12): 1532–1542. doi:10.1176/appi.ajp.2008.08020320.

Brakoulias, Vlasios, Iain E. Perkes, and Emmanouil Tsalamanios. 2018. A Call for Prevention and Early Intervention in Obsessive-Compulsive Disorder. *Early Intervention in Psychiatry* 12 (4): 572–577. doi:10.1111/eip.12535.

Cervin, Matti, Euripedes C. Miguel, Ayşegül Selcen Güler, Ygor A. Ferrão, Ayşe Burcu Erdoğdu, Luisa Lazaro, Sebla Gökçe, et al. 2021. Towards a Definitive Symptom Structure of Obsessive-Compulsive Disorder: A Factor and Network Analysis of 87 Distinct Symptoms in 1366 Individuals. *Psychological Medicine*, 1–13. doi:10.1017/S0033291720005437.

Cookson, Andrew John Boeckldavid. 2017. Encapsulation of lipophilic ingredients in dispersible spray dried powders suitable for inhalation. W O 2017/127641 A1. Patent, filed January 20, 2017, and issued July 27, 2017. https://patentimages.storage.googleapis.com/5c/26/47/6d2cbb98727a3e/AU2017210319A1.pdf.

De Mathis, Maria Alice., Juliana B. Diniz, Ana G. Hounie *et al.*, 2013. Trajectory in Obsessive-Compulsive Disorder Comorbidities. *European Neuropsychopharmacology: The Journal of the European College of Neuropsychopharmacology* 23 (7): 594–601. doi:10.1016/j.euroneuro.2012.08.006.

Dening, Tahnee J., Shasha Rao, Nicky Thomas, and Clive A. Prestidge. 2016. Oral Nanomedicine Approaches for the Treatment of Psychiatric Illnesses. *Journal of Controlled Release: Official Journal of the Controlled Release Society* 223: 137–156. doi:10.1016/j.jconrel.2015.12.047.

Gershkovich, Marina, Brett J. Deacon, and Michael G. Wheaton. 2018. Biomedical Causal Attributions for Obsessive-Compulsive Disorder: Associations with Patient Perceptions of Prognosis and Treatment Expectancy. *Journal of Obsessive-Compulsive and Related Disorders* 18: 81–85. doi:10.1016/j.jocrd.2018.03.001.

Goodman, W. K., Dorothy E. Grice, Kyle A. B. Lapidus, and Barbara J. Coffey. 2014. Obsessive-Compulsive Disorder. *The Psychiatric Clinics of North America* 37 (3): 257–267. doi:10.1016/j.psc.2014.06.004.

Goodman, W. K., L. H. Price, S. A. Rasmussen *et al.* 1989. The Yale-Brown Obsessive Compulsive Scale. I. Development, Use, and Reliability. *Archives of General Psychiatry* 46 (11): 1006–1011. doi:10.1001/archpsyc.1989.01810110048007.

Kumar Prithvi K. R., and Alukal Albert. 2020. Control of mind using nanotechnology. *Journal of Regenerative Biology and Medicine* 2 (4): 1–14.

Mataix-Cols, David, Maria Conceição do Rosario-Campos, and James F. Leckman. 2005. A Multidimensional Model of Obsessive-Compulsive Disorder. *The American Journal of Psychiatry* 162 (2): 228–238. doi:10.1176/appi.ajp.162.2.228.

Nazeer, Ahsan, Finza Latif, Aisha Mondal, Muhammad Waqar Azeem, and Donald E. Greydanus. 2020. Obsessive-Compulsive Disorder in Children and Adolescents: Epidemiology, Diagnosis and Management. *Translational Pediatrics* 9 (Suppl 1): S76–S93. doi:10.21037/tp.2019.10.02.

Osland, Sydney, Paul D. Arnold, and Tamara Pringsheim. 2018. The Prevalence of Diagnosed Obsessive Compulsive Disorder and Associated Comorbidities: A Population-Based Canadian Study. *Psychiatry Research* 268: 137–142. doi:10.1016/j.psychres.2018.07.018.

Pallanti, Stefano, Giacomo Grassi, Elisa Dinah Sarrecchia, Andrea Cantisani, and Matteo Pellegrini. 2011. Obsessive-Compulsive Disorder Comorbidity: Clinical Assessment and Therapeutic Implications. *Frontiers in Psychiatry* 2: 70. doi:10.3389/fpsyt.2011.00070.

Rapinesi, Chiara, Georgios D. Kotzalidis, Stefano Ferracuti, Gabriele Sani, Paolo Girardi, and Antonio Del Casale. 2019. Brain Stimulation in Obsessive-Compulsive Disorder (OCD): A Systematic Review. *Current Neuropharmacology* 17 (8): 787–807. doi:10.2174/1570159X17666190409142555.

Rodriguez-Salgado Beatriz, Helen Dolengevich-Segal, Manuel Arrojo-Romero, *et al.* 2006. Perceived Quality of Life in Obsessive-Compulsive Disorder: Related Factors. *BMC Psychiatry* 6 (1): 20. doi:10.1186/1471-244X-6-20.

Ruscio, A. M., D. J. Stein, W. T. Chiu, and R. C. Kessler. 2010. The Epidemiology of Obsessive-Compulsive Disorder in the National Comorbidity Survey Replication. *Molecular Psychiatry* 15 (1): 53–63. doi:10.1038/mp.2008.94.

Samaridou, Eleni, and Maria José Alonso. 2018. Nose-to-Brain Peptide Delivery—The Potential of Nanotechnology. *Bioorganic & Medicinal Chemistry* 26 (10): 2888–2905. doi:10.1016/j.bmc.2017.11.001.

Srikanth, Maya, and John A. Kessler. 2012. Nanotechnology-Novel Therapeutics for CNS Disorders. *Nature Reviews. Neurology* 8 (6): 307–318. doi:10.1038/nrneurol.2012.76.

Van den Heuvel, Odile A., Guido van Wingen, Carles Soriano-Mas *et al.* 2016. Brain Circuitry of Compulsivity. *European Neuropsychopharmacology: The Journal of the European College of Neuropsychopharmacology* 26 (5): 810–827. doi:10.1016/j.euroneuro.2015.12.005.

Vinzant Nathan, Jamie L. Scholl, Chia-Ming Wu, Trevor Kindle, Ranjit Koodali, and Gina L. Forster. 2017. Iron Oxide Nanoparticle Delivery of Peptides to the Brain: Reversal of Anxiety during Drug Withdrawal. *Frontiers in Neuroscience* 11. doi:10.3389/fnins.2017.00608.

Wu, Monica S., Joseph F. McGuire, Betty Horng, and Eric A. Storch. 2016. Further Psychometric Properties of the Yale-Brown Obsessive Compulsive Scale—Second Edition. *Comprehensive Psychiatry* 66: 96–103. doi:10.1016/j.comppsych.2016.01.007.

Yang, Xiao, Eve McGlynn, Rupam Das, Sergiu P. Paşca, Bianxiao Cui, and Hadi Heidari. 2021. Nanotechnology Enables Novel Modalities for Neuromodulation. *Advanced Materials (Deerfield Beach, Fla.)* 33 (52): e2103208. doi:10.1002/adma.202103208.

Zheng Yage, Ling Xiao, Yinping Xie, Huiling Wang, and Gaohua Wang. 2020. Prevalence and Characteristics of Obsessive-Compulsive Disorder among Urban Residents in Wuhan during the Stage of Regular Control of Coronavirus Disease-19 Epidemic. *Frontiers in Psychiatry* 11: 594167. doi:10.3389/fpsyt.2020.594167.

7 Obsessive-Compulsive Disorder and Histopathology

K. Kalaiarasi† *and Anu Augustine*
Department of Biotechnology, Sree Narayana Guru College, Coimbatore, TN, India

7.1 INTRODUCTION

Obsessive-compulsive disorder is the brain disorder which is characterized by uncontrollable repetitive behavior and reoccurring of thoughts. These random thoughts get stuck in the brain of individuals with OCD. Every human being has OCD behavior. OCD expression is more common in some people, who have thoughts, actions, and habits that don't make sense. If the efforts to ignore or get rid of bothersome thoughts or urges do not work and they keep coming back, this leads to more ritualistic behavior (Barzilay et al., 2022; Tristan et al., 2017; National Institute of Mental Health, 2019; Chiasson et al., 2017).

There are two main components of OCD:

1. Obsession
2. Compulsion

7.1.1 Obsession

Obsession is caused by excessive repetition of unwanted thoughts, which leads to significant anxiety and distress.

Features:

- Fear of one person using the product and others using it.
- I want things to be orderly and symmetrical.
- Unable to control emotions, like aggressive, horrific, and sexual; also includes awareness of blinking, breathing, or other body sensations.

OCD cannot be ignored or controlled, but it can be managed with the right combination of medication and therapy.

7.1.2 Compulsion

Compulsion is a repetitive behavior, or thoughts, where people engage to reduce anxiety and distress. To suppress unwanted thoughts can likewise incorporate maintaining a strategic distance from circumstances that trigger fixations.

---

†This is the first author.

DOI: 10.1201/9781003387893-9

Features:

- **Washing and cleaning:**
    - Washing hands predominantly.
    - Excessive bathing and brushing of teeth.
    - Always trying to clean.
- **Checking:**
    - Checking lock multiple times.
    - Counting steps when walking.
    - Counting flies on the wall.
    - Afraid that something may go wrong.
- **Other features:**
    - Superstitions about certain numbers, colors, or arrangements.
    - Arranging things in proper order till they feel right.
    - Avoiding obsession situations which light triggers.
    - Reassurance is noted by confessing.

7.2 OCD CAUSES

- OCD is still not clearly understood. Few theories may suggest the causes.
- Stress can make symptoms worse; it is mostly common among women compared to men.
- OCD may also be a result of changes in chemical, structural, functional abnormalities in the brain.
- By seeing/watching the family member's behavior, they started learning the behavior, which can lead to obsessive fears and compulsive behavior.
- Several other factors trigger the development of OCD, which may lead to stressful life events, hormonal changes, and changes in personality traits (Brown, 2019; Visvalingam et al., 2022).

7.2.1 Biological Factors

The brain's extremely different pattern of complex structure called nerve cells, also named as neurons, works by sending messages to the brain to perform body function. They communicate with the neurotransmitters that flow so information can be passed. Serotonin is the chemical responsible for OCD. The normal brain is responsible for linking the parts of the brain involved in planning and judgments and then carrying out those functions. In the condition of OCD, the signals are lost. Neurotransmitters are not communicating at a particular time and not responding to the signals and messages.

7.2.2 Genetic Factors

OCD is due to genetic variation, but the specific gene is not identified (Kring et al., 2012). The genetic factors have a major role in the expression of OCD. Studies on genetics will help to identify the specific gene that is present or not.

7.2.3 Environmental Factors

Environmental factors can be a few things:

- Stressors that trigger the condition of OCD.
- Childhood abuse.

- Death of someone close.
- Illness.
- School or work-related issues or changes.

OCD may be due to a specific type of infection caused by the bacteria streptococcus. If not treated, this can lead to the development of mental disorders such as OCD in children.

7.3 SIGNS AND SYMPTOMS OF OCD

Even though obsessive thoughts and compulsive behaviors are there, it does not mean that they have obsessive-compulsive disorder. With OCD, thoughts and behavior cause tremendous distress and take up a lot of time that interferes with daily life and relationships. Some people can have both obsession and compulsion, but few people experience any one of the two (National Institute of Mental Health, 2019).

OCD comes in many forms, but most causes fall into at least one of four general categories.

1. **Contamination:**
   People with OCD may fear contamination, which includes dirt, germs, body fluids, disease contaminants, or chemicals.
2. **Checking:**
   People with OCD have always checks several times the things which are right or wrong.
3. **Symmetry and ordering:**
   People feel the need to have things lined up in a certain way.
4. **Ruminations and intrusive thoughts:**
   May be violent or disturbing.

7.4 SYMPTOMS

The person affected becomes unreasonable with their time management, which affects the normal day-to-day life, school and social life, and all work gets affected. They have repeated unwanted thoughts and get affected by distress or anxiety.

7.4.1 Symptoms in Children

OCD in children is more common in boys than girls (American Academy of Child & Adolescent Psychiatry, 2019; Boileau, 2022).

1. OCD is a condition that causes kids to have unwanted thoughts, feelings, and fears.
2. Kids feel upset, sad, or anxious.
3. To overcome obsession and anxiety, OCD leads kids to do repetitive behavior.
4. They may feel confused and worry that others will tease them about their fears.
5. OCD can cause kids to have trouble concentrating on schoolwork or enjoying activities.
6. Kids may take a much longer time to do everyday tasks, like getting dressed, organizing a backpack, completing homework, or taking a shower.

7.4.2 Symptoms in Teenagers

OCD begins in the teen or young adult years but can be monitored in childhood. Symptoms usually vary throughout their life. There may be chemical structural and functional abnormalities in the brain caused by changes (Chiang and Purdon, 2019). The changes include the following)

1. **Behavioral causes:**
    - Compulsion can be behavioral or thoughts.
    - OCD is a condition which affects kids, bringing unwanted thoughts, mood changes, and new fears.
    - Kids feel anxious.
    - Obsession is relieved by repetitive behavior called compulsion (Cigala et al., 2015).
2. **Risk factors:**
   Factors that may increase the risk of developing or triggering obsessive-compulsive disorder include: age, family history, gender, personality, stressful life events, mental illness, PANDAS (pediatric autoimmune neuropsychiatric disorders associated with *Streptococcus* infections), stress, and drug use.
    - **Age:**
      It is most common in older teens or young adults and begins as early as preschool age and late forties.
    - **Family history:**
      Family members like parents or others with the disorder can increase the likelihood of a person developing OCD.
    - **Gender:**
      In childhood, males are at greater risk of developing OCD. Different symptoms are seen for both women and men. Males mostly complain of obsessions, like sexuality and asymmetry process. And pregnant women have expressed more OCD than general population (Russell et.al., 2013).
    - **Personality:**
      People who score high on the condition of neuroticism will be at greater risk.
    - **Stressful life events:**
      Some events can trigger the intrusive thoughts, rituals, and emotional distress. Being physically or sexually abused would fall into this category (Ahmed et al., 2015).
    - **Mental illness:**
      OCD often may occur in people who suffer from other mental health illnesses. OCD may be due to anxiety, depression, Tourette syndrome, and attention deficit hyperactivity disorder.
    - **PANDAS (pediatric autoimmune neuropsychiatric disorders associated with *Streptococcus* infections):**
      Pediatric autoimmune neuropsychiatric disorders is related to streptococcal infections. These illnesses are seen in children.
    - **Stress:**
      Stress can make OCD symptoms appear and cause various life changes, so they may lose their loved ones, go to the divorce condition, and issues may arise among relationship and also problems in school.
    - **Drug use:**
      There might be chemicals which are taken as a drug that enters our body that may cause chemical changes in the brain which can create a vulnerability to OCD.

People with OCD and their close family members may have a pattern showing that the brain areas which are responsible for stopping habitual behavior are not properly activated (U.S. Food and Drug Administration, 2019; Garrigan et al., 2018).

7.5 FUNCTIONAL CHANGES IN THE BRAIN

OCD is the main disorder which affects about 2–3% of the population. People with the condition of OCD, their close relatives will have abnormal changes in brain areas especially responsible for habitual behavior. The cortical brain region and subcortical brain region comprise a series of functionally relatively separated circuits, which play a different pattern of roles in thoughts and behaviors. Mutation in the serotonin transporter gene leads to OCD. Neurotransmitter and dysregulation of brain neuron both have an important role in OCD. Serotonin is a neurochemical, as well as dopamine and glutamate, present in patients with OCD, where varied chemical imbalance affects or produces changes in the brain—increased level of dopamine in the prefrontal cortex and decrease in serotonin in basal ganglia. Similar to OCD, other neurological diseases such as Parkinson's disease and Tourette's syndrome affect changes in basal ganglia. Basal ganglia plays a role in the pathophysiology of OCD (Maia et al., 1999).

Neuroimaging studies are widely used for the experimental paradigm, which is divided into four categories (Saxena and Rauch, 2000), such as, resting studies, provocation studies, pre-and post-treatment studies, and cognitive activation studies which compare brain activity in patients with OCD (Maia et al., 2008). Many regions including OCD are striatum, anterior cingulate, thalamus, and caudate nucleus (Saxena and Rauch, 2000). Viewing of structural and volumetric abnormalities in the brain of someone with OCD is done by neuroimaging techniques (MRI). People with OCD show a pattern increase in gray matter in the brain and decrease for others. OCD has a reduced measure of fiber density, axonal diameter in white matter. OCD distributed changes in brain structures both in terms of gray matter and white matter tracts.

FMRI is used to monitor patterns of brain activity. Studies are done with structural imaging, CT and MRI, and functional imaging PET, SPECT, fMRI, and MRS (Parmar and Sarkar, 2016).

7.5.1 Functional Magnetic Resonance Imaging (FMRI)

Functional magnetic resonance imaging is used to measure brain activity in the lateral orbitofrontal cortex (OFC), which is located in the frontal lobes. The lateral OFC is involved in decision-making. Hyperactivity in OFC, ACC, is evidence of abnormalities (Parmar and Sarkar, 2016).

OFC includes decision-making, judgments, planning, and reward processing, which is given as evidence by positron emission tomography (PET) and functional magnetic resonance imaging (fMRI) showing significantly more metabolism and activation in the OFC. Harris and Hoehn-Saric (1995) using MRI, fMRI images found no structural differences between controls and those with OCD. The PET scans showed hypermetabolism in the OFC (Kwon et al., 2003).

7.6 BRAIN AREAS INVOLVED IN OCD

The changes in structural and functional patterns shows the differences in PFC (prefrontal cortex), basal ganglia, ACC (anterior cingulated cortex), OFC, and thalamus. PFC's dorsal portion involved is from the frontal lobe. Mostly they are categorized for being involved in cognitive and emotional brain function. The basal ganglia are found deep within the cerebral hemispheres (Purves et al., 2008). The basal ganglia are deep subcortical structures represented by the striatum and perform executive functional behaviors and emotions (Lanciego et al., 2012).

The striatum is one of the major recipients which gives input to basal ganglia. The dysfunction of specific basal ganglia is especially due to the infection of the microorganism *Streptococcus*. The entire frontal cortex sends projections to the basal ganglia through the thalamus. Frontal-striatal circuits' projections are mediated by the excitatory neurotransmitter glutamate (Vera-Estay et al., 2016). The connections take place through two pathways such as direct (inhibitory) pathway and the indirect pathway. Efferent GABA projections into the thalamus. Thalamocortical projections are reported to be excitatory and use aspartate or glutamate as the neurotransmitter (Maia et al., 1999).

Orbitofrontal cortex (OFC) plays a role in learning, emotion, social behaviors, and stops responding to food taste. ACC is important for specific roles like decision-making, which responds only to the presence of conflicts as source, and they are associated with negative emotional states. The anxiety state is observed due to the activation of ACC. Thalamus, a small structure above the brain stem between the cortex and midbrain, functions as a relay motor and sends sensory signals to the cortex. Motor and sensory signals have a role in the regulation of consciousness and alertness. It has a significant role in emotion and memory (Hal Blumenfeld, 2018). More activated in patients with OCD.

The neuroanatomic model of OCD has a standard model with a direct/indirect striatal pathway. Structured event complex (SEC) possesses the following: an error signal in ACC, OFC reward structures, emotional experience, compensatory SEC in basal ganglia, and activation in PFC.

7.7 SEROTONIN

Serotonin is a neurotransmitter which carries the signal between nerve cells that regulates brain functions such as sleep, emotions, mood, and memory. After carrying a message, serotonin is again reabsorbed by the nerve cells, which act as reuptake. SSRIs work by delaying the reuptake, and hence the level of serotonin is freely ready to transfer the next messages to nerve cells. Serotonin level influences our mental health condition, which leads to OCD. hSERT is a gene that acts as a serotonin transporter, which works very fast to collect all the serotonin before it gives a message to the nerve cell. SSRIs will inhibit gene activity and maintain the serotonin level. The serotonergic system in the brain interacts with the cortical and subcortical system (Tork 1990). The system has many receptors (Yohn et al., 2017). The different types of receptors are given in the next section (Alan and Julie, 1999).

7.7.1 Serotonin Receptors

Serotonin receptors are the main group of 5-HT receptor named as 5-hydroxytryptamine receptors. They are the G-protein coupled receptor, ligand-gated ion channels; both are found in the nervous system, central and peripheral. This mediates both inhibitory and excitatory neurotransmission. These receptors are activated by the ligand, and they release many neurotransmitters, such as glutamate, GABA, dopamine, epinephrine/norepinephrine, and acetylcholine and a few hormones, including oxytocin, vasopressin, cortisol. Serotonin receptors are autoreceptors and heteroreceptors, which affect the brain. Receptors such as 5-HT_1, 5-HT_2, 5-HT_3, 5-HT_4, 5-HT_5, 5-HT_6, and 5-HT_7. The 5-HT_1 receptors (5-HT_{1ARs}) are autoreceptors in neurons that play a role in stress reactivity (Yohn et al., 2017). The 5-HT_2 is a heteroreceptor which is expressed in several limbic structures. The 5-HT_4 G-protein coupled receptors increase intracellular cAMP. The 5-HT_{6Rs} are postsynaptic Gαs-coupled heteroreceptors in the striatum. The 5-HT_7 heteroreceptors are present in cortical regions of the brain (Calza et al., 2019).

7.7.2 Serotonin Syndrome

Certain medications can build high levels of serotonin in the body, which can alter the mental state (Rubin et al., 2018). High levels can also be due to a state of anxiety, agitation, high fever, sweating, confusion, tremors, also changes in blood pressure and heart rate. OCD changes the structure of the brain, especially the frontal lobe. These parts of the brain need serotonin to communicate with other cells. The front part of brain are the anterior cingulated cortex, striatum, and deeper parts of the brain contain the thalamus. The whole collection of brain parts are connected by deeper nerves in the center brain.

7.8 DOPAMINES

Interpretation of cognitive results in patients with OCD shows effects on dopamine level in the nucleus, associated with both increased and decreased dopamine signaling. There is increased level in the prefrontal cortex. The anticipation and rewards both can be designed as a key for understanding the behavior changes due to influence of dopaminergic receptors (Vekshina et al., 2017).

7.8.1 Dopaminergic Receptors

The dopaminergic receptors are the class of G-protein coupled receptors present in the central nervous system. They have a role in daily life functions. They are involved in emotions and all reward portions of the central lobe of the brain. Increased levels of dopamine cause rewarded behavior. Five types of dopamine receptors:

D1: memory, attention, impulse control, regulation of renal function, locomotion
D2: locomotion, attention, sleep, memory, learning
D3: cognition, impulse control, attention, sleep
D4: cognition, impulse control, attention, sleep
D5: decision-making, cognition, attention, renin secretion

These five different dopamine receptors can subdivide into two categories, D1 and D5 receptors. They are G-protein coupled receptors which activates adenylyl cyclase leads for the production of cAMP. The protein kinase A are grouped together, and other receptors, D2, D3, D4, are together in another grouping named G inhibitory site. They affect and inhibit the enzyme adenylyl cyclase and can activate protein kinase channels. D2–D4 receptors are specific signaling receptors needed for the survival of human neuron development system (Mishra et al., 2018; Vekshina et al., 2017; Bhatia et al., 2020).

7.9 GLUTAMATE

Usual medication for OCD target neurotransmitters serotonin and dopamine. Recent studies show the involvement of other neurotransmitters, like glutamate, in OCD. High levels of glutamate in CSF brain patterns is seen on people with OCD (Willie, 2014). Glutamate is an excitatory neurotransmitter present in the brain, and they are critical to nerve cells. A high level can lead to neuron damage (Jimenez et al., 2020).

Glutamate is the most abundant excitatory neurotransmitter present in the brain. It is very critical to the communication of nerve cells with one another in practically every circuit present in the nervous system. An abnormally high level of glutamate can lead to neuron damage. The glutamate-modulating therapies medications aimed for normalizing the actions of glutamate in the brain have been explored in medical conditions such as Lou Gehrig's disease and stroke. The hyperglutamatergic state involves the prefrontal brain regions. Glutamate concentration is seen at a decreased level after the treatment (Pittenger et al., 2011).

7.10 SELECTIVE SEROTONIN REUPTAKE INHIBITORS

SSRI is one type of antidepressant drug which is commonly used but has few side effects than other drugs. It inhibits serotonin by blocking the action of serotonin transport. SSRI increases serotonin levels in the brain. SSRIs work by inhibiting the reuptake of serotonin, so the action allows more serotonin available to be taken up by other nerves. SSRI drugs have adverse effects, like nausea, headache, insomnia, constipation, anxiety, and sexual dysfunction. SSRIs function by increasing the level of serotonin in the brain by stopping its reabsorption in the presynaptic cell. The increased level of serotonin in the synaptic cleft binds to the postsynaptic receptor. It has a strong affinity for serotonin transporter and a weak affinity for dopamine transporters (Driesen, 2004). Other agents such as a low dose of antipsychotic medication has been implicated with OCD, which is associated with abnormalities in the cortico-striatal-thalamic-cortical (CSTC) circuitry. OCD is also associated with dysregulation of neurotransmitters within the CSTC, such as serotonin, dopamine, glutamate, and GABA.

We still don't know exactly what the influences of serotonin are on OCD, and some people's symptoms are not improved despite taking SSRIs.

7.10.1 SSRIs Work

SSRIs can treat the condition of depression by increasing levels of serotonin in the brain. SSRIs block the reabsorption (reuptake) of serotonin into neurons. SSRIs are used to treat conditions other than depression, such as anxiety disorders. The SSRIs are approved for treating depression according to the Food and Drug Administration committee.

7.11 NEUROBIOLOGICAL MODEL

Another way OCD can be characterized is by the activity of the cortico-striatal-thalamic-cortical pathway (CSTC), which includes OFC, ACC, basal ganglia (Calza et al., 2019). There are two systems, such as the direct system, which involves direct projections from the striatum to globus pallidus internal, used as substance P as a neurotransmitter with a net effect of excitation present on the thalamus. The indirect system is performing indirect projections from striatum to Gpi via Gpe using enkephalin as a transmitter with the net effect of inhibition on the thalamus (Van Bavel et al., 2015).

7.12 DIRECT AND INDIRECT PATHWAYS OF CSTC

The experience of OCD symptoms comes from binding together SECs in the prefrontal cortex (PFC), the reward signal in reward structures and the OFC, the threshold for activation of SECs in the basal ganglia, emotional relevance by limbic structures, and the error signal by the ACC (Huey et al., 2008; Calza et al., 2019).

7.13 MEGA ANALYSIS

Due to OCD, the cortical thickness decreased in various frontal, parietal, and temporal cortical legions versus controls. The neurobiological state shows the distributed brain networks, resting state, or during cognitive tasks. Neuroimaging studies in OCD have identified hypo connectivity across circuits. Dysconnectivity (in a direction of other circuits) coupling between brain region (Zagaria et al., 2022).

7.14 CONCLUSION

OCD involves having a random thought that gets stuck in the brain which leads to repetitive behavior. Which makes people do things many times. OCD is more common in women than men. Abnormality in the brain is due to some change in chemical, structural and functional modification in the brain. Neurons communicate through chemicals known as neurotransmitters, which pass information. Symptoms in children and teenagers are common only in 2–3% of the population. Serotonin, dopamine, and glutamate play important roles in OCD. So the level of these neurotransmitters have to be kept in control. If changes occur, it should be controlled by SSRIs, which block the action of serotonin transport, which is the best antidepressant drug that can be used for OCD patients.

REFERENCES

Alan, Frazer., & Julie, G. H. (1999). Serotonin, basic neurochemistry: Molecular. *Cellular and Medical Aspects*: 6(13). www.ncbi.nlm.nih.gov/books/NBK20375/

Ahmed, F., Begum, M., Wahab, M. A., & Ahmed, S. K. (2015). Quality of life of patients with obsessive compulsive disorder. *Bangladesh Journal of Psychiatry*: 29(1), 18–22.

American Academy of Child & Adolescent Psychiatry. (2019). *Obsessive-compulsive disorder in children and adolescents*. www.aacap.org/AACAP/Families_and_Youth/Facts_for_Families/FFF-Guide/Obsessive-Compulsive-Disorder-In-Children-And-Adolescents-060.aspx.

Barzilay, S., Fradkin, I., & Huppert, J. D. (2022). Habitual or hyper-controlled behavior: OCD symptoms and explicit sequence learning. *Journal of Behavior Therapy and Experimental Psychiatry*: Jun 1;75, 101723.

Bhatia, A., Lenchner, J. R., & Saadabadi, A. (2020). Biochemistry, dopamine receptors. In *StatPearls*. StatPearls Publishing. www.ncbi.nlm.nih.gov/books/NBK538242/

Boileau, B. (2022). A review of obsessive-compulsive disorder in children and adolescents. *Dialogues in Clinical Neuroscience*: Apr 1.

Brown, A. Y. (2019). *Allscripts EPSi*. Mayo Clinic.

Calza, J., Gursel, D. A., Schmitz-Koep, B., Bremer, B., Reinholz, L., Berberich, G., & Koch, K. (2019). Altered cortico-striatal functional connectivity during resting state in obsessive-compulsive disorder. *Frontiers in Psychiatry*: ***10***, 319.https://doi.org/10.3389/fpsyt.2019.00319

Chiang, B., & Purdon, C. (2019). Have I done enough to avoid blame? Fear of guilt evokes OCD-like indecisiveness. *Journal of Obsessive-Compulsive and Related Disorders*: ***20***, 13–20.

Chiasson, V., Vera-Estay, E., Lalonde, G., Dooley, J. J., & Beauchamp, M. H. (2017). Assessing social cognition: Age-related changes in moral reasoning in childhood and adolescence. *The Clinical Neuropsychologist*: *31*(3), 515–530.

Cigala, A., Mori, A., & Fangareggi, F. (2015). Learning others' point of view: Perspective taking and prosocial behaviour in preschoolers. *Early Child Development and Care*: ***185***(8), 1199–1215.

Driesen, J. L. (2004). *SSRI actions*. www.driesen.com/ssri_actions.htm.

Garrigan, B., Adlam, A. L., & Langdon, P. E. (2018). Moral decision-making and moral development: Toward an integrative framework. *Developmental Review*: ***49***, 80–100.

Hal, Blumenfeld. (2018). *Thalamus anatomy*. Encylopaedia Britannica. www.britannica.com/science/thalamus

Harris, G. J., & Hoehn, Saric, R. (1995). Functional neuroimaging in biological psychiatry. *Advances in Biological Psychiatry*, 113–160.

Huey, E. D., Zahn, R., Krueger, F., Moll, J., Kapogiannis, D., Wassermann, E. M., & Grafman, J. (2008). A psychological and neuroanatomical model of obsessive-compulsive disorder. *The Journal of Neuropsychiatry and Clinical Neurosciences*: ***20***(4), 390–408. https://doi.org/10.1176/jnp.2008.20.4.390

Jimenez-Trevino, L., Gonzalez-Blanco, L., Alvarez-Vazquez, C., Rodriguez-Revuelta, J., & Saiz Martinez, P. A. (2020). Glutamine and new pharmacological targets to treat suicidal ideation. In *Behavioral neurobiology of suicide and self harm* (pp. 179–196). Springer.

Kring, A. M., Johnson, S. L., Davison, G. C., & Neale, J. M. (2012). *Abnormal psychology*, 12th Edition. John Wiley & Sons.

Kwon, J. S., Kim, J. J., Lee, D. W., Lee, J. S., Lee, D. S., Kim, M. S., . . . Lee, M. C. (2003). Neural correlates of clinical symptoms and cognitive dysfunctions in obsessive–compulsive disorder. *Psychiatry Research: Neuroimaging*: ***122***(1), 37–47.

Lanciego, J. L., Luquin, N., & Obeso, J. A. (2012). Functional neuroanatomy of the basal ganglia. *Cold Spring Harbor Perspectives in Medicine*: ***2***(12).

Maia, A. S., Barbosa, E. R., Menezes, P. R., & Miguel Filho, E. C. (1999). Relationship between obsessive-compulsive disorders and diseases affecting primarily the basal ganglia. *Revista do Hospital das Clínicas*: ***54***(6), 213–221.

Maia, T. V., Cooney, R. E., & Peterson, B. S. (2008). The neural bases of obsessive-compulsive disorder in children and adults. *Development and Psychopathology*: ***20***(4), 1251.

Mishra, A., Singh, S., & Shukla, S. (2018). Physiological and functional basis of dopamine receptors and their role in neurogenesis: Possible implication for Parkinson's disease. *Journal of Experimental Neurology*: 12, 11.

National Institute of Mental Health. (2019). *Obsessive-compulsive disorder*. National Institute of Mental Health.

Parmar, A., & Sarkar, S. (2016). Neuroimaging studies in obsessive compulsive disorder: A narrative review. *Indian Journal of Psychological Medicine*: ***38***(5), 386–394. https://doi.org/10.4103/0253-7176.191395

Pittenger, C., Bloch, M. H., & Williams, K. (2011). Glutamate abnormalities in obsessive compulsive disorder: Neurobiology, pathophysiology, and treatment. *Pharmacology & Therapeutics*: ***132***(3), 314–332.

Purves, D., Augustine, G. J., Fitzpatrick, D., Hall, W. C., LaMantia, A.-S., McNamara, J. O., & White, L. E. (Eds.). (2008). *Neuroscience*, 4th Edition. Sinauer Associates, p.857.

Rubin, A. N., Espiridion, E. D., Kattan, M., & Desmarais, E. C. (2018). Serotonin syndrome with atypical hypernatremia. *Cureus*: ***10***(11).

Russell, E. J., Fawcett, J. M., & Mazmanian, D. (2013). Risk of obsessive-compulsive disorder in pregnant and postpartum women: A meta-analysis. *The Journal of Clinical Psychiatry*: ***74***(4), p.18438.

Saxena, S., & Rauch, S. L. (2000). Functional neuroimaging and the neuroanatomy of obsessive-compulsive disorder. *Psychiatric Clinics of North America*: ***23***(3), 563–586.

Tork, I. (1990). Anatomy of the serotonergic system a. *Annals of the New York Academy of Sciences*: ***600***(1), 9–34.

Tristan, Gorrindo, & Ranna, Parekh. (2017). *What is obsessive-compulsive disorder?* The American Psychiatric Association. www.psychiatry.org/patients-families/ocd/what-is-obsessive-compulsive-disorder.

US Food and Drug Administration. FDA permits marketing of transcranial magnetic stimulation for treatment of obsessive compulsive disorder. www.fda.gov/news-events/press-announcements/fda-permits-marketing-transcranial-magnetic-stimulation-treatment-obsessive-compulsive-disorder [Accessed 27 January 2019].

Van Bavel, J. J., FeldmanHall, O., & Mende-Siedlecki, P. (2015). The neuroscience of moral cognition: From dual processes to dynamic systems. ***Current Opinion in Psychology***: ***6***, 167–172.

Vekshina, N. L., Anokhin, P. K., Veretinskaya, A. G., & Shamakina, I. Y. (2017). Heterodimeric D1-D2 dopamine receptors: A review. *Biomeditsinskaya khimiya*: ***63***(1), 5–12.

Vera-Estay, E., Seni, A. G., Champagne, C., & Beauchamp, M. H. (2016). All for one: Contributions of age, socioeconomic factors, executive functioning, and social cognition to moral reasoning in childhood. *Frontiers in Psychology*: ***7***, 227.

Visvalingam, S., Crone, C., Street, S., Oar, E. L., Gilchrist, P., & Norberg, M. M. (2022). The causes and consequences of shame in obsessive-compulsive disorder. *Behaviour Research and Therapy*: Apr 1;151, 104064.

Willie. (2014). *Dopamine imbalance and its effect on compulsive behaviors*. Emotion, Brain, & Behavior Laboratory (EBBL). https://sites.tufts.edu/emotiononthebrain/2014/10/10/dopamine-imbalance-and-its-effect-on-compulsive-behaviors/

Yohn, C. N., Gergues, M. M., & Samuels, B. A. (2017). The role of 5-HT receptors in depression. *Molecular Brain*: ***10***(1), 1–12.

Zagaria, A., Vacca, M., Cerolini, S., Ballesio, A., & Lombardo, C. (2022). Associations between orthorexia, disordered eating, and obsessive–compulsive symptoms: A systematic review and meta-analysis. *International Journal of Eating Disorders*: Mar;55(3), 295–312.

8 OCD
Current Diagnostic and Treatment Challenges

Upasana Sarma, Parthana Sarma, and Aswani T

8.1 INTRODUCTION

Obsessive-compulsive disorder is a highly prevalent chronic condition found worldwide. It is considered an important mental disorder owing to its associated disability levels and the impact it has on the quality of life of an individual if left underdiagnosed or undertreated. The hallmark of classical OCD is the presence of obsessions and/or compulsions. Obsessions are repetitive in nature along with presence of unwanted thoughts, images, and scenarios that are intrusive and often associated with anxiety. Some of the common obsessions and compulsions include repetitive washing, cleaning, counting issues related to symmetry of things, rearrangement, and even hoarding (Stein et al. 2019). As a condition of enormous suffering, OCD affects approximately one person in 40 worldwide. Even though multiple evidence-based treatment strategies have come up in recent times, diagnosis of this common disorder is either often delayed or difficult to assess. Along with this, a suboptimal and delayed treatment add to the mental agony of the affected individuals (Pittenger et al. 2021). With the possibility of hindering one's social growth and development, OCD is listed by WHO as one of the ten most disabling conditions by financial loss and a decrease in quality of life (Brock H et al. 2022).

The *Diagnostic and Statistical Manual of Mental Disorders* (*DSM-5*), currently in its fifth edition, categorizes OCD under obsessive-compulsive and related disorders (Goodman WK et al. 2014). Accurate identification and establishing the OCD diagnosis is the key component of an evidence-based assessment (Rapp et al. 2016). One of the major diagnostic challenges in OCD is the presence of comorbidities, as shown in Figure 8.1. Other psychiatric disorders, which include anxiety, impulse control, trauma-related, panic attacks, substance abuse, social phobia, and personality disorders, are commonly seen amongst patients with OCD. Despite OCD being a common mental illness, most people seek treatment after several years of suffering and hence remain underdiagnosed for most of their lives (Janardhan Reddy YC et al. 2017). As a common heterogeneous psychiatric disorder, OCD includes themes of obsessions, types of rituals, presence or absence of tics, etiology, genetics, and response to pharmacotherapy (Bokor G et al. 2014). Important questions still persist on the disorder's public health significance, appropriate diagnostic classification, and clinical heterogeneity (Ruscio AM et al. 2010). Some studies have even questioned the empirical validity and practical utility of the *Diagnostic and Statistical Manual of Mental Disorders* (Abramowitz JS et al. 2015). Changes in the diagnostic criteria and grouping of these disorders may have significant clinical implications; it is hence subjected to a lot of revisions as well (Van Ameringen et al. M 2014).

Notwithstanding its severity, OCD is also very often diagnosed with delay, which leads to a long latency between onset of the OCD symptoms and the start of proper treatments. The focus on determining the exact age at which OCD symptoms present along with the course of illness aids

DOI: 10.1201/9781003387893-10

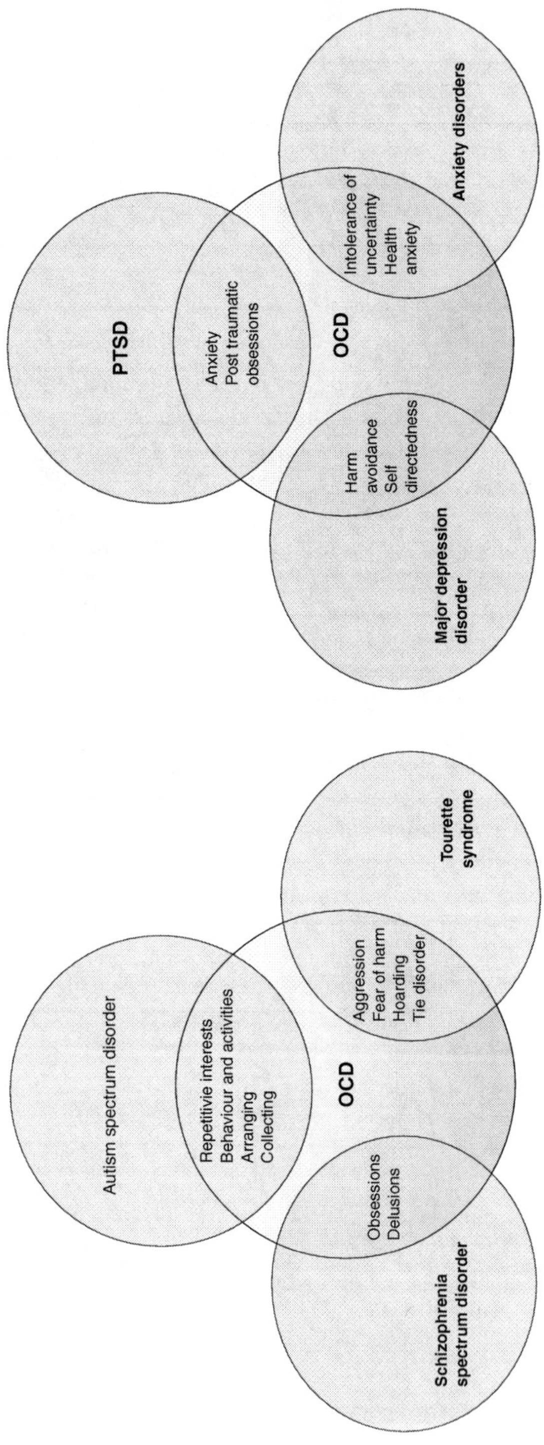

FIGURE 8.1 Comorbid conditions associated with obsessive-compulsive disorder (OCD).

in the overall disease assessment. The difference between early and adult-onset OCD is still not clear-cut, and hence a lot of scrutiny and studies are still under process to give us a clearer and more evidence-backed picture. Crucial steps in a global assessment are an accurate diagnostic formulation. Therefore, the importance of exploring age of onset, family history and involvement, functional assessment, risk assessment, neurocognitive and immunological factors are good considerations for an effective diagnosis in OCD (Pampaloni I et al. 2022). Another major challenge in diagnosis of OCD is nonrecognition and underdiagnosis of symptoms. Though, many scientists regard it as a model psychiatric disease, wherein the symptoms are fascinating but are also seemingly baffling and heterogeneous, which makes diagnosis even more difficult (Robbins TW et al. 2019). A review conducted in Netherlands compared the neurobiological overlap of compulsions in OCD and non-drug-related disorders that share compulsivity, such as pathological gambling, skin-picking, trichotillomania, and compulsive eating. They concluded that compulsivity encompasses many of the Research Domain Criteria (RDoC) constructs in a trans-diagnostic fashion having a common brain circuit dysfunction, which will further help in better diagnosis of OCD (Figee M et al. 2015).

Many individuals have subclinical symptoms years before the development of their OCD, and early treatment may reduce its severity. Through better research there is potential for strategies addressing environmental factors, which may include family factors (Brakoulias V et al. 2017). Several studies have suggested that the genetic influence on OCD is polygenic, with many genes involved which individually exert a relatively small effect on the phenotype. This influence of multiple genes on OCD symptoms has a great implication in diagnostic approach towards the disorder (Krebs G et al. 2014). Additive genetic factors have also been shown to influence OCD. But molecular genetic studies have yielded inconsistent results because of difference in statistical analysis across study models. Further research is needed to investigate the genetic basis of OCD subtypes, such as to properly evaluate early versus late-onset OCD (Taylor S 2013). Advanced research into genetic and biological basis of OCD and its relationship with infections/autoimmunity may help further in diagnosis and understanding the underlying mechanism in OCD and even offer newer and better treatment possibilities (Krebs G et al. 2014).

Some of the common comorbid condition in subjects with obsessive-compulsive disorder (OCD) are bipolar disorder (Saraf G et al. 2017), anxiety along with depression (Klein Breteler J et al. 2021), schizophrenia, amongst many others. Most comorbidity case symptoms are associated with an unfavorable course of the illness; in many cases they are not recognized or treated on time. The situation for patients especially with schizophrenia and obsessive-compulsive disorder (OCD) could improve considerably with more advanced diagnostic techniques (Muller D et al. 2016). Insights into OCD is minimal, and those with preexisting conditions like autism spectrum disorder makes OCD diagnosis complicated. In OCD assessment an essential aspect is to separate the OCD diagnosis from other disorders that may be the cause of the patient's symptomatology instead of OCD (Nazeer A et al. 2020). Moreover, reasonable understanding of the psychological and emotional framework of people with OCD is of utmost importance to avoid the negative effects in life. With the advancement of clinical and technological methods, the detection and thereby treatment of OCD have gained a progressive turn (Kamra et al. 2021). Additionally, various advanced models based on either deep learning or machine learning have been proposed and designed for such early detection of OCD (Kamra et al. 2021). Although these models have proven to be effective, however, due to the heterogenous nature of this mental condition along with its associated uncertainties, accurate detection becomes a challenge. Addressing these challenges is a crucial factor for efficient OCD detection and its subsequent treatment.

The current chapter is an overview on the diagnostic challenges faced by clinicians and researchers in differential diagnosis of OCD along with issues such as delay in diagnosis, nonrecognition of important factors, and effects of comorbidity on diagnosis of the disease. A short summary on challenges of using automation in OCD diagnosis and treatment have also been included.

8.2 DIAGNOSTIC CHALLENGES OF OCD

8.2.1 Differential Diagnosis of OCD

Obsessive-compulsive disorder symptoms often overlap with other disorders like schizophrenia-spectrum disorders (Rasmussen & Parnas 2022), Tourette syndrome (Deeb W et al. 2019), prolonged grief disorder (Szuhany KL et al. 2021), major depressive disorder (Sharma & Sommerdyk 2015), autism spectrum disorders (Paula-Pérez 2013), anxiety or fear (Rozenman M et al. 2017), attention deficit hyperactivity disorder (Brem S et al. 2014), which leads to erroneous over- or under-diagnosis of OCD across different patient age groups. A few simple screening questions can help identify initial OCD symptoms in both children and adolescents and may also lead to early detection and appropriate treatment for the same (Rotberg B et al. 2008).

8.2.1.1 Obsessive-Compulsive Disorder (OCD) and Autism Spectrum Disorders (ASD)

OCD, a common condition characterized by repeated obsessional thoughts and compulsive acts, and ASD, an increasingly recognized neurodevelopmental condition, occurs concurrently with greater frequency in the general population. To explain this occurrence, several factors have been proposed, which includes genetic, anatomic, neurobiological, and psychological ones (Elliott SJ et al. 2021). One study stated that there is a clear diagnostic overlap between ASD and OCD. A better diagnostic difference can be achieved between core symptoms of both the disorders by asking about the target of the interests and routines and comparing it with those of children with OCD (van Steensel FJ et al. 2011). In one of the largest studies of co-occurring OCD and ASD to date, it was found that approximately 25% of young people with OCD also had a diagnosis of ASD, and of those young people who had a diagnosis of ASD, approximately 5% also had a diagnosis of OCD. The severity of OCD symptoms was lower among youth with OCD + ASD compared to those with OCD. This may be due to lesser accuracy at self-reporting by some young people with ASD who may lack deeper insight into their symptoms. In conclusion, this study found that a quarter of young people with OCD also met diagnostic criteria for ASD. This further highlights the need for clinicians working with pediatric OCD to have adequate knowledge of ASD to further ensure accurate detection and diagnosis of these disorders separately (Martin AF et al. 2020). Another study conducted in the United Kingdom found high prevalence of previously undiagnosed ASD in patients with OCD. A greater OCD symptom-severity was found with ASD traits along with poor insight (Wikramanayake WNM et al. 2018). In a recent study, the overlaps among the symptomology of autistic spectrum disorders (ASDs) and obsessive-compulsive disorders (OCDs) was discussed along with the limitations of the *DSM*. It states the importance of flexible ideographic approaches in diagnosing and treating mental health and autism (Edwards DJ et al. 2022). In a comparative study, it stated that obsessions and compulsions are both common in adults with high-functioning ASD and are associated with significant levels of distress, with only somatic obsessions and repeating rituals being more common in the OCD group. These results may have important implications for both clinical practice and research, and proper assessment to distinguish these two types of phenomena is important (Russell AJ et al. 2005). The need to develop tailored assessment and treatment approaches appropriate for individuals with anxiety disorders and obsessive-compulsive disorder (OCD) in autism spectrum disorder (ASD) is much needed amidst all diagnostic challenges (Postorino V et al. 2017).

8.2.1.2 Obsessive-Compulsive Disorder (OCD) and Schizophrenia Spectrum Disorders

Several clinical studies have shown that patients with concurrent schizophrenia and obsessive-compulsive symptoms have more severe psychotic and depressive symptoms, higher suicidality, and lower social functioning (Tezenas du Montcel C et al. 2019). This occurrence causes challenges in the differential diagnosis of OCD, especially in first-contact patients. To add to the problems, the notions of obsession and compulsion are broader and conceptually vague in current diagnostic systems and research instruments (Rasmussen & Parnas 2022). In the current diagnostic criteria, basic concepts of obsession and delusion present in both disorders often overlap, making right diagnosis difficult for either of them. It was further found that a high proportion of clinically diagnosed OCD patients fulfilled diagnostic criteria of a schizophrenia spectrum disorder. There is insufficient differentiation of obsessions from related psychopathological phenomena, and conceptual effort in this domain will further help in proper diagnosis of OCD (Rasmussen AR et al. 2020). Another study has suggested that the presence of prior diagnosis of OCD was associated with an increased risk of developing schizophrenia and schizophrenia spectrum disorders later in life, which is also suggestive of a common etiological pathway for the disorders (Meier SM et al. 2014). The dual diagnosis of the schizo-obsessive disorder makes pharmacological treatment of such patients difficult. Efforts to search for possible specific endophenotypic markers is the need of the hour, which would open new avenues in the knowledge of schizo-obsessive spectrum (Scotti-Muzzi & Saide 2017). Recent studies have focused more on the brain activity in both the diseases, exploring the similarities and differences in brain function between them using near-infrared spectroscopy. A better understanding of the cognitive abilities in both disorders will aid in proper diagnosis (Fu X et al. 2022). In another systemic review involving both children and adolescents, significant GMV abnormalities were seen in multiple brain regions in both OCD and schizophrenia. A closer look at such studies also reveals presence of heterogenous results (based on different regions of the brain) which can help differentiate between both disorders (Liu J et al. 2022).

8.2.1.3 Challenges of Distinguishing OCD from Tourette Syndrome

A commonly seen co-occurring combination is OCD with Tourette syndrome. Both are overlapping conditions, but diagnosis and clinical intervention have different courses. OCS in TS is observed in up to 90%, while tics in OCD up to 37% of the times. Hence proper assessment needs high clinical experience, careful planning, and ongoing evaluation to differentiate, diagnose, and treat both disorders (Rothenberger & Roessner 2019). A cohort study setting in Yale Child Study Center tic and OCD outpatient specialty clinic concluded that obsessive-compulsive disorder symptoms in children with TS became more severe at a later age and were more likely to persist than tic symptoms (Bloch MH et al. 2006). Several evidence-based studies have explored the role of immune dysregulation in the pathophysiology of obsessive-compulsive disorder (OCD) and Tourette syndrome. Recent advances in neuroimaging and animal model work are creating new opportunities to decipher the mechanism behind both disorder, which in turn will aid in better diagnosis of the same (Frick L et al. 2016). The impacts of tic severity and OCD on internalizing (e.g., anxiety) and externalizing (e.g., aggression) psychopathology has also been explored to improve differential diagnosis of both disorders (Pollak Y et al. 2009). In a more recent study, differences of the clinical spectrum of OCD in patients with TS compared to those with "pure OCD" without tics was illustrated. The revised version of the Yale Global Tic Severity Scale (YGTSS-R) was recommended to assess tics, while for the assessment of OCD, the Yale-Brown Obsessive-Compulsive Disorder Scale (Y-BOCS) was recommended (Szejko & Müller-Vahl 2021).

8.2.1.4 Challenges of Distinguishing OCD from PTSD

Obsessive-compulsive disorder (OCD) and post-traumatic stress disorder (PTSD) share the basic clinical feature of anxiety, which makes diagnosis difficult for clinicians (D'Urso G et al. 2018). Following traumatic events in one's life, it becomes increasingly difficult to interpret and differentiate OCD from PTSD. Emergence of post-traumatic obsessions have been reported from several studies conducted in recent times. A case study conducted on veterans who had served in the Second Lebanon War initially found symptoms related to the trauma which later became generalized and an independent form of OCD (Leah Fostick et al. 2012). Acute and chronic symptoms of both PTSD and OCD appear to be closely related to each other, often mistaken to be the same phenomenon. However, a closer look at the symptomatology reveals certain patterns of obsessional thoughts common in OCD (Yehuda Sasson et al. 2005). OCD + PTSD is equally as heterogeneous as OCD and PTSD alone, making assessment and diagnosis difficult. Hence, differential diagnosis and specialized treatment can be a complex where rates of co-occurring OCD within PTSD populations are even higher (Pinciotti CM et al. 2022). More and more studies support the need to consider PTSD symptoms in the assessment and treatment of OCD (Ojserkis R et al. 2017).

8.2.1.5 Challenges of Distinguishing OCD from Major Depressive Disorder

Many disorders with anxiety issues often overlap because of universal dimensions of distress or negative affectivity, a shared genetic predisposition, and a common neurobiology (Goodwin GM 2015). Studies to examine the dimensional personality profile associated with OCD have shown results wherein comorbid depressive symptoms and hoarding obsessions and compulsions were significantly associated with high harm avoidance scores. These results support the existence of a dimensional personality profile associated with OCD (Alonso P et al. 2008). A comprehensive study conducted in early 2000 stated that OCD and major depression share similar personality deviations on harm avoidance, self-directedness, and co-cooperativeness, but low novelty-seeking in OCD is a distinct characteristic difference from major depression which can help in diagnostic approaches (Kusunoki K et al. 2000). A longitudinal twin study compared two hypotheses, whether depression is a functional consequence of OCD or whether these disorders share a common genetic etiology, and the findings showed that OCD symptoms and depressive symptoms co-occur primarily due to shared genetic factors rather than environmental effects (Bolhuis K et al. 2014).

8.2.1.6 Challenges of Distinguishing OCD from Anxiety Disorders

Several studies have indicated that anxiety plays a significant role in the clinical picture of OCD, and severity level can be as high as in generalized anxiety disorder (Citkowska-Kisielewska A et al. 2019). Significant overlapping of symptoms makes differential diagnosis a challenge in such cases. Since both are early onset disorders, diagnostic challenges arises especially in young patients in the developmental stage of life, and intolerance of uncertainty (IU) is one such shared phenomenology (Gillett CB et al. 2018). A much older study highlighted that obsessive-compulsive disorder is mostly characterized by recurrent unwanted but irresistible thoughts and acts of repetition wherein absence of preexisting psychosis or depression is observed (Cameron OG 1985). Diagnostic uncertainty of anxiety subgroups like health anxiety (HA) is often overlooked, and hence overlaps is seen as part of OCD (Villadsen A et al. 2017). A recent study found pediatric OCD with HA symptoms is characterized by more anxiety symptoms and results in a more heterogeneous OCD symptom profile (Duholm CS et al. 2022). Even illness anxiety disorder and obsessive-compulsive disorder (OCD) often co-occur, making differential diagnosis challenging (Knowles KA et al. 2022).

A summarized findings of overlapping symptoms in differential diagnosis of OCD is shown in Table 8.1, where column one and two showcase the specific symptoms of comorbid conditions. Column three denotes the overlapping symptoms along with references in the last column.

8.2.2 Delay in Diagnosis of Early Onset OCD

Early onset mental disorder is a major public health concern as it mainly affects the young working population of a nation. Diagnostic challenges and treatment delay makes this scenario even more difficult (Vaingankar JA et al. 2013). Diagnosing early onset OCD comes with its own set of challenges. A particular study conducted on young children assigned "provisional OCD" to 12% of the total study population as per *DSM 4* as they could not achieve the necessary information required to establish a full OCD diagnosis. Many times, young children are unable to express their symptoms as per requirement leading to incorrect or delayed diagnosis (Garcia AM et al. 2009). Recent studies have denoted early onset OCD as a precise subtype of OCD itself. Moving from a behavioral-based to a biological-based classification of mental disorders in the future is crucial from a diagnostic and treatment point of view (Grassi G et al. 2021). Understanding the distinction between early and late onset will lead to better diagnosis in future investigations (Taylor S 2011).

TABLE 8.1
Summary of Overlapping Symptoms in Differential Diagnosis of OCD

OCD Symptoms	Comorbid Conditions	Overlapping Symptoms	Reference
Obsessions, compulsions, intrusive thoughts (Stein et al. 2019)	Autism spectrum disorders	Repetitive interests, behavior, and activities, arranging, collecting	Jacob S 2009
	Schizophrenia spectrum disorders	Obsessions and delusions	Rasmussen AR et al. 2020
	Tourette syndrome	Aggression, fear of harm, hoarding, tic disorder	Diniz JB et al. 2006
	PTSD	Anxiety, post-traumatic obsessions	Leah Fostick and Zohar 2012
	Major depression disorder	Harm avoidance, self-directedness	Kusunoki K et al. 2000
	Anxiety disorders	Intolerance of uncertainty (IU), health anxiety (HA)	Gillett CB et al. 2018; Villadsen A et al. 2017

8.3 USE OF AUTOMATION IN OCD DIAGNOSIS

In recent years, the automatic detection of OCD through deep learning and machine learning methods have gained huge momentum (Kalmady et al. 2022). Studies have highlighted the significance of using automated processes to detect OCD efficiently. Specifically, identifying the key biomarkers involving neural changes at the functional, structural, and behavioral levels is of utmost importance for accurate diagnosing of mental illness (G.-D. Liu et al. 2022). In this regard, several research involving brain signals for detection of OCD have been reported so far. For instance, Hu et al. (2019) illustrated that patients suffering from certain mental conditions can be segregated from healthy people by employing functional magnetic resonance imaging (fMRI). The studies exploited the information contained in fMRIs to build reliable models for prediction. Nevertheless, these predictions have been conducted on a small subject pool of less than 100. The designed model thus has insufficient data for both cross validation and learning. This results in skewed predictions by the model, thereby limiting its generality. Therefore, getting the required dataset size for the correct prediction can be a huge challenge in automatic detection of OCD. In this regard, Takagi et al. (Takagi et al. 2017) performed OCD prediction by considering dataset with more than 100 subjects. The average accuracies have been reported in the range of approximately 75.5%, which is comparatively low for detection of mental illness. Similarly, Kalmady et al. conducted a study for OCD prediction by employing a machine learning–based model aided by neurobiological feature set (Kalmady et al. 2022). The study used resting-state functional magnetic resonance imaging for the purpose of generalization, reliability, and interpretability. A framework named EMPaSchiz (ensemble algorithm with multiple parcellations for schizophrenia prediction) based on multiparcellation ensemble learning has been employed for the automatic prediction of OCD. The dataset consisted of 350 subjects. Further, transfer learning has been employed to locate the effective underlying neural indications for OCD diagnosis. The results indicated an improved performance of approximately 80%. Although the proposed model performs comparatively better, nonetheless, the study suffers from several challenges. For example, as in most psychiatric studies, the ground truth employed in the model might be not suitably defined, leading to faulty results. Further, the specificity of the reported results is also unclear. The use of a homogenous dataset can also pose hindrance if the model is tested on heterogenous subjects (mixed population, ethnicity, gender, etc.). In another study by Kamra (2022), a predictive model has been introduced to automatically diagnose mental conditions pertaining to bipolar disorder, obsessive-compulsive disorder, and schizophrenia. The model has been tested on the general disease dataset, and the results indicated the highest accuracy of approximately 96%. The major issue in this reported result is the generalization, as the model has been tested only on a single dataset. The power of the model in predicting OCD and other related mental illness over variability remains to be highlighted. Additionally, studies have also focused on detecting specific traits of OCD, such as aggressiveness, as reported by Madanan and Sayed (2022). The study explored the diagnosis of deep obsessive-compulsive disorder (DOCD) among children. For the purpose, a novel Inception-V3 transfer learning CNN has been proposed. The results indicated that aggression level among children can be detected by employing data augmentation with improved performance. Although these works have demonstrated suitable OCD prediction and diagnosis of other related mental illness, there are certain issues which need to be highlighted, as discussed in the next section.

8.3.1 Challenges in Automatic OCD Detection

A major challenge is the requirement of higher computational power and configurations to design models in deep learning. Further, since reliable models can only be built with larger datasets, the storage of such data also becomes an issue. Also, the time consumption of running such models also needs to be properly considered (G.-D. Liu et al. 2020).

The stages of building a predictive model are highly dependent on the feature extraction processes. Current studies are primarily based on shallow learning algorithms, making it difficult to be used for datasets having higher dimensional feature set. Moreover, the present models are mostly based on supervised learning with homogenous datasets, making it a challenge for generalization. To reliably detect OCD from other mental illness, unsupervised models need to be designed along with factors such as ensemble learning, migration learning, and multi-view learning needs to be considered (G.-D. Liu et al. 2020).

Another crucial challenge in automatic OCD prediction is the heterogeneity of the datasets employed. To suitably understand the underlying neural activities and brain region activations pertaining to OCD, the models need to accurately recognize these symptoms aided with proper clinical association. However, the relation between identifying the neuropathological activity and clinical diagnosis of OCD remains indistinct and weak, leading to improper predictive models. This occurs as several brain regions and neural activities concerned with other mental illness can also be associated with OCD. Such feature fusions can lead to incorrect detection among people (Kalmady et al. 2022).

Most of the studies reviewed here have generally focused on categorizing the type of mental illness of the subject. However, the studies failed to take into account the inconsistencies of the various symptoms (Sarkar et al. 2022). Such issues are a major challenge in any automatic OCD detection. The summarized evaluations have been highlighted in Table 8.2, where first and second columns denote the authors and key findings, respectively. The last column displays the challenges.

8.4 TREATMENT CHALLENGES OF OCD

OCD was previously thought to be untreatable, but the last decade has seen several treatment options come up for patients suffering from this disorder. Despite these therapeutic advances, a lot of challenges lie ahead in treating OCD (Hezel & Simpson 2019). A case study on suicidality of a male patient with OCD highlights the necessity of a thorough and individualized approach to treatment for maximum effect (Ross BM et al. 2018). One of the most effective and first-line treatment strategy for OCD has been prolonged administration of selective serotonin reuptake inhibitors (SSRIs) combined with cognitive-behavioral therapy (CBT) (Del Casale A et al. 2019). However, a significant proportion of patients do not respond satisfactorily to first-choice treatments and are regarded as treatment-resistant OCD. Addition of antipsychotics to SSRIs has shown improved results but

TABLE 8.2
Summarized Overview of Challenges of Using Automation in OCD Diagnosis

Authors	Key Findings	Challenges
Kalmady et al. (2022)	Designed EMPaSchiz leading to a prediction performance of 80.3% accuracy for OCD diagnosis.	Used homogeneous dataset.
(Kamra 2022)	Presented a model for predicting psychological conditions such as OCD, bipolar disorder, and schizophrenia.	Method not generalized.
(Madanan & Sayed 2022)	Employed novel model called Inception-V3 transfer learning CNN for prediction of aggression levels for DOCD.	Heterogeneity of features not considered.
(Erguzel et al. 2015)	Employed quantitative EEG features to discriminate OCD and trichotillomania (TTM) using machine learning.	Number of subjects considered is less.
(Clemmensen et al. 2022)	Introduced an improved indicator of OCD severity considering gender and age.	Subject pool is very limited.
(Sekaran & M. 2020)	Revealed the significant gene markers of OCD with active drug responsiveness.	
(Srividya et al. 2018)	Used defined machine learning models to categorize psychological conditions including depression, OCD, etc. of a target group.	Model is strictly based on feature set obtained, which are fused.

Note: EMPaSchiz: ensemble algorithm with multiple parcellations for schizophrenia prediction; DOCD: deep obsessive-compulsive disorder.

remains to be studied extensively (Albert U et al. 2018). The proven second-line treatments include the tricyclic clomipramine and the addition of low-dose neuroleptic medications. However, treatment refractory OCD remains a profound challenge for clinicians across the world (Pittenger & Bloch 2014). A long latency period and high doses of medication are factors that pose difficulties for patients (Mavrogiorgou & Hegerl 1999). Treatment augmentation for resistant patients also lead to adverse side effects (Bandelow B 2008). Clear understanding of mechanism of psychological, pharmacological, and surgical approaches to treat OCD are under a lot of review and ongoing debates (Robbins TW et al. 2019). Another major challenge in treating OCD patients is family members' perceptions of the link between OCD and the person. Unhelpful family member perceptions leads to the severity of OCD being unrecognized, creating treatment challenges (Pedley R et al. 2017). Multiple barriers have been identified as to why people delay treatment for OCD. Some of them are stigma, internal/cognitive factors, fear of criminalization, concern about the reaction of health professionals (Robinson KJ et al. 2017). Even unacceptable taboo thoughts comprising of sexual, religious, and impulsive aggressive obsessions and mental rituals lead to greater levels of hostility among patients and difficulty in treatment of affected individuals (Brakoulias V et al. 2013). Due to the heterogenous nature of OCD both biologically and clinically, patients show different responses to similar pharmacological treatment. The need to divide OCD patients into subgroups for future studies will yield better outcomes (Sumitani S et al. 2006).

8.5 CONCLUSION

OCD patients require a step-by-step hierarchical diagnosis and treatment. As a heterogenous condition with multiple subtypes, OCD differential diagnosis need high-level precision and clarity in understanding the core symptoms for individual subtypes. Co-occurring symptoms need to be addressed and prioritized in new diagnostic approaches. A better understanding of symptomatic, cognitive, and neural patterns in OCD will help mitigate challenges faced by clinicians worldwide. As a common but underdiagnosed psychiatric disorder, OCD requires early intervention and appropriate treatment, which can lead to reduced symptom severity, improve quality of life, and minimize the functional disability in affected individuals. The influence of multiple genes on OCD symptoms is also an area of interest that needs attention of researchers. The detailed understanding of the psychobiology and genomics of OCD will lead to sharper constructs of OCD and its overlapping diseases. Diagnostic classification system of mental health disorders also needs to look deeper into the genetic basis of interrelated disorders. Understanding environmental triggers, OCD subtypes, and their pathophysiology will help locate the genes responsible for it. The focus needs to be on a diagnosis based on biological validity. Delay and incorrect diagnosis lead to lifelong impact on development and function.

REFERENCES

Abramowitz JS, Jacoby RJ. Obsessive-compulsive and related disorders: A critical review of the new diagnostic class. Annu Rev Clin Psychol. 2015;11:165–186. doi: 10.1146/annurev-clinpsy-032813-153713. Epub 2015 Jan 2. PMID: 25581239.

Albert U, Marazziti D, Di Salvo G, Solia F, Rosso G, Maina G. A systematic review of evidence-based treatment strategies for obsessive- compulsive disorder resistant to first-line pharmacotherapy. Curr Med Chem. 2018;25(41):5647–5661. doi: 10.2174/0929867325666171222163645. PMID: 29278206.

Alonso P, Menchón JM, Jiménez S, Segalàs J, Mataix-Cols D, Jaurrieta N, Labad J, Vallejo J, Cardoner N, Pujol J. Personality dimensions in obsessive-compulsive disorder: Relation to clinical variables. Psychiatry Res. 2008 Jan 15;157(1–3):159–168. doi: 10.1016/j.psychres.2006.06.003. Epub 2007 Sep 27. PMID: 17900706.

Bandelow B. The medical treatment of obsessive-compulsive disorder and anxiety. CNS Spectr. 2008 Sep;13(9 Suppl 14):37–46. doi: 10.1017/s1092852900026924. PMID: 18849910.

Bloch MH, Peterson BS, Scahill L, Otka J, Katsovich L, Zhang H, Leckman JF. Adulthood outcome of tic and obsessive-compulsive symptom severity in children with Tourette syndrome. Arch Pediatr Adolesc Med. 2006 Jan;160(1):65–69. doi: 10.1001/archpedi.160.1.65. PMID: 16389213; PMCID: PMC2291298.

Bokor G, Anderson PD. Obsessive-compulsive disorder. J Pharm Pract. 2014 Apr;27(2):116–130. doi: 10.1177/0897190014521996. Epub 2014 Feb 27. PMID: 24576790.

Bolhuis K, McAdams TA, Monzani B, Gregory AM, Mataix-Cols D, Stringaris A, Eley TC. Aetiological overlap between obsessive-compulsive and depressive symptoms: A longitudinal twin study in adolescents and adults. Psychol Med. 2014 May;44(7):1439–1449. doi: 10.1017/S0033291713001591. Epub 2013 Aug 7. PMID: 23920118; PMCID: PMC3959155.

Brakoulias V, Perkes IE, Tsalamanios E. A call for prevention and early intervention in obsessive-compulsive disorder. Early Interv Psychiatry. 2018 Aug;12(4):572–577. doi: 10.1111/eip.12535. Epub 2017 Dec 14. PMID: 29239120.

Brakoulias V, Starcevic V, Berle D, Milicevic D, Moses K, Hannan A, Sammut P, Martin A. The characteristics of unacceptable/taboo thoughts in obsessive-compulsive disorder. Compr Psychiatry. 2013 Oct;54(7):750–757. doi: 10.1016/j.comppsych.2013.02.005. Epub 2013 Apr 12. PMID: 23587527.

Brem S, Grünblatt E, Drechsler R, Riederer P, Walitza S. The neurobiological link between OCD and ADHD. Atten Defic Hyperact Disord. 2014 Sep;6(3):175–202. doi: 10.1007/s12402-014-0146-x. Epub 2014 Jul 14. PMID: 25017045; PMCID: PMC4148591.

Brock H, Hany M. Obsessive-compulsive disorder. In: *StatPearls* [Internet]. Treasure Island, FL: StatPearls Publishing, 2022 Aug 15; 2022 Jan–. PMID: 31985955.

Cameron OG. The differential diagnosis of anxiety. Psychiatric and medical disorders. Psychiatr Clin North Am. 1985 Mar;8(1):3–23. PMID: 3887337.

Citkowska-Kisielewska A, Rutkowski K, Sobański JA, Dembińska E, Mielimąka M. Anxiety symptoms in obsessive-compulsive disorder and generalized anxiety disorder. Psychiatr Pol. 2019 Aug 31;53(4):845–864. English, Polish. doi: 10.12740/PP/105378. Epub 2019 Aug 31. PMID: 31760413.

Clemmensen LKH, Lønfeldt NN, Das S, Lund NL, Uhre VF, Mora-Jensen ARC, Pretzmann L, Uhre CF, Ritter M, Korsbjerg NLJ, Hagstrøm J, Thoustrup CL, Clemmesen IT, Plessen KJ, Pagsberg AK. Associations between the severity of obsessive-compulsive disorder and vocal features in children and adolescents: Protocol for a statistical and machine learning analysis. JMIR Research Protocols. 2022;11(10):e39613. https://doi.org/10.2196/39613

Deeb W, Malaty IA, Mathews CA. Tourette disorder and other tic disorders. Handb Clin Neurol. 2019;165:123–153. doi: 10.1016/B978-0-444-64012-3.00008-3. PMID: 31727209.

Del Casale A, Sorice S, Padovano A, Simmaco M, Ferracuti S, Lamis DA, Rapinesi C, Sani G, Girardi P, Kotzalidis GD, Pompili M. Psychopharmacological treatment of obsessive-compulsive disorder (OCD). Curr Neuropharmacol. 2019;17(8):710–736. doi: 10.2174/1570159X16666180813155017. PMID: 30101713; PMCID: PMC7059159.

Diniz JB, Rosario-Campos MC, Hounie AG, Curi M, Shavitt RG, Lopes AC, Miguel EC. Chronic tics and Tourette syndrome in patients with obsessive-compulsive disorder. J Psychiatr Res. 2006 Sep;40(6):487–493. doi: 10.1016/j.jpsychires.2005.09.002. Epub 2005 Nov 14. PMID: 16289552.

Duholm CS, Højgaard DRMA, Skarphedinsson G, Thomsen PH, Rask CU. Health anxiety symptoms in pediatric obsessive-compulsive disorder: Patient characteristics and effect on treatment outcome. Eur Child Adolesc Psychiatry. 2022 Aug;31(8):1317–1328. doi: 10.1007/s00787-021-01774-x. Epub 2021 Apr 16. PMID: 33861384.

D'Urso G, Mantovani A, Patti S, Toscano E, de Bartolomeis A. Transcranial direct current stimulation in obsessive-compulsive disorder, posttraumatic stress disorder, and anxiety disorders. JECT. 2018 Sep;34(3):172–181. doi: 10.1097/YCT.0000000000000538. PMID: 30095684.

Edwards DJ. Going beyond the DSM in predicting, diagnosing, and treating autism spectrum disorder with covarying alexithymia and OCD: A structural equation model and process-based predictive coding account. Front Psychol. 2022 Sep 6;13:993381. doi: 10.3389/fpsyg.2022.993381. PMID: 36148114; PMCID: PMC9485626.

Elliott SJ, Marshall D, Morley K, Uphoff E, Kumar M, Meader N. Behavioural and cognitive behavioural therapy for obsessive compulsive disorder (OCD) in individuals with autism spectrum disorder (ASD). Cochrane Database Syst Rev. 2021 Sep 3;9(9):CD013173. doi: 10.1002/14651858.CD013173.pub2. PMID: 34693989; PMCID: PMC8543671.

Erguzel TT, Ozekes S, Sayar GH, Tan O, Tarhan N. A hybrid artificial intelligence method to classify trichotillomania and obsessive compulsive disorder. Neurocomputing. 2015;161:220–228. https://doi.org/10.1016/j.neucom.2015.02.039

Figee M, Pattij T, Willuhn I, Luigjes J, van den Brink W, Goudriaan A, Potenza MN, Robbins TW, Denys D. Compulsivity in obsessive-compulsive disorder and addictions. Eur Neuropsychopharmacol. 2016 May;26(5):856–868. doi: 10.1016/j.euroneuro.2015.12.003. Epub 2015 Dec 11. PMID: 26774279.

Frick L, Pittenger C. Microglial dysregulation in OCD, tourette syndrome, and PANDAS. J Immunol Res. 2016;2016:8606057. doi: 10.1155/2016/8606057. Epub 2016 Dec 7. PMID: 28053994; PMCID: PMC5174185.

Fu X, Quan W, Liu L, Li T, Dong W, Wang J, Tian J, Yan J, Liao J. Similarities and differences in brain activation between patients with schizophrenia and obsessive-compulsive disorder: A near-infrared spectroscopy study. Front Psychiatry. 2022 Apr 26;13:853428. doi: 10.3389/fpsyt.2022.853428. PMID: 35558422; PMCID: PMC9086627.

Garcia AM, Freeman JB, Himle MB, et al. Phenomenology of early childhood onset obsessive compulsive disorder. J Psychopathol Behav Assess. 2009;31(2):104–111. doi:10.1007/s10862-008-9094-0

Gillett CB, Bilek EL, Hanna GL, Fitzgerald KD. Intolerance of uncertainty in youth with obsessive-compulsive disorder and generalized anxiety disorder: A transdiagnostic construct with implications for phenomenology and treatment. Clin Psychol Rev. 2018 Mar;60:100–108. doi: 10.1016/j.cpr.2018.01.007. Epub 2018 Feb 2. PMID: 29426573.

Goodman WK, Grice DE, Lapidus KA, Coffey BJ. Obsessive-compulsive disorder. Psychiatr Clin North Am. 2014 Sep;37(3):257–267. doi: 10.1016/j.psc.2014.06.004. Epub 2014 Jul 23. PMID: 25150561.

Goodwin GM. The overlap between anxiety, depression, and obsessive-compulsive disorder. Dialogues Clin Neurosci. 2015 Sep;17(3):249–260. doi: 10.31887/DCNS.2015.17.3/ggoodwin. PMID: 26487806; PMCID: PMC4610610.

Grassi G, Cecchelli C, Mazzocato G, Vignozzi L. Early onset obsessive-compulsive disorder: The biological and clinical phenotype. CNS Spectr. 2021 Feb 1:1–7. doi: 10.1017/S1092852921000122. Epub ahead of print. PMID: 33517936.

Hezel DM, Simpson HB. Exposure and response prevention for obsessive-compulsive disorder: A review and new directions. Indian J Psychiatry. 2019 Jan;61(Suppl 1):S85–S92. doi: 10.4103/psychiatry.IndianJPsychiatry_516_18. PMID: 30745681; PMCID: PMC6343408.

Hu X, Zhang L, Bu X, Li H, Li B, Tang W, Lu L, Hu X, Tang S, Gao Y, Yang Y, Roberts N, Gong Q, Huang X. Localized connectivity in obsessive-compulsive disorder: An investigation combining univariate and multivariate pattern analyses. Front. Behav. Neurosci. 2019;13. www.frontiersin.org/articles/10.3389/fnbeh.2019.00122

Jacob S, Landeros-Weisenberger A, Leckman JF. Autism spectrum and obsessive-compulsive disorders: OC behaviors, phenotypes and genetics. Autism Res. 2009;2(6):293–311. doi:10.1002/aur.108

Janardhan Reddy YC, Sundar AS, Narayanaswamy JC, Math SB. Clinical practice guidelines for obsessive-compulsive disorder. Indian J. Psychiatry. 2017 Jan;59(Suppl 1):S74–S90. doi: 10.4103/0019-5545.196976. PMID: 28216787; PMCID: PMC5310107.

Kalmady SV, Paul AK, Narayanaswamy JC, Agrawal R, Shivakumar V, Greenshaw AJ, Dursun SM, Greiner R, Venkatasubramanian G, Reddy YCJ. Prediction of obsessive-compulsive disorder: Importance of neurobiology-aided feature design and cross-diagnosis transfer learning. Biol Psychiatry Cogn Neurosci Neuroimaging. 2022;7(7):735–746. https://doi.org/10.1016/j.bpsc.2021.12.003

Kamra V. Machine learning and natural language processing based cognitive disease prediction model for prediction of psychological diseases. J. Comput. Sci. 2022;13(2):6.

Kamra V, Kumar P, Mohammadian M. Natural language processing enabled cognitive disease prediction model for varied medical records implemented over ML techniques. *2021 3rd International Conference on Signal Processing and Communication (ICPSC)*. 2021, 494–498. https://doi.org/10.1109/ICSPC51351.2021.9451785

Klein Breteler J, Ikani N, Becker ES, Spijker J, Hendriks G. Comorbid depression and treatment of anxiety disorders, OCD, and PTSD: Diagnosis versus severity. J. Affect Disord. 2021 Dec 1;295:1005–1011. doi: 10.1016/j.jad.2021.08.146. Epub 2021 Sep 3. PMID: 34706408.

Knowles KA, Jakes KS, Olatunji BO. Obsessive-compulsive disorder and illness anxiety: Examining commonalities and comorbidity. J. Cogn Psychother. 2022 Aug 24:JCP-2022–0027.R1. doi: 10.1891/JCP-2022-0027. Epub ahead of print. PMID: 36002282.

Krebs G, Heyman I. Obsessive-compulsive disorder in children and adolescents. Arch Dis Child. 2015 May;100(5):495–499. doi: 10.1136/archdischild-2014-306934. Epub 2014 Nov 14. PMID: 25398447; PMCID: PMC4413836. (twice)

Kusunoki K, Sato T, Taga C, Yoshida T, Komori K, Narita T, Hirano S, Iwata N, Ozaki N. Low novelty-seeking differentiates obsessive-compulsive disorder from major depression. Acta Psychiatr Scand. 2000 May;101(5):403–405. doi: 10.1034/j.1600-0447.2000.101005403.x. PMID: 10823301.

Leah Fostick NN, Zohar J. Acute obsessive compulsive disorder (OCD) in veterans with posttraumatic stress disorder (PTSD). World J. Biol. Psychiatry. 2012;13(4):312–315. DOI: 10.3109/15622975.2011.607848

Liu J, Wen F, Yan J, Yu L, Wang F, Wang D, Zhang J, Yan C, Chu J, Li Y, Li Y, Cui Y. Gray matter alterations in pediatric schizophrenia and obsessive-compulsive disorder: A systematic review and meta-analysis of voxel-based morphometry studies. Front Psychiatry. 2022 Mar 2;13:785547. doi: 10.3389/fpsyt.2022.785547. PMID: 35308883; PMCID: PMC8924120.

Liu GD, Li YC, Zhang W, Zhang L. A brief review of artificial intelligence applications and algorithms for psychiatric disorders. Engineering. 2020;6(4): 462–467. https://doi.org/10.1016/j.eng.2019.06.008

Liu J, Wang H. Analysis of educational mental health and emotion based on deep learning and computational intelligence optimization. Front. Psychol. 2022;13:898609. https://doi.org/10.3389/fpsyg.2022.898609

Madanan M, Sayed BT. Designing a deep learning hybrid using CNN and inception V3 transfer learning to detect the aggression level of deep obsessive compulsive disorder in children. Int. J. Biol. Biomed. 2022;16:207–220. https://doi.org/10.46300/91011.2022.16.27

Martin AF, Jassi A, Cullen AE, Broadbent M, Downs J, Krebs G. Co-occurring obsessive-compulsive disorder and autism spectrum disorder in young people: Prevalence, clinical characteristics and outcomes. Eur Child Adolesc Psychiatry. 2020 Nov;29(11):1603–1611. doi: 10.1007/s00787-020-01478-8. Epub 2020 Feb 1. PMID: 32008168; PMCID: PMC7595977.

Mavrogiorgou P, Hegerl U. Die medikamentöse Behandlung der Zwangsstörung. Mit dem richtigen Mittel und etwas Geduld helfen Sie vielen Patienten [Drug treatment of obsessive-compulsive disorder. With proper drugs and some patience many patients can be helped]. MMW Fortschr Med. 1999 Sep 30;141(39): 32–34, 37. German. PMID: 10897982.

Meier SM, Petersen L, Pedersen MG, Arendt MC, Nielsen PR, Mattheisen M, Mors O, Mortensen PB. Obsessive-compulsive disorder as a risk factor for schizophrenia: A nationwide study. JAMA Psychiatry. 2014 Nov;71(11):1215–1221. doi: 10.1001/jamapsychiatry.2014.1011. PMID: 25188738.

Muller D, Schirmbeck F, de Haan L. Aanbevelingen voor diagnostiek en behandeling van dwangsymptomen bij schizofrenie [Recommendations regarding diagnosis and treatment of obsessive-compulsive symptoms in schizophrenia]. Tijdschr Psychiatr. 2016;58(7):529–534. Dutch. PMID: 27397805.

Nazeer A, Latif F, Mondal A, Azeem MW, Greydanus DE. Obsessive-compulsive disorder in children and adolescents: Epidemiology, diagnosis and management. Transl Pediatr. 2020 Feb;9(Suppl 1):S76–S93. doi: 10.21037/tp.2019.10.02. PMID: 32206586; PMCID: PMC7082239.

Ojserkis R, Boisseau CL, Reddy MK, Mancebo MC, Eisen JL, Rasmussen SA. The impact of lifetime PTSD on the seven-year course and clinical characteristics of OCD. Psychiatry Res. 2017;258:78–82.

Pampaloni I, Marriott S, Pessina E, Fisher C, Govender A, Mohamed H, Chandler A, Tyagi H, Morris L, Pallanti S. The global assessment of OCD. Compr Psychiatry. 2022 Oct;118:152342. doi: 10.1016/j.comppsych.2022.152342. Epub 2022 Aug 6. PMID: 36007341.

Paula-Pérez I. Differential diagnosis between obsessive compulsive disorder and restrictive and repetitive behavioural patterns, activities and interests in autism spectrum disorders. Rev Psiquiatr Salud Ment. 2013 Oct–Dec;6(4):178–186. English, Spanish. doi: 10.1016/j.rpsm.2012.07.005. Epub 2012 Sep 13. PMID: 23084795.

Pedley R, Bee P, Berry K, Wearden A. Separating obsessive-compulsive disorder from the self. A qualitative study of family member perceptions. BMC Psychiatry. 2017 Sep 7;17(1):326. doi: 10.1186/s12888-017-1470-4. PMID: 28882118; PMCID: PMC5590215.

Pinciotti CM, Fontenelle LF, Van Kirk N, Riemann BC. Co-occurring obsessive-compulsive and posttraumatic stress disorder: A review of conceptualization, assessment, and cognitive behavioral treatment. Journal of Cognitive Psychotherapy. 2022; 36(3):207–225.

Pittenger C, Bloch MH. Pharmacological treatment of obsessive-compulsive disorder. Psychiatr Clin North Am. 2014 Sep;37(3):375–391. doi: 10.1016/j.psc.2014.05.006. Epub 2014 Jul 24. PMID: 25150568; PMCID: PMC4143776.

Pittenger C, Brennan BP, Koran L, Mathews CA, Nestadt G, Pato M, Phillips KA, Rodriguez CI, Simpson HB, Skapinakis P, Stein DJ, Storch EA. Specialty knowledge and competency standards for pharmacotherapy for adult obsessive-compulsive disorder. Psychiatry Res. 2021;300:113853. https://doi.org/10.1016/j.psychres.2021.113853

Pollak Y, Benarroch F, Kanengisser L, Shilon Y, Ben-Pazi H, Shalev RS, Gross-Tsur V. Tourette syndrome-associated psychopathology: Roles of comorbid attention-deficit hyperactivity disorder and obsessive-compulsive disorder. J Dev Behav Pediatr. 2009 Oct;30(5):413–419. doi: 10.1097/dbp.0b013e3181ba0f89. Erratum in: J Dev Behav Pediatr. 2011 Nov–Dec;32(9):691. Benpazi, Hilla [corrected to Ben-Pazi, Hilla]. PMID: 19827221.

Postorino V, Kerns CM, Vivanti G, Bradshaw J, Siracusano M, Mazzone L. Anxiety disorders and obsessive-compulsive disorder in individuals with autism spectrum disorder. Curr Psychiatry Rep. 2017 Oct 30;19(12). doi: 10.1007/s11920-017-0846-y. PMID: 29082426; PMCID: PMC5846200.

Rapp AM, et al. Evidence-based assessment of obsessive–compulsive disorder. J. Cent. Nerv System Disease. 2016;8:13–29. doi: 10.4137/JCNSD.S38359.

Rasmussen AR, Nordgaard J, Parnas J. Schizophrenia-spectrum psychopathology in obsessive-compulsive disorder: An empirical study. Eur Arch Psychiatry Clin Neurosci. 2020 Dec;270(8):993–1002. doi: 10.1007/s00406-019-01022-z. Epub 2019 May 25. PMID: 31129700; PMCID: PMC7599137.

Rasmussen AR, Parnas J. What is obsession? Differentiating obsessive-compulsive disorder and the schizophrenia spectrum. Schizophr Res. 2022 May;243:1–8. doi: 10.1016/j.schres.2022.02.014. Epub 2022 Feb 23. PMID: 35219003.

Robbins TW, Vaghi MM, Banca P. Obsessive-compulsive disorder: Puzzles and prospects. Neuron. 2019 Apr 3;102(1):27–47. doi: 10.1016/j.neuron.2019.01.046. PMID: 30946823.

Robinson KJ, Rose D, Salkovskis PM. Seeking help for obsessive compulsive disorder (OCD): A qualitative study of the enablers and barriers conducted by a researcher with personal experience of OCD. Psychol Psychother. 2017 Jun;90(2):193–211. doi: 10.1111/papt.12090. Epub 2017 Apr 10. PMID: 28397354.

Ross BM, Lee JD, Espiridion ED. Suicidality in a 27-year-old male with obsessive compulsive disorder. Cureus. 2018 Dec 31;10(12):e3804. doi: 10.7759/cureus.3804. PMID: 30868018; PMCID: PMC6402744.

Rotberg B, Zalsman G, Weizman A. [Obsessive-compulsive disorder in children and adolescents]. Harefuah. 2008 Apr;147(4):324–329, 374, 373. Hebrew. PMID: 18686815.

Rothenberger A, Roessner V. Psychopharmacotherapy of obsessive-compulsive symptoms within the framework of tourette syndrome. Curr Neuropharmacol. 2019;17(8):703–709. doi: 10.2174/1570159X16666180828095131. PMID: 30152283; PMCID: PMC7059153.

Rozenman M, Peris T, Bergman RL, Chang S, O'Neill J, McCracken JT, Piacentini J. Distinguishing fear versus distress symptomatology in pediatric OCD. Child Psychiatry Hum Dev. 2017 Feb;48(1):63–72. doi: 10.1007/s10578-016-0653-4. PMID: 27225633; PMCID: PMC5860879.

Ruscio AM, Stein DJ, Chiu WT, Kessler RC. The epidemiology of obsessive-compulsive disorder in the national comorbidity survey replication. Mol Psychiatry. 2010 Jan;15(1):53–63. doi: 10.1038/mp.2008.94. Epub 2008 Aug 26. PMID: 18725912; PMCID: PMC2797569.

Russell AJ, Mataix-Cols D, Anson M, Murphy DG. Obsessions and compulsions in Asperger syndrome and high-functioning autism. Br J Psychiatry. 2005 Jun;186:525–528. doi: 10.1192/bjp.186.6.525. PMID: 15928364.

Saraf G, Paul I, Viswanath B, Narayanaswamy JC, Math SB, Reddy YC. Bipolar disorder comorbidity in patients with a primary diagnosis of OCD. Int J Psychiatry Clin Pract. 2017 Mar;21(1):70–74. doi: 10.1080/13651501.2016.1233344. Epub 2016 Sep 20. PMID: 27646489.

Sarkar A, Mallik SS, Roy K. A survey of machine learning based methods for the diagnosis of mental health a survey of machine learning based methods for the diagnosis of mental health. Asia-Pacific Journal of Management and Technology (AJMT). 2022;3(2):Article 2.

Scotti-Muzzi E, Saide OL. Schizo-obsessive spectrum disorders: An update. CNS Spectr. 2017 Jun;22(3): 258–272. doi: 10.1017/S1092852916000390. Epub 2016 Sep 27. PMID: 27669819.

Sekaran K, M., S. Predicting drug responsiveness with deep learning from the effects on gene expression of obsessive–compulsive disorder affected cases. Comput. Commun. 2020;151:386–394. https://doi.org/10.1016/j.comcom.2019.12.049

Sharma V, Sommerdyk C. Obsessive-compulsive disorder in the postpartum period: Diagnosis, differential diagnosis and management. Womens Health (Lond). 2015 Jul;11(4):543–552. doi: 10.2217/whe.15.20. Epub 2015 Aug 6. PMID: 26246310.

Srividya M, Mohanavalli S, Bhalaji N. *Behavioral modeling for mental health using machine learning algorithms | SpringerLink*. 2018. https://link.springer.com/article/10.1007/s10916-018-0934-5

Stein DJ, Costa D, Lochner C, Miguel EC, Reddy Y, Shavitt RG, van den Heuvel OA, Simpson, HB. Obsessive-compulsive disorder. Nat. Rev. Dis. Primers. 2019;5(1):52. https://doi.org/10.1038/s41572-019-0102-3

Sumitani S, Ueno S, Ishimoto Y, Taniguchi T, Tomotake M, Motoki I, Yamauchi K, Ohmori T. [Clinical features of patients with obsessive-compulsive disorder showing different pharmacological responses]. Seishin Shinkeigaku Zasshi. 2006;108(12):1282–1292. Japanese. PMID: 17334085.

Szejko N, Müller-Vahl KR. Challenges in the diagnosis and assessment in patients with tourette syndrome and comorbid obsessive-compulsive disorder. Neuropsychiatr Dis Treat. 2021 Apr 30;17:1253–1266. doi: 10.2147/NDT.S251499. PMID: 33958867; PMCID: PMC8096634.

Szuhany KL, Malgaroli M, Miron CD, Simon NM. Prolonged grief disorder: Course, diagnosis, assessment, and treatment. Focus (Am Psychiatr Publ). 2021 Jun;19(2):161–172. doi: 10.1176/appi.focus.20200052. Epub 2021 Jun 17. PMID: 34690579; PMCID: PMC8475918.

Takagi Y, Sakai Y, Lisi G, Yahata N, Abe Y, Nishida S, Nakamae T, Morimoto J, Kawato M, Narumoto J, Tanaka SC. A neural marker of obsessive-compulsive disorder from whole-brain functional connectivity. Scientific Reports. 2017;7(1):Article 1. https://doi.org/10.1038/s41598-017-07792-7

Taylor S. Early versus late onset obsessive-compulsive disorder: Evidence for distinct subtypes. Clin Psychol Rev. 2011 Nov;31(7):1083–1100. doi: 10.1016/j.cpr.2011.06.007. Epub 2011 Jul 1. PMID: 21820387.

Taylor S. Molecular genetics of obsessive–compulsive disorder: A comprehensive meta-analysis of genetic association studies. Mol Psychiatry. 2013;18:799–805. https://doi.org/10.1038/mp.2012.76

Tezenas du Montcel C, Pelissolo A, Schürhoff F, Pignon B. Obsessive-compulsive symptoms in schizophrenia: An up-to-date review of literature. Curr Psychiatry Rep. 2019 Jul 1;21(8). doi: 10.1007/s11920-019-1051-y. PMID: 31263973.

Vaingankar JA, Rekhi G, Subramaniam M, Abdin E, Chong SA. Age of onset of life-time mental disorders and treatment contact. Soc Psychiatry Psychiatr Epidemiol. 2013 May;48(5):835–843. doi: 10.1007/s00127-012-0601-y. Epub 2012 Oct 18. PMID: 23076588.

Van Ameringen M, Patterson B, Simpson W. DSM-5 obsessive-compulsive and related disorders: Clinical implications of new criteria. Depress Anxiety. 2014 Jun;31(6):487–493. doi: 10.1002/da.22259. Epub 2014 Mar 10. PMID: 24616177.

van Steensel FJ, Bögels SM, Perrin S. Anxiety disorders in children and adolescents with autistic spectrum disorders: A meta-analysis. Clin Child Fam Psychol Rev. 2011 Sep;14(3):302–317. doi: 10.1007/s10567-011-0097-0. PMID: 21735077; PMCID: PMC3162631.

Villadsen A, Thorgaard MV, Hybel KA, Jensen JS, Thomsen PH, Rask CU. Health anxiety symptoms in children and adolescents diagnosed with OCD. Eur Child Adolesc Psychiatry. 2017 Feb;26(2):241–251. doi: 10.1007/s00787-016-0884-8. Epub 2016 Jun 29. PMID: 27357512.

Wikramanayake WNM, Mandy W, Shahper S, Kaur S, Kolli S, Osman S, Reid J, Jefferies-Sewell K, Fineberg NA. Autism spectrum disorders in adult outpatients with obsessive compulsive disorder in the UK. Int J Psychiatry Clin Pract. 2018 Mar;22(1):54–62. doi: 10.1080/13651501.2017.1354029. Epub 2017 Aug 11. PMID: 28705096.

Yehuda Sasson, Dekel Sharon, Nacasch Nitza, Chopra Miriam, Zinger Yaffa, Amital Daniella, Zohar Joseph. Posttraumatic obsessive–compulsive disorder: A case series. Psychiatry Research. 2005;135(2):145–152, ISSN 0165–1781, https://doi.org/10.1016/j.psychres.2004.05.026.

9 An Outlook on Various Diagnostic Approaches of OCD and its Types

Nethravathy V,[a,*,†] *Sowmya Kumar,*[b] *and Mrinalini UG*[c]

[a] Department of Biotechnology, SoAS, REVA University, Bengaluru, KA, India

[b] Department of Life Science, Mount Carmel College, Bengaluru, KA, India

[c] Department of Biotechnology and Microbiology, Bangalore University, Bengaluru, KA, India

9.1 INTRODUCTION

OCD is a condition characterized by obsessions (repeating unreasonable desires or beliefs) and compulsions (behaviors) that interfere with a person's quality of life. The *Diagnostic and Statistical Manual of Mental Disorders, Fifth Edition*, now identifies OCD as a mental disorder (*DSM-5*). OCD is a chronic condition that affects people from early infancy through adulthood, ranging from overt OCD to obsessive-compulsive features. OCD is frequently underdiagnosed and is frequently detected in OCD sufferers with little understanding. Some neurodevelopmental problems, such as OCD and autism spectrum disorder, are more likely to emerge in families with older parents. The top ten human disorders according to the World Health Organization include OCD. OCD can start in childhood, and about eight in ten people who develop it do so before the age of 18. Based on theories of genesis, the research looks for different OCD subtypes (Nazeer A *et al.* 2020). According to research, there are numerous underlying etiologic variables that might cause OCD to develop in people of all ages. These encompass connected problems with psychological, neurological, viral, and immunological roots. Psychological causes can be complex and vast and may be overlapping with other behavioral issues of non-OCD conditions. A complex process in which the OCD individual learns or is motivated to identify what are viewed as faults in their thought patterns in order to cope with ingrained anxiety is described by many psychological and neurological models of genesis. The impact of strict regulations and overly protective parents, the belief that one's acts are never acceptable to others, and other negative childhood experiences are still being contested, especially in light of twentieth- and twenty-first-century approaches to understanding the central nervous system (Doreen LY *et al.* 2018).

Due to alterations in the central nervous system that cause neurological dysfunctions, OCD is recognized as a complicated, diverse condition. These studies on the brain used MRIs, whole-brain VBM, diffusion tensor imaging, and brain scans (DTI). Due to alterations in the central nervous system that cause neurological dysfunctions, OCD is recognized as a complicated, diverse

*Corresponding author.

†This is the first author.

DOI: 10.1201/9781003387893-11

condition. As demonstrated by the story of a child who developed OCD after contracting acute disseminated encephalomyelitis (ADEM), a disease that attacked the cerebral white matter, the involvement of the immune system and infection is still being explored. People with OCD may notice inflammatory damage to their monocytes, which is altered by the effects of OCD treatment (Rozenman M *et al.* 2017).

The microbiome-gut-brain axis, in which gastrointestinal (GI) bacteria influence the brain to affect human behavior via GI inflammation and immunological consequences, is another fascinating field of investigation in the etio-pathogenesis of psychiatric disorder. Depression, OCD, PANS, PANDAS, and other diseases can be caused by or affected by the presence of different GI microorganisms, which are influenced by different internal and external variables (stressors). Group A beta-hemolytic *Streptococcus* and *Bacteroidetes* are two bacteria that are being studied (i.e., *Bacteroides, Odoribacter*, and *Oscillospira*) to understand the influence of microbiome in OCD conditions (Wu Monica S *et al.* 2018).

Genetic contributions to OCD have been analyzed to understand the genetic etiology. The complexities of genetic contributions to OCD and other neuropsychiatric/neurodevelopmental disorders continue to be unraveled by researchers. One report noted deletions at two chromosomes (18q22.1 and 13q12.3-q13.1) in a person with OCD along with Tourette syndrome and dysmorphism. Contrary to dizygotic twins, which have a concordance rate of 0.22, monozygotic twins have a concordance rate of 0.57. First-degree relatives are two times more likely to have OCD than the average person. If childhood OCD is present, this risk is multiplied by ten. Serotonin transporter polymorphism 5-HTTLPR and HTR2A polymorphism are two examples of those that have been linked to OCD. The glutamate transporter (epithelial and neuronal) gene, SLC1A1, and its possible connection to OCD are the subject of research (Janecka M *et al.* 2018).

In this report, Tourette syndrome (TS), pediatric obsessive-compulsive disorder, adult obsessive-compulsive disorder are discussed with respect to its epidemiology, etiology, and differential diagnosis in detail.

9.2 TOURETTE SYNDROME

The French physician Jean Marc Gaspard Itard identified the Marquis of Dampierre as having Tourette syndrome (TS) in 1825 and wrote the first account of a person who has the tics and behaviors that define TS (Walusinski O and Féray J-C 2020). While a student at the Hospital de La Salpêtrière, Gilles de La Tourette reported the pathology as a disorder marked by multiple tics, including the involuntary use or inappropriate repetition of profane words (coprolalia) and sounds, words, or phrases from another person (echolalia), according to the report of Itard itself. It wasn't until 1884 that this pathology was given the name syndrome of Gilles de La Tourette (TS)(Andreas H *et al.* 2018).

Multiple motor tics and at least one phonic tic is present in Tourette syndrome (TS), a disorder that usually develops in childhood and lasts for more than a year but can fluctuate over time (Jones *et al.* 2022). Tics are frequently mild and unnoticeable, but they can also cause physical pain, academic and professional difficulties, and social impairment in certain people. Tics in TS may be comorbid with an array of disorders, such as anxiety, obsessive-compulsive disorder (OCD), and attention deficit hyperactivity disorder (ADHD). A portion of the risk is influenced by genetics, and the pathophysiology emerges to be significantly influenced by the hyperactivity of the dopaminergic circuitry. The first step in management is to educate people about tics (Emilija V *et al.* 2018). There are numerous well-known and developing therapeutic methods that may be taken into account when tics negatively impact the quality of life.

Sometimes the TS seemed to be a curse, and whoever possessed it was destined to exhibit strange actions for the remainder of his life. It is still recognized today as a condition that makes its sufferers struggle to integrate and occasionally makes them inappropriate in numerous circumstances. Children and teenagers that suffer from the disease are frequently discriminated against and pose obstacles to the psychosocial development of children. This illness may cause symptoms to become persistent, which may result in the emergence of other personality disorders.

TS is a neuropsychiatric disorder that manifests in childhood and is marked by the affected individuals' psychological and social commitment, negatively impacting the carriers' and their families' lives. When children are 3 years old, behavioral disorders, most frequently ADHD, start to appear. In contrast, the severity of OCD symptoms peaks in late adolescence, up to many years after the inception of tics. Even when TS is in remission, both of these psychological comorbidities frequently last until adulthood (Ueda K and Black KJ 2021).

9.3 EPIDEMIOLOGY

The first motor tic episode often appears between the ages of 4–6 years. With a 3–1 frequency, TS is more common in boys than in girls (Black KJ *et al.* 2016). The motor tics are typically followed by the phonic ones. Primitive simplex noises like coughing and sneezing are sometimes initially misdiagnosed as "allergies" in practice. According to the literature, TS manifestations are most severe between the ages of 10 and 12. After becoming 21 years old, the majority of patients will experience a full or almost full remission of their disease.

As per studies that included both identified and undiagnosed TS children, 0.6%, or 1 in every 162 children, are thought to have TS (Knight T *et al.* 2012). One in every 333 (0.3%) children in the United States aged 3 to 17 have been diagnosed with TS, according to a CDC study that used parent-reported data; this equates to around 174,000 children between 2016 and 2019 (Bitsko RH *et al.* 2022). This shows that around 50% of children with TS may not be diagnosed.

9.4 PATHOPHYSIOLOGY

The clinical characteristics of TS include simple and/or complex physical tics, as well as simple and complex verbal tics that severely impede social or other important functioning domains or cause significant distress (Eddy CM *et al.* 2011). The motor and limbic systems become disinhibited in TS, which is most likely the outcome of a problem with the cortico-striatal-thalamic-cortical (mesolimbic) circuit. But the situation is still uncertain exactly where anatomically tics are expressed (Augustine F and Singer HS 2019).

The concepts of neurochemically based anomalies for TS have been developed because of many reports of clinical trials, imaging studies, and individual sample research. Dopaminergic dysfunction, originally considering the observation that dopamine receptor-blocking medications known as neuroleptics were most effective in reducing tics, is the most well-known neurochemical explanation of TS (Leckman JF and Bloch MH 2007). Positron emission tomography (PET) studies have not revealed any dopaminergic hyperinnervation, despite some investigations noting aberrations in dopamine transporter combining ability and enlarged cortical and striatal dopamine receptors. The pathophysiology of this syndrome is thought to involve several different neurotransmitters, including dopamine, glutamate, GABA, serotonin, noradrenergic, cholinergic system, serotonergic system, and histaminergic pathways, despite significant progress in understanding the precise cellular and molecular causes of TS (Hienert M *et al.* 2018). These research areas have helped the therapeutic approach to the control of TS and have also laid the groundwork for future advancements in therapeutic tactics. The pathogenesis of TS is thus thought to involve a number of neurotransmitters (Katz TC *et al.* 2022).

Dopaminergic pathways are connected to a variety of movement disorders in addition to the tics produced by TS. The substantia nigra's pars compacta (PC), which receives input from ascending D pathways, is involved in controlling the striatum's output. The excitability of pyramid cells is directly regulated by dopaminergic fibers that emerge from the ventral tegmental area (VTA) in frontal areas, and it is also indirectly regulated by synapses on interneurons (Leckman JF *et al.* 2010). The striatum receives glutaminergic system input, which inhibits medium spiny neuron (MSN) projection through direct pathways on D1 R and indirect pathways on D2 R. In TS there is an upsurge in binding of the vesicular monoamine transporter (VMAT). Cortical dopamine may play a substantial role in the production of tics in TS since the dopamine receptor (DR) binding potential reduces subcortically and cortically. Dopamine active transporter (DAT) binding increases in the neostriatum, while D 1 and D 2 release increases in the ventral striatum (VS). In the thalamus and cortex, the additional striatal D2 receptors are reduced. The various D antagonists are used to prevent certain or specified DR tics from being produced in TS (Nolan S *et al.* 2020).

There is evidence that there is a strong genetic basis for TS. Monozygotic twins are more frequently concordant for the presence of TS in up to 53% of instances, and any tics in up to 77% of cases compared to dizygotic twins (up to 8% concordant for TS and 23% concordant for any tics). Although it is obvious that inherited factors have a substantial influence, the phenotypic can vary and is not always connected to full-blown TS (Ueda K and Black KJ 2021).

9.5 DIAGNOSTICS

There is currently no accurate method or blood test to recognize and diagnose Tourette's syndrome. The patient's history of different signs and symptoms is typically used to make the diagnosis. Tic disorders are identified and diagnosed using a thorough history taking down the start of childhood, family history, detection of a wide variety of behavioral and motor manifestations, and neurological testing (Jones KS *et al.* 2022). When combined, verbal and motor tics appear at some time in the patient's history and persist for more than a year. Additionally, the onset of tic disorder must last for at least 18 years, and tics are not caused by drugs, other substances, or the persistence of other medical conditions.

The presence of comorbidities is not the requirement for correctly diagnosing tic conditions. Tics differ from other movement disorders by having a number of unique characteristics, such as a tendency to wax and wane in severity, a tendency to be connected to sensory stimuli, temporary suppression, and a tendency for movement characteristics to change over time. Tic severity often peaks in children between 10 and 12 years old, and in most cases, symptoms will lessen as they approach puberty. The impact of tics rarely persists into puberty (Liu Z-S *et al.* 2020).

Tics, stereotypies, chorea, dyskinesia, myoclonia, and synkinesis are just a few of the movements that are related to the TS diagnosis. The term "stereotypies behaviors" refers to basic, erratic, and random behaviors in addition to aimless, repetitive, and seemingly voluntary movements. Dyskinesia is a result of protracted twisting movements that are interspersed with prolonged bouts of muscle tension. The basic, fleeting, shock-like muscle contractions that cause an individual's muscles or groups of muscles to change are indicated by myoclonic movements. Synkinesis describes unintentional motions that are connected to precise deliberate actions, such as opening the corner of the mouth while closing just one eye (Ramteke A and Lamture Y 2022).

9.6 TREATMENT

Since the early 1990s, TS has been thoroughly described, and both the pathophysiology of the condition and its treatment has advanced. Since TS was formerly thought to be simply a psychiatric condition, only psychotherapy techniques, particularly psychodynamic ones, were employed in its care. The period of pharmacotherapy for the disorder's symptoms started when haloperidol, a representative of classical antipsychotics, was brought into the treatment of tics in the early 1960s of the previous centuries. Deep brain stimulation (DBS) and behavioral therapy (BT) became crucial as the disorder's nature was more understood (Sarkhel S *et al.* 2020).

The patient, along with his or her family, teachers, and fellow pupils should all receive socio-education about the benignity of tics as the first line of treatment. Neuropsychological therapies may aid in boosting self-esteem and easing depression brought on by a disorder. The most widely used and supported treatments appear to be habit reversal training (HRT), exposure and response prevention (ER), and comprehensive behavioral intervention (CBIT) (Cen SS *et al.* 2020). HRT includes five essential strategies: generalization of abilities, development of a competitive reaction, contingency management, and awareness training (McGuire JF *et al.* 2015). In ER, a type of cognitive behavioral therapy, patients confront their concerns and stop using their escape response. CBIT is a prevalent form of structured therapy that teaches clients how to recognize their tics and provides them with specific behavioral techniques for tic reduction (Fründt O *et al.* 2017).

The second line of treatment, medication, should be regarded as the following four reasons by a physician: The following are the outcome of tics on patients: (a) subjective discomfort (such as pain or damage); (b) long-term social issues (such as bullying or social isolation); (c) emotional issues (such as depression); and (d) functional interference (e.g., impairment of academic achievements). The antipsychotics haloperidol, pimozide, and risperidone appear to be the most efficient medications. Tiapride and sulpiride are often used in several nations. The dualistic neuroleptic aripiprazole has gained prominence recently. It differs from conventional antipsychotics in that it effectively treats tic disorders while having a markedly lower frequency of drowsiness and metabolic adverse effects. Cannabis or other medications that affect the cannabinoid system may be a useful alternative for treating TS, according to the preliminary findings (Novotny M *et al.* 2018).

The final therapeutic option is neurosurgery, particularly DBS. A number of sites, including the thalamus, globus pallidus internal segment, anterior limb internal capsule/nucleus accumbens, and numerous targets have been studied for DBS and GPi DBS as surgical approaches for treating medication-refractory TS. The patient should be informed of potential risks related to neurosurgery as well as its own failure (bleeding and infections at the site of the electrode application). Even its own stimulating effect has certain drawbacks. Sedation, abulia, weariness, apathy, sexual dysfunction, and vision abnormalities are possible side effects for some patients (Lozano AM *et al.* 2019).

9.7 OCD IN ADULTS

OCD is considered a leading cause of disability world over (WHO 2008). Earlier regarded as an untreatable brain disorder has now been considered as highly manageable condition with the advent of new psychological and pharmacological therapies. Among adults, males and females are equally vulnerable to OCD. In men, GAD and tic is often associated with hoarding symptoms, whereas in women, body dysmorphic disorder, nail biting, post-traumatic stress disorder (PTSD), social anxiety, and skin picking are prevalent. Risk factors such as childhood personality, environmental factors and neurodevelopmental factors, abuse during prenatal development are predisposing factors for development of OCD in adults (Lack CW 2012).

Substantial comorbidity is associated OCD. Comorbid disorders are often associated with 75% of adult patients. The most common comorbidity in adults associated with OCD are depression, alcohol abuse, and social anxiety. Phase specific comorbidities are often associated with OCD. Certain patterns of comorbidity are associated with different primary obsessive-compulsive disorder. Other comorbidities such as eating disorders, attention deficit hyperactive disorder (ADHD), disruptive disorders, autism spectrum disorder (ASD), schizophrenia are often associated with OCD, which assists in designing viable treatment strategy (Stein DJ *et al.* 2019). Genetic studies with respect to OCD and comorbidities has also been studied. Genome-wide associated studies (GWA) show a correlation between multiple variations along the genome and disease and has diagnostic value (Gordon 2018, National Advisory Mental Health Council workshop in genomics 2019).

9.8 DIAGNOSIS

Diagnosis of OCD is highly complicated as other non-OCDs are also having similar characteristics. Diagnosis plays a very important role in any clinical condition. In case of OCD diagnosis, a thorough medical history (including family history), physical examination, diagnostic interviews, and psychological assessments are used.

The first step in diagnosis and management of OCI is comprehensive assessment. Rationale behind this assessment is, it aids in making appropriate diagnosis, assessing symptom severity, gaining an insight into obsessive-compulsion symptoms, assessing severity of symptoms, and in choosing appropriate treatment target. For an extensive diagnosis of OCD, well-trained clinicians who conduct personal interviews of patients and a personalized treatment and diagnose the hallmark if OCD. Severities of symptoms are assessed by aid of validated instruments. Structured interviews over free-form interviews are advantageous as they are objective and have psychometric properties (Fineberg N *et al.* 2020).

9.9 TREATMENT

Psychological and pharmacological treatments backed by research evidence are available for treatment of OCD in adults. Serotonin reuptake inhibitor (SRI) therapy has proven to have a marked effect in adult patients (Eddy KT Dutra). Residual impairments and symptoms are unavoidable, and multiple studies have shown a relapse in the condition of OCD. Psychological therapies, SSRI, and tricyclic clomipramine are the most preferred method of treatment (Keeley ML *et al.* 2008). Cognitive behavioral therapy is a psychological treatment of choice for adults and children, especially with exposure and response prevention (EX/RP). The therapy duration is between 12 and 16 sessions. The treatment begins with complete assessment of triggers of obsession and assessment of the extent of distress the patient undergoes when prevented from performing compulsion. Similar to difference in response to CBT therapy, there is a difference amongst population to pharmacological treatment (Lack CW 2012).

9.10 CHILD OCD

About 4% of the general population suffers from obsessive-compulsive disorder (OCD), which is characterized by a variety of compulsions and obsessions. According to the Mayo Clinic, it can have a modest to severe impact on a person's quality of life.

Obsessions are described in the *DSM-4-TR* as habitual recurring, continual thoughts, notion, or inclination and images that are felt as interfering and improper and that significantly increase apprehension or uncertainty; compulsions are described as repetitive actions (such as cleansing hands repeatedly, systematic arrangements, and checking) or mental acts (such as repeating words aloud, praying, counting). Pathologies like eating disorders, body dysmorphic disorders, and impulse control disorders like trichotillomania and paraphilias are not included in these symptoms (Boileau B 2011).

9.11 EPIDEMIOLOGY

Lifetime prevalence estimates for pediatric OCD range from 0.25% to 2–3%. It is a chronic illness. OCD is frequently linked to serious impairments in peer interactions, family functioning, and academic performance. Early onset OCD often manifests at an average age of 10.3 years, ranging from 7.5 to 12.5 years. However, OCD can sometimes appear very young. In a sample of 58 children, the mean age of onset was 4.95 years, and in a study from Turkey, OCD is mentioned in children as young as 2.5 years old. The frequency of subclinical OC symptoms was reported to range from 7% and 25% in various epidemiological investigations (Brezinka V *et al.* 2020).

9.12 ETIOLOGY

9.12.1 Psychological Causes

In order to manage their ingrained anxiety, people with OCD learn or are pushed to identify what they perceive to be flaws in their thought processes, according to a variety of psychological and neuropsychological etiology models. In light of twentieth- and twenty-first-century approaches to studying the central nervous system, the significance of overly protective parenting styles with strict rules, a person's perspective that their actions are never acceptable to others, and various other adverse childhood experiences is still being debated (Chung EH *et al.* 2020).

9.12.2 Environmental Factors

For CO-OCD, there are no definite environmental risk factors. Many patients talk about how their symptoms started after a stressful triggering incident. Although negative childhood events were linked to the existence of psychiatric comorbidities in people with OCD, such as depression, a recent retrospective analysis revealed no evidence for an association between OCD and those experiences. Some statistics point to a potential link between greater recalled rates of prenatal problems and older fathers (Wu Y *et al.* 2012). Pediatric autoimmune disorder associated with streptococcus (PANDAS) is the name given to an early onset, abrupt type of OCD that has been linked to streptococcal infection. The severity of this association is debatable, and there is only weak evidence for causation (Westwell-Roper C *et al.* 2019). According to the description of Sydenham's chorea, a neuropsychiatric consequence of rheumatic fever, these patients are believed to develop postinfectious autoimmunity that targets the basal ganglia. Additionally, more recent epidemiological data point to a stronger correlation between OCD and non-streptococcal pharyngitis than with other anxiety disorders. It has been suggested that the more inclusive term pediatric acute neuropsychiatric syndrome (PANS), which is unrelated to streptococcal infection, encompasses all instances of acute-onset OCD or reduced food intake with concomitant neuropsychiatric symptoms. PANS/PANDAS may affect 5% or less of kids who visit general pediatric OCD (POC) outpatient clinics (Thomsen PH 1995).

9.12.3 Genetics

While the underlying molecular mechanisms are still unknown, meta-analyses of family research indicate heritability between 45% and 65%[2829] higher than in patients with adult-onset illness and higher than most other anxiety and mood disorders in children. A first-degree relative of a proband may have an OCD risk of up to 10–17%, as opposed to 2–3% in controls. Segregation analysis, linkage studies, candidate gene studies, and genome-wide association studies, among other molecular genetic research, point to a complicated condition for which numerous gene variants probably confer vulnerability, each with a tiny effect. Single nucleotide polymorphisms in genes related to glutamatergic transmission and synaptic function have been identified in genetic association studies, albeit none have yet attained genome-wide significance. OCD susceptibility may also be influenced by genes that have an impact on serotonergic and dopaminergic function (Jaspers-Fayer F *et al.* 2017).

9.12.4 Pathophysiological Factors

Functional neuroimaging evidence support the impairment of the orbitofrontal cortex, basal ganglia, thalamus, and interconnecting networks, despite the contradictory results of structural research. It's significant to note that various symptom dimensions might be mediated by several but overlapping brain systems [3738]. These might include a dorsolateral "cognitive" circuit and a ventromedial "emotion" circuit. According to one theory, continuous thoughts of danger and threat activate the orbitofrontal cortex, which in turn triggers actions aimed at removing the perceived threat. Growing evidence points to a malfunctioning reward circuitry in addition to the amygdala and prefrontal cortex's roles in the regulation of fear and anxiety linked to various symptom dimensions (Calaprice D *et al.* 2017).

9.12.4.1 Immune Dysfunction

Data from investigations in people with "classic" OCD reveal a link with abnormal innate immune activation, notwithstanding the lack of evidence for causal pathogenic autoantibodies in CO-OCD, including PANDAS. This includes lower than anticipated immunoglobulin levels as well as altered levels of systemic cytokines such as tumor necrosis factor- and interleukin-1. Furthermore, a recent positron emission tomography imaging research in people with OCD, many of whom also had CO, revealed an increased volume of translocator protein-18 distribution across cortico-striato-thalamo-cortical circuits, suggesting extensive microglial activation. Changes in cellular and soluble inflammatory indicators are either linked epiphenomena, a result of illness progression, or represent underlying pathogenesis (Melin K *et al.* 2018).

9.12.4.2 Diagnosis

Before a proper diagnosis can be made, OCD frequently goes unnoticed for a long period of time. Delays in diagnosis in young individuals may be caused by shame and attempts to hide symptoms, lack of awareness, and/or difficulties telling the difference between actual OCD and normal rituals during growth. Even though OCD is frequently accompanied by a specific set of symptoms (such as excessive washing and frequent checking), the condition is very variable; two OCD sufferers may present with completely distinct symptom profiles, which can make diagnosis difficult (Scahill L *et al.* 1997).

OCD should be taken into consideration in the differential diagnosis of all children and adolescents because of the high prevalence and large cost associated with delayed treatment displaying nervousness or repetitious habits (Storch EA *et al.* 2011). Since young people with obsessions and compulsions frequently do not report their symptoms on their own or may downplay the severity of their symptoms (especially violent or sexual obsessions), a systematic approach to assessment is essential and should be based on history collected from the patient, their family members, and closely associated ones (Foa EB *et al.* 2010).

Differential diagnostic criteria include various assessment processes.

1. *Diagnostic and Statistical Manual of Mental Disorders-5* diagnostic criteria
2. Pediatric acute neuropsychiatric syndrome/pediatric autoimmune disorder associated with *Streptococcus* diagnostic criteria

9.13 CHALLENGES ASSOCIATED WITH DIAGNOSIS

OCD is strikingly heterogeneous with entirely nonoverlapping symptom profiles between two individuals. Detection in adolescent is challenging due to attempts to conceal symptoms or reflect embarrassment and poor insight or difficulty differentiating true OCD from regular behavioral changes during development. Sexual obsessions are common among young people, and they may be reluctant to disclose them during diagnosis. OCD diagnosis is challenging in differentiating between psychosis (such as hallucinations and thought disorder) and OCD. Both are common in adolescents. OCD symptoms can easily be confused with delusions, leading to inappropriate diagnosis and treatment. Aggressive obsessions such as a fear of being harmed can appear similar to paranoia. Hence, careful examination is critical in the assessment of the disease.

9.14 TREATMENT

For the treatment of pediatric OCD, two approaches have a strong body of research behind them: CBT (cognitive-behavior therapy) including exposure with response prevention (E/RP) and selective serotonin reuptake medications (SSRIs). CBT is a reasonably quick-acting treatment for pediatric

OCD, often requiring 12 to 20 weekly sessions (Krebs G and Heyman I 2015). The primary therapy approach is E/RP, in which the young person gradually confronts their dreaded situations (like touching unclean door handles) and abstains from performing compulsions (like handwashing) in an effort to neutralize their anxiety or feared consequence. The young person is advised to wait until their anxiety naturally subsides before practicing the identical E/RP activity repeatedly until it completely vanishes (i.e., habituation).

CBT is an effective treatment for pediatric OCD, according to randomized controlled trials (RCTs). When given in a developmentally appropriate manner, the treatment can be helpful for kids as young as 3 years old and is linked to a 40–65% reduction in symptoms, which has been demonstrated to be maintained up to an 18-month follow-up, suggesting that they are reasonably long-lasting. Fortunately, similar results have been seen in community clinics (i.e., settings other than research settings), indicating that CBT regimens work well in everyday clinical practice (Watson HJ and Rees CS 2008).

The effectiveness of CBT versus SSRI medication in treating pediatric OCD has only been explicitly examined in one study to date. According to this study, combined CBT and SSRI treatment produced better results than either sertraline or CBT alone in terms of symptom reduction. The same group has more recently looked at how much CBT helps young people who are taking SRIs for OCD (Lewin AB *et al.* 2014).

9.15 CONCLUSION

There are still a lot of unresolved questions even though more than 130 years have elapsed since the OCD in 1885. Changes in brain tissue have an unknown specific origin. Similar to this, there is yet no therapy that is highly effective, focused, and secure. There are still a number of hot-button issues surrounding OCD therapy, and the absence of comparative or therapeutically applicable studies is their common denominator. These factors make it unclear whether psychotherapy or medication should be the first course of treatment, or whether using cannabis in TS treatment is appropriate or not. Apart from the treatment, diagnosis is highly challenging as OCD is heterogeneous and affects people of all age groups. Lack of biochemical or metabolic markers is a drawback in the accurate diagnosis of OCD. The effective therapy is possible only by accurate assessment of the symptoms. Research should emphasize on the synergistic evaluation of biochemical-marker-based diagnosis, immunodiagnostic, and molecular correlations.

REFERENCES

Augustine F, Singer HS. Merging the pathophysiology and pharmacotherapy of tics. Tremor Other Hyperkinet Mov. 2019;8:595.

Bitsko Rebecca H, Claussen Angelika H, Lichstein Jesse, Black Lindsey I, Jones Sherry Everett, Danielson Melissa L, Hoenig Jennifer M, Davis Jack Shane P, Brody Debra J, Gyawali Shiromani, Maenner Matthew J, Warner Margaret, Holland Kristin M, Perou Ruth et al. Mental health surveillance among children – United States, 2013–2019. MMWR Suppl. 2022;71(2):1–42.

Black Kevin J et al. Provisional tic disorder: What to tell parents when their child first starts ticking. F1000Res. 2016;5:696.

Boileau B. A review of obsessive-compulsive disorder in children and adolescents. Dialogues Clin Neurosci. 2011;13(4):401–11. doi: 10.31887/DCNS.2011.13.4/bboileau. PMID: 22275846; PMCID: PMC3263388.

Brezinka V, Mailänder V, Walitza S. Obsessive compulsive disorder in very young children—a case series from a specialized outpatient clinic. BMC Psychiatry. 2020;20:366. https://doi.org/10.1186/s12888-020-02780-0

Calaprice D, Tona J, Parker-Athill EC, Murphy TK. A survey of pediatric acute-onset neuropsychiatric syndrome characteristics and course. J Child Adolesc Psychopharmacol. 2017;27:607–18.

Cen SS, Yu J, Wang Q, et al. Multidisciplinary telemedicine care for tourette syndrome: Minireview. Front Neurol. 2020;11:573576.

Chung Estefani Hee, Chou Jesse, Brown Kelly A. Neurodevelopmental outcomes of preterm infants: a recent literature review. Transl Pediatr (Neurodevelopmental and Neurobehavioral Disorders in Children). 2020 February 22;9(Suppl 1).

Eddy Clare M, Rickards Hugh E, Cavanna Andrea E. Treatment strategies for tics in Tourette syndrome. Ther Adv Neurol Disord. 2011;4(1):25–45.

Fineberg Naomi A, Hollander Eric, Pallanti Stefano, Walitza Susanne, Grünblatt Edna, Dell'Osso Bernardo Maria, Albert Umberto, Geller Daniel A, Brakoulias Vlasios, Janardhan Reddy YC, Arumugham ShyamSundar, Shavitt Roseli G, Drummond Lynne, Grancini Benedetta, De Carlo Vera, Cinosi Eduardo, Chamberlain Samuel R, Ioannidis Konstantinos, Rodriguez Carolyn I, Garg Kabir, Castle David, Van Ameringen Michael, Stein Dan J, Carmi Lior, Zohar Joseph, Menchon Jose M. Clinical advances in obsessive-compulsive disorder: A position statement by the International College of Obsessive-Compulsive Spectrum Disorders. Int Clin Psychopharmacol. 2020 July;35(4):173–93. doi: 10.1097/YIC.0000000000000314

Foa EB, Coles M, Huppert JD, Pasupuleti RV, Franklin ME, March J. Development and validation of a child version of the obsessive compulsive inventory. Behav Ther. 2010;41:121-32.

Fründt O, Woods D, Ganos C. Behavioral therapy for Tourette syndrome and chronic tic disorders. Neurol Clin Pract. 2017;7(2):148–56.

Hartmann Andreas et al. Tourette syndrome research highlights from 2017. F1000Res. 2018;7:1122.

Hienert Marius, Gryglewski Gregor, Stamenkovic Mara, Kasper Siegfried, Lanzenberger Rupert. Striatal dopaminergic alterations in Tourette's syndrome: A meta-analysis based on 16 PET and SPECT neuroimaging studies. Transl Psychiatry 2018;8(1):143.

Janecka M, Hansen SN, Modabbernia A, Browne HA, Buxbaum JD, Schendel DE, Reichenberg A, Parner ET, Grice DE. Parental age and differential estimates of risk for neuropsychiatric disorders: Findings from the Danish birth cohort. J Am Acad Child and Adolesc Psychiatry. 2019;58(6):618–27. https://doi.org/10.1016/j.jaac.2018.09.447

Jaspers-Fayer F, Han SH, Chan E, McKenney K, Simpson A, Boyle A, et al. Prevalence of acute-onset subtypes in pediatric obsessive-compulsive disorder. J Child Adolesc Psychopharmacol 2017;27:332–41.

Jones Kimberly S, Saylam Ezgi, Ramphul Kamleshun. Tourette syndrome and other tic disorders. Statpearls. 2022.

Katz Tamar C, Bui Thanh Hoa, Worhach Jennifer, Bogut Gabrielle, Tomczak Kinga K. Tourettic OCD: Current understanding and treatment challenges of a unique endophenotype. Front. Psychiatry 2022;13.

Keeley ML, Storch EA, Merlo LJ, Geffken GR. Clinical predictors of response to cognitive-behavioral therapy for obsessive–compulsive disorder. Clin. Psychol. Rev. 2008;28(1):118–30. https://doi.org/10.1016/j.cpr.2007.04.003

Knight T, Steeves T, Day L, Lowerison M, Jette N, Pringsheim T. Prevalence of tic disorders: A systematic review and meta-analysis. Pediatr Neurol. 2012;47(2):77–90.

Krebs G, Heyman I. Obsessive-compulsive disorder in children and adolescents. Arch Dis Child. 2015 May;100(5):495–9. doi: 10.1136/archdischild-2014-306934. Epub 2014 Nov 14. PMID: 25398447; PMCID: PMC4413836.

Lack CW. Obsessive-compulsive disorder: Evidence-based treatments and future directions for research. World J. Psychiatry. 2012;2(6):86–90. https://doi.org/10.5498/wjp.v2.i6.86

Leckman James F, Bloch Michael H. Tourette's syndrome. In Molecular Neurology, edited by Stephen G. Waxman, 385–408. Academic Press, 2007.

Leckman James F et al. Neurobiological substrates of Tourette's disorder. J Child Adolesc Psychopharmacol. 2010;20(4):237–47.

Lewin AB, Park JM, Jones AM, et al. Family-based exposure and response prevention therapy for preschool-aged children with obsessive-compulsive disorder: A pilot randomized controlled trial. Behav Res Ther. 2014;56:30–8.

Lewis Yael Doreen, Mann Tal Gilon, Enoch-Levy Adi, Dubnov-Raz Gal, Gothelf Doron, Weizman Abraham, Stein Daniel. Obsessive–compulsive symptomatology in female adolescent inpatients with restrictive compared with binge–purge eating disorders. Eur. Eat. Disord. Rev. 2018;27:224–35.

Liu Z-S, Cui Y-H, Sun D, Lu Q, Jiang Y-W, Jiang L, Wang J-Q, Luo R, Fang F, Zhou S-Z, Wang Y, Cai F-C, Lin Q, Xiong L, Zheng Y, Qin J. Current status, diagnosis, and treatment recommendation for tic disorders in China. Front. Psychiatry 2020;11:774.

Lozano AM, Lipsman N, Bergman H, et al. Deep brain stimulation: Current challenges and future directions. Nat Rev Neurol. 2019;15(3):148–60.

McGuire JF, Ricketts EJ, Piacentini J, Murphy TK, Storch EA, Lewin AB. Behavior therapy for tic disorders: An evidenced-based review and new directions for treatment research. Curr Dev Disord Rep. 2015;2(4):309–17.

Melin K, Skarphedinsson G, Skärsäter I, Haugland BS, Ivarsson T. A solid majority remit following evidence-based OCD treatments: A 3-year naturalistic outcome study in pediatric OCD. Eur Child Adolesc Psychiatry. 2018;27:1373–81.

Nazeer A. et al. Obsessive-compulsive disorder in children and adolescents: Epidemiology, diagnosis and management. Transl Pediatr. 2020;9(Suppl 1):S76–S93. doi: 10.21037/tp.2019.10.02

Nolan Suzanne O, Zachry Jennifer E, Johnson Amy R, Brady Lillian J, Siciliano Cody A, Calipari Erin S. Direct dopamine terminal regulation by local striatal microcircuitry. J Neurochem. 2020;155:475–93.

Novotny M, Valis M, Klimova B. Tourette syndrome: A mini-review. Front Neurol. 2018;9:139.

Ramteke Anshuta, Lamture Yashwant. Tics and tourette syndrome: A literature review of etiological, clinical, and pathophysiological aspects. Cureus 2022;14(8):e28575.

Rozenman M, Peris T, Bergman RL, et al. Distinguishing fear versus distress symptomatology in pediatric OCD. Child Psychiatry Hum Dev. 2017;48:63–72. https://doi.org/10.1007/s10578-016-0653-4

Sarkhel S, Singh OP, Arora M. Clinical practice guidelines for psychoeducation in psychiatric disorders general principles of psychoeducation. Indian J Psychiatry. 2020;62(Suppl 2):S319–S323.

Scahill L, Riddle MA, McSwiggin-Hardin M, Ort SI, King RA, Goodman WK, et al. Children's yale-brown obsessive compulsive scale: Reliability and validity. J Am Acad Child Adolesc Psychiatry. 1997;36:844–52.

Stein DJ, Costa DLC, Lochner C, Miguel EC, Janardhan Reddy YC, Shavitt RG, Simpson HB. Obsessive–compulsive disorder. Nat Rev Dis Primers. 2019;5(1):52. https://doi.org/10.1038/s41572-019-0102-3

Storch EA, Park JM, Lewin AB, Morgan JR, Jones AM, Murphy TK, et al. The Leyton obsessional inventory-child version survey form does not demonstrate adequate psychometric properties in American youth with pediatric obsessive-compulsive disorder. J Anxiety Disord. 2011;25:574–8.

Thomsen PH. Obsessive-compulsive disorder in children and adolescents: A study of parental psychopathology and precipitating events in 20 consecutive Danish cases. Psychopathology. 1995;28:161–7.

Ueda Keisuke, Black Kevin J. A comprehensive review of tic disorders in children. J Clin Med. 2021;10(11):2479.

Veljkovic Emilija, Xia Wenhao, Phillips Blaine, Wong Ee Tsin, Ho Jenny, Alberto Oviedo Alberto, Hoeng Julia, Peitsch Manuel. Tourette syndrome. In Nicotine and Other Tobacco Compounds in Neurodegenerative and Psychiatric Diseases, edited by Wenhao Xia, Blaine Phillips, Ee Tsin Wong, Jenny Ho, Alberto Oviedo, Julia Hoeng, Manuel Peitsch, Emilija Veljkovic, 37–42. Academic Press, 2018.

Walusinski O, Féray J-C. The Marquise de Dampierre identified at last, the first described clinical case of Gilles de la Tourette syndrome. Rev Neurol. 2020;176(10):754–62.

Watson HJ, Rees CS. Meta-analysis of randomized, controlled treatment trials for pediatric obsessive-compulsive disorder. J Child Psychol Psychiatry. 2008;49:489–98.

Westwell-Roper Clara, Evelyn Stewart S. Challenges in the diagnosis and treatment of pediatric obsessive–compulsive disorder. Indian J Psychiatry. 2019 January;61(Suppl 1):p S119–S130. doi: 10.4103/psychiatry.IndianJPsychiatry_524_18

Wu Monica S, et al. Comparing OCD-affected youth with and without religious symptoms: Clinical profiles and treatment response. Compr Psychiatry. 2018;86:47–53. doi:10.1016/j.comppsych.2018.07.009.

Wu Y, Liu X, Luo H, Deng W, Zhao G, Wang Q, et al. Advanced paternal age increases the risk of schizophrenia and obsessive-compulsive disorder in a Chinese Han population. Psychiatry Res. 2012;198:353–9.

10 Pathophysiology of Obsessive-Compulsive Disorder

Santhosh Kumar Yasam,[a,†] *Dhasarathdev Srinivasan,*[a,†]
C. Thangam,[b,†] *Langeswaran Kulanthaivel,*[c,†]
A. Billy Asir,[a,†] *and Gowtham Kumar Subbaraj*[a,*]

[a] Faculty of Allied Health Sciences, Chettinad Hospital and Research Institute, Chettinad Academy of Research and Education (Deemed to be University), Old Mahabalipuram Road (OMR), Kelambakkam-603 103, Tamil Nadu, India

[b] Department of Pharmacology, KSR Institute of Dental Science and Research, Tiruchengode, Namakkal-637215, Tamil Nadu, India

[c] Department of Biotechnology, Science campus, Alagappa University, Karaikudi-630001, Tamil Nadu, India

10.1 INTRODUCTION

The presence of intrusive, unwanted thoughts, images, desires, and urge-neutralizing repetitive activities (such as examining) or mental actions (such as praying) is what distinguishes OCD from other mental health conditions (compulsions) (Rasmussen and Eisen, 1994). These include obsessions, which are undesired but persistent thoughts that repeat again, and compulsions to act, which are urges to repeat portions of behavior in a ritualistic, stereotyped order (Graybiel and Rauch, 2000). The term "OCD" refers to a group of symptoms that influence both cognition and motor activity (Moheb et al., 2019). In general, some compulsive behaviors are exhibited as a reaction to a particular fixation in an attempt to counterbalance the stress and symptoms of that OCD, while others are found as a result of other factors (Paula-Pérez, 2013). The most frequent compulsive behaviors are cleaning, ordering, straightening, and checking (repeatedly doing behavior in reaction to obsessive self-doubts about whether it was completed). OCD patients have uncontrollable routines, behaviors, and habit.

According to the circumstances, larger diagnostic cluster, there was a shift away from the focus being placed on anxiety as the primary symptom and toward the focus being placed on repetitive behaviors or mental processes (Foa et al., 2008).

*Corresponding author: Dr. Gowtham Kumar Subbaraj, Faculty of Allied Health Sciences, Chettinad Hospital and Research Institute, Chettinad Academy of Research and Education (Deemed to be University), Old Mahabalipuram Road (OMR), Kelambakkam-603 103, Tamil Nadu, India.
Email: gowtham_phd@yahoo.com
†Equal first authors.

DOI: 10.1201/9781003387893-12

The following are significant findings concerning these symptoms for the neurobiologist:

- The person with OCD is usually aware that the obsessive thoughts and compulsions are not rational; however, despite great efforts, he or she has been unable to get them under control.
- Most of the time OCD is not uncommon. They typically have good or indeed ordinary contents.
- Despite differences in individual symptoms, there is a notable crossed continuity in the themes of the OCD.
- One group of patients experiences compulsions more often than obsessions, another group more frequently than both. In other words, OCD can manifest as either largely an executive-behavioral disease or primarily a cognitive-affective condition.
- The compulsions and obsessions may last for several hours. For instance, a "checker" checks, but because they are unable to be certain, they must check again.

These observations indicate that repetitive actions and thoughts that coincide with other things in conscious awareness can be triggered by brain circuits in OCD, and they also indicated that the trigger can be caused by situations that would otherwise be harmless (American Psychiatric Association, 1995). OCD has an early beginning and is commonly connected to treat chronic illnesses. Early start of OCD affects boys more than girls and is connected to motor nervous tic conditions; it is frequently unresponsible traditional therapy (Baxter et al., 1988) Adult OCD, around age 20, affects a large number of women and is more treatable. Notably, 1–3% of the worldwide population has OCD. Patients have a 60–70% lifetime incidence of major depressive illness. The link between OCD and OC-spectrum illnesses has had the most influence on OCD neurobiology, while 90% of Tourette syndrome (TS) patients have obsessive-compulsive (OC) behavior (Breiter and Rauch, 1996).

Due to their frequent co-occurrence and phenomenological similarities, OCD and TS may be genetically and pathophysiologically connected. Other OC illnesses include body dysmorphic disorder and trichotillomania (compulsive hair pulling), obsessions with appearance, and ritualistic attempts to change it (Saxena et al., 1998). A similar strategy was to establish a new category that would gather together all OCRD conditions, including olfactory disorder hypochondriasis, hoarding, OCD, and repetitive bodily movement behaviors (Stein et al., 2016).

Evidence of similar physiological and epidemiological traits is the focus of this new diagnostic category, as well as common traits of intrusive thoughts and repeated activities (Okada et al., 2015).

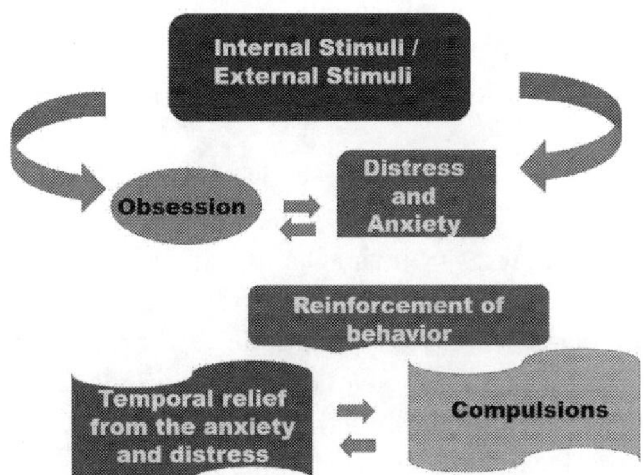

FIGURE 10.1 Factors responsible for obsessive-compulsive disorder.

OCD being a part of the OCRD cluster, which emphasizes tedious behaviors, is unable to acknowledge two things: first, the fully functioning connection between compulsions and obsessions and, second, the manifestation heterogeneity of OCD, which have a number of psychological and neurological mechanisms in common (Hu et al., 2016).

OCD is assumed to feature intrusive thoughts about potential risks, followed by repetitive compulsive behaviors or rituals meant to counteract such threats (Heatherton and Wagner, 2011). There is no doubt that compulsions provide a momentary relief from discomfort, but they also strengthen the cycle, making it stronger over time. This obsessive-compulsive cycle appears to be accompanied by two psychological mechanisms, a cognitive mechanism characterized by executive dysfunction and a psychological process marked by constant worry about danger and high levels of anxiety. Due to the high emotional arousal that patients have experienced, the *DSM-IV* grouping has been used to define OCD as an anxiety disorder. The presence of pervasive anxiety symptoms has been the defining characteristic of this group (Calkins et al., 2013). In contrast, anxiety is characterized by an affective-motivational imbalance between defensive and appetitive processes, as evidenced in the predisposition for disgust, shame, and guilt associated with OCD. It has been shown, therefore, that a dysregulation at higher level emotional processing exists due to the presence of an affective-motivational imbalance (Baxter et al., 1988). There is a wealth of proof that executive functioning problems present in OCD. Different components of executive functioning that are often linked with OCD are relevant: working memory, response inhibition, and cognitive flexibility (Breiter and Rauch, 1996).

Exposure and response/ritual prevention are two treatments used in first-line psychotherapy treatment for OCD in order to address each of the processes present in OCD. A response and ritual prevention approach is different from exposure because it focuses on improving inhibitory control and improving cognitive flexibility, while exposure involves subjecting patients to stressful conditions in order to restrict their emotional expression (Saxena et al., 1998). OCD has been linked to a variety of diverse pathophysiological processes at the levels of immunology, neurochemistry, and neuroanatomical (Saxena and Rauch, 2000). To explain the pathophysiology of OCD, a number of neuroanatomical theories were put forth, all of which agreed that there were malfunctioning circuits in the cortico-striatal-thalamus-cortex (CSTC). The CSTC's contribution to self-regulation enabled OCD to be reclassified as a disease of reaction inhibition rather than as an anxiety disorder (Richards et al., 2020).

Despite this, further studies have shown functional problems in more extensive brain networks. In OCD, frontal-subcortical circuits are widely acknowledged to play a vital part (Whiteside et al., 2004). The purpose of this paper is to discuss evidence that suggests that OCD may be caused by abnormal brain functioning, which could explain the disorder. Deficits in thinking, planning, and problem-solving, as well as in working memory and the regulation of negative emotions (including fear, defensiveness, disdain, guilt, and humiliation).

Figure 10.2 provides a depiction of the original and modified groupings of OCSD and OCRD disorders, including notation of other disorders considered by some as part of a compulsive-impulsive spectrum group of disorders.

The first step in the presentation is a timeline which shows the progress of brain research on OCD starting with PET and SPECT at the start, all the way to today's symptom provocation fMRI (functional magnetic resonance imaging) which is being carried out (Del Casale et al., 2011). Next, we discuss the most up-to-date research on alterations in functional connectivity in the OCD brain via resting state paradigms. Then researchers looked at how the brain's functional structure changed in response to different emotional activities (anger/defensiveness, disgust, guilt and shame, working memory, cognitive flexibility, and inhibitory control) (Thorsen et al., 2015). The initial functional investigations on OCD found elevated metabolism in areas connected to "CSTC loops," including the putamen, caudate, thalamus, and orbitofrontal cortex. In order to study the extent of glucose metabolism in this area, positron emission tomography (PET) was used, which measures the amount of glucose in the brain (Mataix et al., 2004). Among the regions associated with OCD symptoms, according to early studies using symptom provocation paradigms and still utilizing crude fMRI

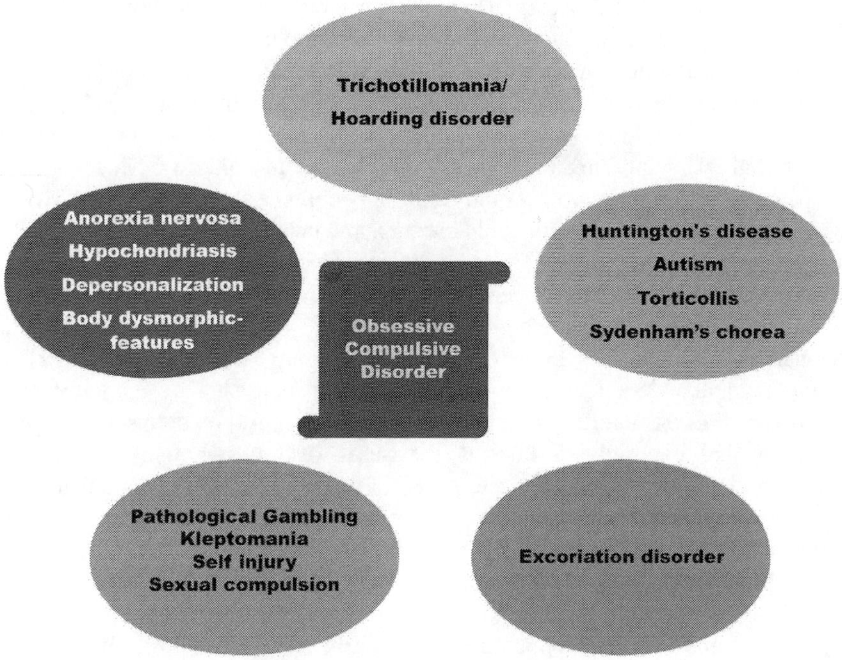

FIGURE 10.2 Signs and symptoms associated with obsessive-compulsive disorder.

techniques, are the anterior cingulate, the amygdala and caudate, the putamen, the globus pallidus, the temporal cortex, the insular cortex, the medial orbital gyrus, and the insular cortex (Murayama et al., 2013).

The thalamus, caudate nucleus, orbitofrontal cortex, anterior cingulate, and other areas in the brain showed signs of hyperactivation in the opening stories assessments of brain functions investigations of OCD (Tian et al., 2016). However, a few years later, the authors were aware that while there is evidence of overactive CSTC areas, there were still a number of discrepancies in the results of other investigations. There was a major publication in the field of functional neuroimaging out in 2004, which was one of the first comprehensive quantitative meta-analyses of neuroimaging investigations to be released. It included controlled investigations performed between 1987 and 2003 using SPECT (single-photon emission computed tomography) and PET (Single-PET). There has been a possible alteration in the circuits that drive working memory, cognitive flexibility, and other executive areas involving the inhibition of motor functions in the thalamus, caudate nucleus, and orbitofrontal cortex as a result of elevated metabolism in these regions (Harrison et al., 2009).

An evaluation of fMRI stands for functional magnetic resonance imaging and PET investigations using rasterization quantitative meta-analyses based on symptom provocation paradigms (Sakai et al., 2011). The eight research that made up this meta-analysis discovered more evidence in favor of the hyperactivation of CSTC circuits. Other areas, including additionally active regions were the precuneus, left superior temporal gyrus, and dorsolateral prefrontal cortex. This shows decreased mental imagery, visual-spatial memory (precuneus), and memory monitoring (superior temporal gyrus) (Kang et al., 2013).

Most of the results that were presented were confirmed, demonstrating that the CSTC circuits were only one part of an even more complex picture that also involved other parts of the brain, such as the parietal, temporal, hippocampal, and even the cerebellum. Recent research has shown that successful treatment for OCD patients results in a normalization of the patient's brain activity, specifically a reduction in activity in the thalamus, temporal, and occipital cortices as well as the caudate, orbitofrontal, and cingulated areas (Beucke et al., 2014).

According to Mataix-Cols, the discrepancy in the range of responses in individuals with distinct subtypes of OCD exhibited in various symptom provocation paradigms, along with the results of the research previously described, may be a result of the discrepancy between the subtypes actually exhibiting the symptoms (Hou et al., 2012). A collection of multisymptomatic OCD patients, along with controls as healthy individuals was examined to test this theory. The authors examined a group of patients with multiple symptoms and healthy controls to determine whether inducing different symptoms (specifically contamination) would result in different patterns of brain activity (such as aggressiveness and checking, aggression and hoarding, and unpleasant control). In order to achieve this, they used a sample of patients with OCD exhibiting multiple symptoms to extract the data (Hou et al., 2014; Posner et al., 2014). It has been observed that in OCD patients exposed to a condition in which washing symptoms are triggered, there are increases in activity in both anterior cingulate and orbitofrontal gyri on both sides, as well as in the left caudate, left middle frontal gyrus, and left dorsal anterior cingulate gyrus. The left middle frontal gyrus, left caudate, and left dorsal anterior cingulate gyri, as well as the anterior cingulate and orbitofrontal gyri on both sides, show increased activity in OCD patients exposed to a situation that causes washing symptoms.

Individuals with hoarding OCD had higher levels of activity in the right orbitofrontal, fusiform, and left precentral/superior frontal gyri. Brain responses to a specific symptom-provoking paradigm were also found to have significant positive links to symptomatic features (Ping et al., 2013). In recent years, Murayama et al. (2013) demonstrated that distinction between the activation of the brain in individuals with OCD, washers, and healthy controls by using a symptom provocation task. Comparing checkers and washers revealed that while washers showed significantly higher levels of activation in more widespread brain regions such as the right cerebellum, the right posterior cingulate cortex, the right medial frontal gyrus, the left middle temporal gyrus, and the inferior occipital gyrus, checkers showed significantly lower levels of activation (compared to normal controls) of the left caudate and left anterior cingulate. The harshness of washers and checkers was found to be significantly correlated having both the left anterior cingulate cortex and the right orbitofrontal cortex active, both the left and right orbitofrontal cortex, anterior cingulate cortex, respectively. Once more, there appeared to be a connection between particular OCD clinical characteristics and particular brain regions. According to the results, checkers exhibit a different degree of activity in regions associated with emotional processing and control, whereas washers show a heightened brain response in CSTC regions primarily tied to reaction inhibition (Abe et al., 2015).

Researchers have only lately begun using resting-state OCD studies to examine OCD functional connections. The resting-state functional studies use functional magnetic resonance imaging (fMRI) to examine the simultaneous activity of many brain areas at the same time during the resting state. These analyses are conducted in an effort to determine the level of functional dependency between the different brain areas (Chen et al., 2016).

Based on preliminary findings of functional resting-state fMRI studies, it appears that those with OCD may have altered connectivity in the CSTC circuits due to changes related to these studies, which is ongoing. CSTC loops are considered to be one of the most functionally active loops in OCD, according to one of the earliest resting state studies (Jhunng et al., 2014). It is interesting to note, however, that the ventral caudate loop was found to be more connected with the orbitofrontal cortex, while the dorsal caudate/putamen loop was found to be less connected to the lateral prefrontal cortex in the CSTC's ventral loop (dorsal loop). Association of increased connectivity indicates a discordant network structure in the pathway for reduced connection, inhibition of emotion, and inhibitory control indicates decreased connectivity pathway for working memory and cognitive flexibility. OCD is associated with increased communication between the ventral striatum and ventral caudate (Fox et al., 2005). Recent studies have found that the regions of the ventral loop CSTC that were found to be abnormally active during an inhibitory response test in patients with OCD have better connectivity while they are at rest, following their abnormal hyperactivity during the inhibitory response test (Buckner et al., 2008). Without any mediating effect from other striate regions, the orbitofrontal cortex has an outsized effect on the ventral striate in people with obsessive-compulsive disorder. Several factors may contribute to the emotional dysregulation seen in OCD. One of the most important factors is an overactive orbitofrontal cortex, which is associated with difficulties in balancing reward and punishment contingencies within a limbic system. Left dorsal caudate putamen houses both the limbic loop and the sensory-motor loop, but research shows that it is functionally less connected to the left supplementary motor region in contrast to enhanced communication between the inferior parietal lobule, the front prefrontal cortex, and the right dorsal caudate (Du, 2010; Andrews-Hanna, 2012).

It is possible that this disparity in connection between the dorsal and ventral CSTC loops is due to differences in OCD subtypes. An investigation was exhibited to observe the association between OCD patients who did not exhibit contamination/washing symptoms, healthy controls, and those whose contamination/washing symptoms were the most pronounced. Studies have shown that people with OCD share a common alteration in the connection between their dorsal horn and temporal cortex (Jang et al., 2010). Increased communication between the dorsal striate and insula can be found in the CSTC's foundational region, which is especially important for those with OCD, who experience a preoccupation with germs and the need to constantly clean. Enhanced both long-distance and short-distance connections between the orbitofrontal cortex and putamen in untreated OCD, a long way from CSTC (precentral and superior temporal areas) (Whitfield-Gabrieli and Ford, 2012).

Default mode network (DMN) is a network that is activated in a resting state and links the posterior cingulate, precuneus, and medial prefrontal cortex and parietal lobes. Core nodes of the DMN can also be found in the dorsolateral prefrontal cortex and the medial temporal lobe, resulting in an expanded DMN subsystem. Particular attention has been paid to the default mode network, one of these networks. The medial prefrontal cortex, precuneus, posterior cingulate, and areas of the parietal cortex form the network known as the DMN, which is active while the brain is at rest. Extensive DMN subsystems with core nodes are also present in the medial temporal lobe and dorsolateral prefrontal cortex. The three DMN subsystems were also found to be connected differently in terms of connectivity patterns (Cheng et al., 2013). Several resting state networks in OCD patients have been studied. Improved connectivity between a number of areas of the brain, including the frontal cortex and cingulate, precuneus, thalamus is one example of the confirmed altered connectivity of the top-down control network (Beucke et al., 2014).

Changes in the cortico-striatal network and the DMN were subsequently validated. Relationship between the fronto-parietal network (FPRN) and default mode network (DMN) system involved attention and executive processes (Zhang et al., 2011). Compared to healthy people, people with OCD have stronger connections between many fronto-parietal seeds and DMN areas. Examples of regions with enhanced connectivity with the anterior insula include the dorsomedial prefrontal cortex, parahippocampus, left posterior inferior parietal lobe, posterior cingulate cortex, and precuneus. According to these studies, OCD may disrupt multiple resting state networks at the same time or at different times, potentially leading to a wide range of psychological problems. There was a similar trend in the connection of DMN nodes (Hou et al., 2013).

There are clear discrepancies in the trends of variance in connection findings, and this may be due to the fact that researchers employ a wide range of image capturing and processing techniques. There is growing evidence that the CSTC is connected to other brain areas as well as to regions outside of the CSTC, has changed. Specifically, there is evidence that the connections between the dorsal and ventral CSTC loops have changed, and that the dorsal and ventral CSTC loops are no longer as tightly coupled as they once were (Göttlich et al., 2014).

OCD patients often struggle with the various inhibitory control tasks, such as interference control, action restrict (suppression of a prepotent reaction), and action cancel (inhibition of an already initiated response, such as a stop signal). The aberrant functioning of the CSTC loop was observed in numerous studies examining a variety of inhibitory control tasks (Zhu et al., 2015). The temporal and posterior cingulate regions were more active than usual during interference control, while the temporal and CSTC loop regions were less active than normal during action constraint (left cerebellum and posterior cingulate). Alterations in the lateral dorsal prefrontal cortex (DLPFC) brain network during cognitive switch task may explain the increased adaptability of the CSTC loop (van Velzen et al., 2014). The cerebellum, posterior cingulate, and temporal cortex are all engaged in these tasks because they require inhibitory regulation.

Patients with OCD may compensate for deficient activation during inhibitory control tasks via CSTC loops or temporal-parietal networks increasing activity in the posterior cingulate and cerebellum. The CSTC orbitofrontal brain loop appears to be critical for inhibitory control. Over-monitoring of unsuccessful or challenging trials was associated with the anterior cingulate loop, another part of the CSTC system hypothesized to be involved in emotional processing control. OCD patients showed increased activity in the dorsal anterior cingulate compared to healthy controls during incorrect trials on an interference control task (Hughes et al., 2013).

This increased activity stands in severe contrast to the diminishing parietal, occipital, parietal-temporal, and frontal-temporal subcortical areas that all showed activity in OCD patients performing the shifting task. The frontal-parietal control network and the corticospinal (CSTC) loop were less active during the inhibitory control task (Maltby et al., 2005). The CSTC loops' abnormal activation is defined by overactive or underactive CSTCs that are compensated for by stimulation of more posterior neural circuits. Impaired inhibitory control tasks have been linked to functional abnormalities in a number of brain areas (such as interference control, cancellation, and display restraint).

People with OCD are also cited as experiencing difficulties to change their way of thinking as an example of executive dysfunction as well. The ability to modify one's way of thinking and acting in response to new circumstances is known as cognitive flexibility. One way to evaluate this ability is by looking at how easily paradigms can be changed (using different rules to complete the same task) (Fitzgerald et al., 2005). OCD patients often struggle when required to shift gears or reorder their priorities. Likely due to the lack of dorsal loop activation in the CSTC, these findings were further elaborated, previously cited by showing that OCD patients had lower activity in the subcortical, frontal, temporal, parietal, and occipital regions areas when switching tasks (Morein-Zamir et al., 2016). When the switch minus repetition condition was performed, OCD patients had decreased activity levels in the prefrontal and subcortical areas than controls, and they made a lot more mistakes when moving between tasks. When you switch between different task types, it appears as though the ventromedial/anterior loops as well as the dorsolateral cingulate loops of the CSTC are inactive.

Traditionally, flexibility has been associated with cognitive abilities, while emotional regulation seems to be important for regulating emotions and has the potential to resolve conflict when work shifts from one state to another. Consistent with the foregoing, the investigators found that severe depressive disorder patients and obsessive-compulsive disorder exhibited abridged activity in the prefrontal cortex (Adelbratt and Strang, 2000). However, there was more activation of the putamen in OCD patients than in major depressive patients or healthy controls.

It is well recognized that OCD patients have significantly impaired working memory, which is defined as the capacity to store and update information in real time over the internet. A study conducted by Van Der Wee et al. in the area of OCD working memory functional magnetic resonance imaging (fMRI) processes has become the first study of its kind. Researchers here compared females with OCD to healthy controls using the spatial n-back test, which was progressively more difficult for each group. Since both groups showed similar patterns of activation, independent authors have come to the conclusion that OCD does not seem to interfere with the usual fronto-parietal working memory network due to the memory load (Daros et al., 2014). The increased activity in the anterior cingulate was thought to be a result of the constant internal conflict that characterizes people with obsessive-compulsive disorder. Some of the previously mentioned studies suggest that the working memory deficits experienced by OCD patients may be attributable to a lack of activity in brain regions that are common to both OCD patients and controls. Comparable levels of behavior are seen in both groups (Cardoner et al., 2011). Previous findings were supported by the fact that both groups used similar task-specific brain regions. It's possible that OCD sufferers' increased activity in these regions is a result of an attempt to make up for underlying weaknesses in working memory. A study found that the right insula, left gyrus, and right cuneus were less active than the cuneus, left superior temporal gyrus, and left dorsolateral prefrontal cortex during working memory tasks. This lends credence to the hypothesis that OCD patients' exhibit altered brain activity in identical brain regions when doing working memory tasks, as was found in a prior study. Despite the fact that there were no notable variations between OCD patients and healthy controls in terms of brain activation for low cognitive working memory demands, there was a distinct pattern of activations throughout load levels. Patients with OCD showed increased activity in the dorsal anterior cingulate (Koek et al., 2016).

Along with other anxiety disorders, OCD's pathophysiology stresses the emotional tendency toward fear reaction and the corresponding activation of the protective system. Examining the participant's reaction to pictures of people's faces expressing fear is one technique to explore the neural networks in the brain tasked for processing fear (Thiel et al., 2014). OCD patients' amygdala reaction to fear stimuli versus pleasant and neutral faces. Amygdala activity for fear faces did not vary between OCD and controls. OCD sufferers and healthy controls participated in an emotional face matching experiment (Husted et al., 2006; Banca et al., 2015). There seems to be an activation of the amygdala as well as the visual cortex, the fusiform gyrus, the hippocampal formation, when fear faces are matched, the premotor cortex, the lateral prefrontal cortex, and the orbitofrontal cortex respond.

In all OCD symptom dimensions, there is greater amygdala activity. Therefore, using symptom-related paradigms rather than merely using generic unpleasant stimuli may make the generalization of amygdala activity across symptoms clearer (Hübner et al., 2016). It has been shown that OCD sufferers have distinct brain responses to fundamental emotional images that are intended to arouse both the appetitive (pleasant stimuli) and defensive (painful stimuli) systems in contrast to healthy controls. According to the findings of a study, when OCD sufferers are confronted with unpleasant stimuli, they often activate a large network of neural connections that include the temporal, parietal, and subcortical nuclei (Hyman and Pedrick, 2010). Interestingly, in contrast, there was a significant decrease in brain activity in the posterior visual processing area of the brain when pleasant images were presented to people with OCD.

OCD symptoms were all correlated with feelings of disgust, except for hoarding. A study by Whitton et al. (2015) revealed that, as compared to those without OCD and those with other anxiety disorders (non-OCD sample), OCD patients were probable to be disgusted when confronted with disgusting stimuli, such as bodily waste, when compared to both those without OCD and those with other anxiety disorders (non-OCD sample). The majority of people suffering from OCD with contamination preoccupations prefer disgusting, threatening, and neutral images to healthy controls (Basile et al., 2011). When disgust-inducing stimuli were presented to OCD patients, right insula activity significantly increased, left parahippocampal activity increased, lower frontal gyrus on the left activity increased, activity in the left inferior occipital region and left posterior cingulated gyrus both increased. There are four types of emotional images used in a sample of OCD patient with a matching control group. Typically those images used are disgusting, evoke a sense of fear, are affectively neutral, or are related to symptoms. It has been found that recent research using real-time MRI has shown an increased performance of OCD patients whose symptoms have improved when they are taught to reduce anterior insula activity in the presence of repulsive stimuli, leading to better symptoms of contamination obsessions and obsessive washing (Pincus and Morley, 2001).

There were significant correlations between harm and symmetry (unacceptable thoughts and contamination) with shame propensity, compared to other symptoms of OCD. The results of a study was associated with that of a group of healthy controls, along with social anxiety, body dysmorphic disorder, and a healthy control group (Nicki, 2001). Researchers found that patients with body dysmorphic disorder expressed higher levels of shame relating to their bodies than patients with generalized anxiety. However, these feelings of shame were less prevalent in generalized anxiety patients. It was found that the patients who were suffering from obsessions were more likely to feel shame due to these feelings. A study on the emotional processing of shame and guilt has been added to the OCD literature. It was found that OCD sufferers and healthy control subjects showed unique patterns of brain responses to neutral, shameful, and guilt-provoking events. OCD sufferers showed, in the same circumstance, there was increased limbic, temporal, midbrain, and sublobar activity and decreased middle frontal gyrus and inferior parietal lobe activation. The temporal sulcus, limbic system, and frontal lobe all showed a considerable rise. Activation factors in OCD patients when faced with guilt-related situations. However, there was a decrease in anterior cingulate activity in OCD patients.

An obsession may be motivated by a central feeling of pathological guilt, which is why OCs tends to obsess about things like scrupulosity, aggression, and sexual contamination, as well as by the fact that most compulsions (like cleaning, checking, repeating, counting, and arranging) are motivated by the same feeling (Dalsgaard et al., 2020). A sample of OCD sufferers and healthy controls was taken to differentiate between their brains' reactions to three main types of guilt: Altruistic guilt, deontological guilt, and the two additional control emotions of grief and rage, as well as a control emotion. In addition to noticing that the anterior cingulate was less active, the scientists also found that the insula and precuneus also showed a reduction in activity as a result of the deontological guilt stimuli (Ashburn and Staats, 1999). As compared to the control stimuli, the anterior cingulate, insula, and precuneus showed a reduction in activity when the deontological guilt stimuli were present. In OCD patients, there was no discernible difference between the activation caused by sad or angry stimuli as compared to altruistic guilt (Del Casale et al., 2019).

It was 100 years ago that obsessive-compulsive disorder (OCD) was first identified as a psychiatric disease. A pathognomonic sign of this condition (compulsions) is the persistence, difficult, nonsensical thoughts and urges as well as repeated, deliberate actions. There are many patients who suffer from this condition, who struggle to overcome their irrational and excessive thoughts and actions (Chamberlain et al., 2005). A population-based study conducted in several parts of the country and abroad has estimated that between 1 and 3% of the population suffers from OCD. There is no doubt that despite the fact that this condition can affect people of varying ages, the most vulnerable period is from infancy to early adulthood when it comes to the condition. The World Health Organization states that OCD is one of the topmost ten medical conditions affecting social, occupational, and academic performance the most (Kurlan, 1998). There are many cases in which patients suffer from chronic or episodic illnesses, and their family members also experience a high level of stress. In spite of the fact that medications and behavioral treatment may be active in relieving some of the symptoms of the condition, the majority of patients suffer from chronic or recurring conditions, and a cure is rare (Barrett et al., 2008).

OCD neurocircuitry models are among the most well-established of all neuropsychiatric diseases, and a growing body of research suggests that circuits in the cortico-striatal-thalamic-cortical (CSTC) play a significant role in disease expression. OCD patients have striatal damage, according to magnetic resonance imaging and spectroscopic investigations, and these patterns of activity are thought to be the result of a breakdown in striate-thalamic inhibition. Neuropharmacological theories about the origin of OCD and the underlying genes have been radically revised in light of research suggesting that the serotonergic system regulates OCD symptoms (Black et al., 1992). This has resulted in a considerable change in the neuropharmacological theories about the origins and genes of OCD as a result. Neuropharmacological theories about the origin of OCD and the underlying genes have been drastically revised in light of research suggesting that the serotonergic system regulates OCD symptoms. This has significantly shifted neuropharmacological hypotheses about the genetics of OCD's origin (Rucci et al., 2003).

Several neurotransmitter systems inside CSTCs have been linked to OCD due to their anatomical placements or roles within CSTC circuits. Some examples of these systems are the cholinergic, endogenous opioid, glutamate, and GABA pathways. The pathophysiology of OCD can only be understood by learning more about its underlying genetic causes. The disorder is probably genetic, and streptococcal infection is the only major environmental possibility besides inheritance in some cases of OCD (Walitza et al., 2010). Understanding pathogenic reasons is crucial to designing effective treatments, and gene discovery is at the heart of this endeavor.

Since the early twentieth century, clinical researchers have speculated about the possible influence of genetics on the onset of obsessive-compulsive disorder. One of the earliest records of obsessional neurosis in English literature comes from the Maudsley Hospital in London, which cared for 50 patients. Researchers at Hopkins found that OCD affected 37% of patients' parents and 21% of their siblings. Two research with remarkably comparable findings (Lenane et al., 1990).

More than 15 familial OCD studies have been undertaken. The results of a vast majority of this research provide belief to the idea that the condition can be transmitted from generation to generation. Results showed that parents of OCD probands had a considerably greater age-adjusted morbidity risk of "broadly defined OCD" (16% versus 3%). Both classic OCD and obsessive-compulsive disorder (OCD) with subsyndromal symptoms are included in this broad definition (Althoff et al., 2005).

According to research out of the University of Windsor, both OCD and subthreshold OCD carry similar morbidity risks (18% vs. 4%), and first-degree relatives of OCD sufferers are at greater threat of developing the disorder than psychiatrically healthy controls (10% vs. 1.9%) (Nestadt et al., 2000). According to the Hopkins OCD family research, 7.8 siblings of those with early onset OCD had OCD, making the prevalence of OCD 17.9% higher than that of other mental disorders, including bipolar disorder and panic disorder. In contrast, the frequency of OCD among controls' relatives was 2.7% (Nestadt et al., 2000). It has been shown that young, inexperienced probands are more likely to have a high family loading, and a case study of children and adolescents who presented with OCD at a young age indicated that first-degree relatives of the probands had an average prevalence rate of 7% to 15% (Cryan and Mombereau, 2004).

Relatives of the proband who developed OCD before the age of 19 were shown to be at an elevated risk. According to the Hopkins OCD family study, 13.8% of relatives of probands under the age of 18 had OCD, compared to 0% of probands after the age of 18. The early onset age has been successfully utilized to subclassify conditions, including schizophrenia, breast cancer, and Alzheimer's disease, and there are multiple examples of monozygotic twin pairs who share symptoms of obsessive-compulsive disorder. A number of monozygotic twins with obsessive-compulsive symptoms have been reported since Lewis' initial report of the illness in 1936 (Sutcliffe et al., 2005). The concordance rate for identical twins was 80% for MZ twins and 50% for DZ twins in one study. In their investigation, Carey and Gottesman discovered that MZ and DZ twins were 87% concordant, leading them to estimate heritability at 80%. Seventy-eight percent of MZ twin pairs were identical. Additional research employing questionnaire diagnosis found just a weak heritability. There has been a lack of research into the adoption of OCD (Zalsman et al., 2002). Recent research on the heritability of OCD has pointed to a candidate gene as the disease's origin. Some new evidence suggests that the dominant transmission theory is correct. Researchers discovered that a mixed inheritance model best characterized the inheritance patterns of 52 families where at least two members had at least one form of OCD; a single locus inheritance, no inheritance, and polygenic inheritance are all ruled out. The sporadic and environmental models were disproved by segregation analysis of the Hopkins OCD family study data, but not the Mendelian dominant or codominant models (Ruzilawati et al., 2020).

Due to the serotonergic system's wealth of information regarding the disease's pathophysiology and pharmacology, it has been the primary focus of the research (Alsobrook et al., 2002). As a result of considering these factors, candidate genes have been selected. There has been study into the role of serotonin receptors like 5HT2C receptors, 5HT1-D beta serotonin receptors, 5HT2A serotonin receptors, and 5HTTLPR serotonin transporters; two families have found evidence that OCD is associated with co-occurring diseases (anorexia nervosa, Asperger's syndrome/autism), and a rare gain-of-function mutation in the serotonin. These probands and their relatives may be affected by a "double-hit" effect of two polymorphisms, a mutation in the serotonin transporter promoter polymorphism's coding area and an allele that is more frequently expressed, resulting in extraordinary severity and treatment resistance (Zhou et al., 2021). Two other research conducted within the past few years on OCD households support this finding. However, it's possible that this variant only affects a tiny fraction of families. Goldman and co-workers discovered that the L(a) allele is unusually passed on to people with obsessive-compulsive disorder. But most recent case-control and family-based investigations, neither serotonin transporter nor tryptophan hydroxylase nor serotonin 1B, 2A, or 1D-beta receptor polymorphisms have been linked to any correlations (Tang et al., 2014). Strong evidence has been found in several research to support the theory that OCD is caused by a repeat in the dopamine receptor type 4 gene, whereas others have found either weak or inconsistent evidence in support of this theory (not nearly statistically significant). Except for people who have OCD, the dopamine D2 receptor is not linked to obsessive-compulsive disorder. However, no connection was found between the dopamine transporter and the dopamine receptor DRD3 (McCarthy and Hirschhorn, 2008). In particular, males have been discovered to carry an allele of catechol-O-methyltransferase (COMT) that has been linked to OCD. The OCD patients have been observed to have a higher frequency of heterozygous genotypes, albeit this discovery has not been confirmed. COMT has been linked to obsessive-compulsive disorder in women, but not in men (Hamshere et al., 2009). An analogous study found a connection between monoamine oxidase A (MAO-A) and OCD in male individuals but not in female participants. Researchers have recently discovered a number of genes that may play a role in OCD, including brain-derived neurotrophic factor (BDNF), glutamate (NMDA) subunit receptor, GABA type B receptor, oligodendrocyte glycoprotein 2, and myelin oligodendrocyte glycoprotein 2 (Cooper et al., 2022).

Despite the fact that obsessive-compulsive disorder (OCD) has an etiological diversity that makes it difficult to find genetic determinants (Chabane et al., 2005). In the context of a more general description of the condition, a number of approaches have been devised to distinguish between groups with etiologically similar characteristics. The age at onset variable has been found to be useful in the clinical classification of individuals in terms of their genetic risk (Naj et al., 2014). According to the childhood OCD family study and the Hopkins OCD family study, there appears to be an inverse association between the age when an individual first developed OCD symptoms and the risk that their relatives may develop OCD symptoms in the future. In the field of OCD, there is currently no clinical or other type of variable which has been well-supported (Mahjani et al., 2022). Family studies indicate that those with OCD before the age of 18 form a particularly close group of individuals (Pauls et al., 1986).

Using genetics as an outcome metric, the Hopkins team found that having an AAO before the age of 18 was preferable to the norm. The previously mentioned linkage study utilized a covariate-based linkage analysis, in which age of onset was included as a covariate.

In a large population-based sample, they identified three distinct forms of OCD: "simple," "tic-related," and "affective-related." Different phenotypic groupings can be shown by the presence of coexisting mental problems (such as anxiety, tics, and affective disorders). Subtypes of OCD may have a genetically uniform population, allowing for more accurate detection of genetic variation (Samuels et al., 2007; Dinecola, 2018). The connection between OCD and TS is acknowledged in the tic-related section. There is substantial genetic overlap between OCD and TS (Tourette syndrome). It is not known whether or not OCD has a hereditary foundation in TS families, as it does in all other individuals (Marincowitz et al., 2021). Multiple studies have shown that OCD sufferers are clinically distinct from OCD sufferers who don't have this particular compulsion. It has been shown, for example, that relatives of OCD probands who hoard are more prone to develop the disorder themselves, as shown by the Hopkins OCD Family Study (Pinto et al., 2008). Our team used a variety of statistical methods on Yale-Brown Obsessive-Compulsive Scale (Y-BOCS) data, and we were able to narrow the symptoms of OCD down to four or five dimensions that may be useful for categorizing sufferers. A dichotomous factor analysis was carried out on the 16 YBOCS symptom categories in the Johns Hopkins OCD Family Study, and it resulted in the discovery of four symptom factors: Hoarding disorders; aggressive, sexual, and religious fixation; contamination and compulsive cleaning disorders. Among the many factors shown to have a bearing on when symptoms first appeared, how well patients responded to treatment, and whether or not they had co-occurring disorders was the level of sibling-sibling association (Alexander et al., 2017). The OCD Collaborative Genetics Study found that 418 OCD patients (251 affected sibling pairs) shared common dimensions based on principal component analysis of the YBOCS symptom specification categories. This data suggested that hoarding was the most strongly heritable factor between siblings. Using the YBOCS symptom checklist items of individual OCGS participants as the basis for the exploratory dichotomous factor analysis, they found that the OCGS participants had very similar dimensions (including a taboo factor) and that there were significant sibling-sibling correlations, particularly when it comes to hoarding (Muiños-Gimeno et al., 2009). The basic framework of a child or teenager is similar to that of an adult. According to a study conducted by the American Sociological Association, first-degree relatives of OCD patients who exhibit hoarding behaviors had a higher prevalence of hoarding behaviors and had distinct clinical characteristics that were distinct from those observed across the entire OCD population (Dozier and Ayers, 2017).

One-LOD support has been found in a 10.9Mb area in 14q31–32. This region contains a number of possible genes, some of which include the Estrogen Receptor 2 (ESR2), the Neurexin 3 (NRXN3), the Voltage-Gated Potassium Channel Subfamily H Member 5 (KCNH5), the Voltage-Gated Potassium Channel Subfamily K Member 10 (KCNK10), and others (Grados et al., 2003). In order to ensure that only those with OCD hoarding, regardless of whether they also had any of the other symptoms of OCD, were included in the analyses, analyses were performed again, this time using the phenotypic result for OCD hoarding as the phenotypic result (Hill, 2004). In research that stratified the X chromosome, the linkage signal for compulsive hoarding was fairly high (indicating that the disease's genes may be found on either chromosome) (Menzies et al., 2007).

In the genetics of OCD, it has been shown that intermediate phenotypes can potentially be exploited as phenotypic tactics. Evidence for the heritability of OCD symptoms has been shown in multiple researches (Sterneckert et al., 2014). Researchers have found that even in the absence of direct genetic link, relatives of people with OCD score abnormally high on these tests. There has been some study in recent years linking particular facets of cognition to the emergence of OCD. Treatments have tried to target both the clinical elements of OCD (obsessions and compulsions) and the brain areas associated with them (Kho et al., 2012). It is obvious that the patient has trouble putting an end to repetitive activities given the executive task "set shifting," which is meant to assess the patient's apparent difficulties in stopping. Cognitive treatments have been used to treat OCD by focusing on both its compulsions and obsessions as well as the brain areas connected to it. Attempts have been made to prove this using the executive task "set shifting," which evaluates a patient's apparent inability to refrain from performing a series of similar actions (Machalińska et al., 2012). The Iowa Gambling Task attempts to replicate the conditions under which real-world decisions are made, providing further evidence of the frontal lobe's susceptibility to dysfunction. Many people with OCD find it challenging to see the task through to its conclusion, which suggests that this may be yet another endophenotype of OCD. Environmental factors (such as trauma and infection) and gene-by-environment interaction may both show a part in the development of OCD, as is the case with other neuropsychiatric diseases. It's also likely that genetic elements contribute to the disorder's onset and progression.

Thanks to recent developments in analytical and molecular genetics, scientists now have a number of tools at their disposal for pinpointing the genetic causes of disease. Researching complicated genetic illnesses is far more challenging than researching Mendelian disorders (De Haan et al., 2015). Studies focusing on linkages are the gold standard for diagnosing rare diseases with high penetrance and substantial influence genes. Diseases with less severe consequences are less amenable to linkage studies because of the impracticality of detecting small-effect genes through linkage analysis. Kho et al. (2012) report that while association research has moved forward, the technical, informatics, and statistical underpinnings for whole genome analysis were built as a result of advances made within association research (Pritchard and Cox, 2002). To conduct genome-wide association studies, "the technical, informatic, and statistical underpinnings have been built for whole genome analyses," Genome-wide association has been used for quite some time, and there have been several recent cases demonstrating its usefulness. However, it is worth mentioning that a genome-wide association research was able to uncover an area on a chromosome that had not previously been related with Hirschsprung Disease, despite the fact that this ailment is well-studied and has eight known mutations. In a study, scientists examined SNPs (single nucleotide polymorphisms) in people with AMD (age-related macular degeneration) and healthy controls. The CFH gene has been shown to be within an area that has been associated with AMD, leading researchers to risk that it may play a role in the disease's progression. Based on these data, the researchers hypothesized that the CFH gene was significantly linked to AMD. Two WGAs are being done for OCD right now, and the results of these studies may help provide insight on what causes OCD. The associational method isn't perfect; it does have certain limitations (Pauls, 2008). Negative linkage-disequilibrium values, for instance, do not always point to a very influential region of the genome in terms of genes. Therefore, it may be difficult, if not impossible, to assess the level of LD in a given location due to the physical separation of the SNPs employed therein. Thus, it is feasible that disease-causing genes will be missed even at the maximum density when using non-random SNPs over the entire genome. In addition, LD may be beneficial for finding illness loci with common variations that affect susceptibility, but it is not likely to be useful for finding disease loci with many rare susceptibility variants. Because of this, LD can be used to pinpoint a wide variety genetic variation that contributes to disease susceptibility. It is important to note that OCD is a highly complex hereditary condition whose underlying causes are mostly unknown. Traditional linkage methods are unlikely to have found the vast majority of these genetic determinants, although there are likely to be a number of common genetic impacts of minor significance. Researchers lack a firm grasp of the biochemical underpinnings of OCD, making it tough for them to narrow their genetic search for the disorder's cause down to specific metabolic pathways. Although the gene-based association studies are crucial to understanding OCD. Several studies have found evidence linking potential genes to the pathophysiology of OCD. Finding the causal genetic variants in OCD will require a number of approaches, some of which are already being used and others that have not been tried before. Some of these methods include expanding our knowledge of clinical and cognitive phenotypes, exploring from an epigenetic angle, and employing deep genomic sequencing. Some methods are already in use to discover relevant genetic differences in OCD, whereas others have not been investigated yet. Investigating epigenetics and making use of deep genetic sequencing methods are two examples of these strategies, both of which have the potential to improve our understanding of clinical and cognitive phenotypic traits (Aubin-Horth and Renn, 2009). Discovering the environmental and genetic elements that contribute to the development of obsessive-compulsive disorder will undoubtedly have far-reaching implications for those living with the condition. Once the underlying genetic and environmental causes are understood, we may work toward developing effective medicines and implementing preventative measures (Baum, 1995).

10.2 CONCLUSION

Multiple variants in numerous genes are increasingly important to the threat of developing OCD. As a result, it is similar to many complex medical and neuropsychiatric disorders. In addition, while the details have yet to be clarified, it seems that both environmental and genetic factors play an important role in the risk of developing OCD. In this sense, OCD also resembles other complex disorders. This has made it difficult to identify and confirm disease alleles, which has made it difficult to determine and quantify how a specific genetic variant affects the risk of developing OCD in an individual, which has made it difficult to identify and confirm disease alleles. Likewise, OCD treatment guidelines have yet to take genetic data into account and are based solely on symptoms and clinical observations. There are numerous genetic studies that suggest that glutamatergic signaling pathways may be potential new therapeutic targets, and this merits further investigation. The use of modern genotyping and sequencing technologies in larger cohorts of affected individuals, as has been demonstrated with other commonly occurring, complex neuropsychiatric disorders like autism and schizophrenia, shows great promise in advancing the field towards the identification, validation, and quantification of specific gene variants and biologic mechanisms that may be associated with these disorders. It is evident that the pace of genetic discoveries in OCD is speeding up rapidly, and it is reasonable to expect that the pace of genetic discoveries will accelerate in the future and that new interventions will be developed as time goes by.

REFERENCES

Abe Y, Sakai Y, Nishida S, Nakamae T, Yamada K, Fukui K, Narumoto J. Hyper-influence of the orbitofrontal cortex over the ventral striatum in obsessive-compulsive disorder. European Neuropsychopharmacology. 2015;25(11):1898–905.

Adelbratt S, Strang P. Death anxiety in brain tumour patients and their spouses. Palliative Medicine. 2000 Sep;14(6):499–507.

Alexander SP, Striessnig J, Kelly E, Marrion NV, Peters JA, Faccenda E, Harding SD, Pawson AJ, Sharman JL, Southan C, Davies JA. The Concise Guide to PHARMACOLOGY 2017/18: Voltage-gated ion channels. British Journal of Pharmacology. 2017;174:S160–94.

Alsobrook JP, Zohar AH, Leboyer M, Chabane N, Ebstein RP, Pauls DL. Association between the COMT locus and obsessive-compulsive disorder in females but not males. American Journal of Medical Genetics. 2002;114(1):116–20.

Althoff RR, Faraone SV, Rettew DC, Morley CP, Hudziak JJ. Family, twin, adoption, and molecular genetic studies of juvenile bipolar disorder. Bipolar Disorders. 2005;7(6):598–609.

American Psychiatric Association. Diagnostic and statistical manual of mental disorders, 4th Edition: Primary care version. Washington DC: American Psychiatric Association; 1995.

Andrews-Hanna JR. The brain's default network and its adaptive role in internal mentation. The Neuroscientist. 2012 (3):251–70.

Ashburn MA, Staats PS. Management of chronic pain. The Lancet. 1999;353(9167):1865–9.

Aubin-Horth NA, Renn SC. Genomic reaction norms: Using integrative biology to understand molecular mechanisms of phenotypic plasticity. Molecular Ecology. 2009;18(18):3763–80.

Banca P, Voon V, Vestergaard MD, Philipiak G, Almeida I, Pocinho F, Relvas J, Castelo-Branco M. Imbalance in habitual versus goal directed neural systems during symptom provocation in obsessive-compulsive disorder. Brain. 2015;138(3):798–811.

Barrett PM, Farrell L, Pina AA, Peris TS, Piacentini J. Evidence-based psychosocial treatments for child and adolescent obsessive–compulsive disorder. Journal of Clinical Child & Adolescent Psychology. 2008;37(1):131–55.

Basile B, Mancini F, Macaluso E, Caltagirone C, Frackowiak RS, Bozzali M. Deontological and altruistic guilt: Evidence for distinct neurobiological substrates. Hum Brain Mapp. 2011; 32(2):229–39.

Baum F. Researching public health: Behind the qualitative-quantitative methodological debate. Social Science & Medicine. 1995;40(4):459–68.

Baxter Jr LR, Schwartz JM, Mazziotta JC, Phelps ME, Pahl JJ, Guze BH, Fairbanks L. Cerebral glucose metabolic rates in nondepressed patients with obsessive-compulsive disorder. The American Journal of Psychiatry. 1988 Dec 1;145(12):1560–3.

Beucke JC, Sepulcre J, Eldaief MC, Sebold M, Kathmann N, Kaufmann C. Default mode network subsystem alterations in obsessive–compulsive disorder. The British Journal of Psychiatry. 2014 Nov;205(5):376-82.

Black DW, Noyes R, Goldstein RB, Blum N. A family study of obsessive-compulsive disorder. Archives of General Psychiatry. 1992;49(5):362–8.

Breiter HC, Rauch SL. Functional MRI and the study of OCD: From symptom provocation to cognitive-behavioral probes of cortico-striatal systems and the amygdala. Neuroimage. 1996;4(3):S127–S38.

Buckner RL, Andrews-Hanna JR, Schacter DL. The brain's default network: Anatomy, function, and relevance to disease. Annals of the New York Academy of Sciences. 2008;24(1):1–38.

Calkins AW, Berman NC, Wilhelm S. Recent advances in research on cognition and emotion in OCD: A review. Current Psychiatry Reports. 2013;(5):1–7.

Cardoner N, Harrison BJ, Pujol J, Soriano-Mas C, Hernandez-Ribas R, López-Solá M, Real E, Deus J, Ortiz H, Alonso P, Menchon JM. Enhanced brain responsiveness during active emotional face processing in obsessive compulsive disorder. The World Journal of Biological Psychiatry. 2011;12(5):349–63.

Chabane N, Delorme R, Millet B, Mouren MC, Leboyer M, Pauls D. Early-onset obsessive-compulsive disorder: A subgroup with a specific clinical and familial pattern?. Journal of Child Psychology and Psychiatry. 2005;46(8):881–7.

Chamberlain SR, Blackwell AD, Fineberg NA, Robbins TW, Sahakian BJ. The neuropsychology of obsessive compulsive disorder: The importance of failures in cognitive and behavioural inhibition as candidate endophenotypic markers. Neuroscience & Biobehavioral Reviews. 2005;29(3):399–419.

Chen Y, Juhás M, Greenshaw AJ, Hu Q, Meng X, Cui H, Ding Y, Kang L, Zhang Y, Wang Y, Cui G. Abnormal resting-state functional connectivity of the left caudate nucleus in obsessive-compulsive disorder. Neuroscience Letters. 2016 Jun 3;623:57–62.

Cheng Y, Xu J, Nie B, Luo C, Yang T, Li H, Lu J, Xu L, Shan B, Xu X. Abnormal resting-state activities and functional connectivities of the anterior and the posterior cortexes in medication-naive patients with obsessive-compulsive disorder. PLoS One. 2013 Jan;8(6):e67478.

Cooper SE, van Dis EA, Hagenaars MA, Krypotos AM, Nemeroff CB, Lissek S, Engelhard IM, Dunsmoor JE. A meta-analysis of conditioned fear generalization in anxiety-related disorders. Neuropsychopharmacology. 2022:1–0.

Cryan JF, Mombereau C. In search of a depressed mouse: Utility of models for studying depression-related behavior in genetically modified mice. Molecular Psychiatry. 2004;9(4):326–57.

Dalsgaard S, McGrath J, Østergaard SD, Wray NR, Pedersen CB, Mortensen PB, Petersen L. Association of mental disorder in childhood and adolescence with subsequent educational achievement. JAMA Psychiatry. 2020;77(8):797–805.

Daros AR, Zakzanis KK, Rector NA. A quantitative analysis of facial emotion recognition in obsessive–compulsive disorder. Psychiatry Research. 2014;215(3):514–21.

De Haan S, Rietveld E, Stokhof M, Denys D. Effects of deep brain stimulation on the lived experience of obsessive-compulsive disorder patients: In-depth interviews with 18 patients. PLoS One. 2015;10(8):e0135524.

Del Casale A, Kotzalidis GD, Rapinesi C, Serata D, Ambrosi E, Simonetti A, Pompili M, Ferracuti S, Tatarelli R, Girardi P. Functional neuroimaging in obsessive-compulsive disorder. Neuropsychobiology. 2011 Jun 21;64(2):61–85.

Del Casale A, Sorice S, Padovano A, Simmaco M, Ferracuti S, Lamis DA, Rapinesi C, Sani G, Girardi P, Kotzalidis GD, Pompili M. Psychopharmacological treatment of obsessive-compulsive disorder (OCD). Current Neuropharmacology. 2019;17(8):710–36.

Dinecola CM. Predictors of Graduate Students' Self-Efficacy for Working with Persons with Autism Spectrum Disorder and Their Families. Louisiana State University and Agricultural & Mechanical College; 2018.

Dozier ME, Ayers CR. The etiology of hoarding disorder: A review. Psychopathology. 2017;50(5):291–96.

Du X. Development of a Mathematical Model of Neurosensory Gating. University of Houston; 2010.

Fitzgerald KD, Welsh RC, Gehring WJ, Abelson JL, Himle JA, Liberzon I, Taylor SF. Error-related hyperactivity of the anterior cingulate cortex in obsessive-compulsive disorder. Biological Psychiatry. 2005;57(3):287–94.

Foa EB, Chrestman KR, Gilboa-Schechtman E. Prolonged exposure therapy for adolescents with PTSD emotional processing of traumatic experiences, therapist guide. Oxford University Press; 2008.

Fox MD, Snyder AZ, Vincent JL, Corbetta M, Van Essen DC, Raichle ME. The human brain is intrinsically organized into dynamic, anticorrelated functional networks. Proceedings of the National Academy of Sciences. 2005;102(27):9673–8.

Göttlich M, Krämer UM, Kordon A, Hohagen F, Zurowski B. Decreased limbic and increased Fronto-parietal connectivity in unmedicated patients with obsessive-compulsive disorder. Human Brain Mapping. 2014 5(11):5617–32.

Grados MA, Walkup J, Walford S. Genetics of obsessive-compulsive disorders: New findings and challenges. Brain and Development. 2003;25:S55–61.

Graybiel AM, Rauch SL. Toward a neurobiology of obsessive-compulsive disorder. Neuron.2000;28(2):343–7.

Hamshere ML, Gordon-Smith K, Forty L, Jones L, Caesar S, Fraser C, Hyde S, Tredget J, Kirov G, Jones I, Craddock N. Age-at-onset in bipolar-I disorder: Mixture analysis of 1369 cases identifies three distinct clinical sub-groups. Journal of Affective Disorders. 2009;116(1–2):23–9.

Harrison BJ, Soriano-Mas C, Pujol J, Ortiz H, López-Solà M, Hernández-Ribas R, Deus J, Alonso P, Yücel M, Pantelis C, Menchon JM. Altered corticostriatal functional connectivity in obsessive-compulsive disorder. Archives of General Psychiatry. 2009;66(11):1189–200.

Heatherton TF, Wagner DD. Cognitive neuroscience of self-regulation failure. Trends in Cognitive Sciences. 2011;15(3):132–9.

Hill EL. Evaluating the theory of executive dysfunction in autism. Developmental Review. 2004;24(2):189–233.

Hou J, Song L, Zhang W, Wu W, Wang J, Zhou D, Qu W, Guo J, Gu S, He M, Xie B. Morphologic and functional connectivity alterations of corticostriatal and default mode network in treatment-naïve patients with obsessive-compulsive disorder. PLoS One. 2013 (12):e83931.

Hou J, Wu W, Lin Y, Wang J, Zhou D, Guo J, Gu S, He M, Ahmed S, Hu J, Qu W. Localization of cerebral functional deficits in patients with obsessive–compulsive disorder: A resting-state fMRI study. Journal of Affective Disorders. 2012;138(3):313–21.

Hou JM, Zhao M, Zhang W, Song LH, Wu WJ, Wang J, Zhou DQ, Xie B, He M, Guo JW, Qu W. Resting-state functional connectivity abnormalities in patients with obsessive–compulsive disorder and their healthy first-degree relatives. Journal of Psychiatry and Neuroscience. 2014 Sep 1;39(5):304–11.

Hu X, Liu Q, Li B, Tang W, Sun H, Li F, Yang Y, Gong Q, Huang X. Multivariate pattern analysis of obsessive–compulsive disorder using structural neuroanatomy. European Neuropsychopharmacology. 2016;26(2): 246–54.

Hübner C, Wiesendahl W, Kleinstäuber M, Stangier U, Kathmann N, Buhlmann U. Facial discrimination in body dysmorphic, obsessive-compulsive and social anxiety disorders. Psychiatry Research. 2016; 236:105–11.

Hughes ME, Johnston PJ, Fulham WR, Budd TW, Michie PT. Stop-signal task difficulty and the right inferior frontal gyrus. Behavioural Brain Research. 2013;256:205–13.

Husted DS, Shapira NA, Goodman WK. The neurocircuitry of obsessive–compulsive disorder and disgust. Progress in Neuro-Psychopharmacology and Biological Psychiatry. 2006;30(3):389–99.

Hyman BM, Pedrick C. The OCD Workbook: Your Guide to Breaking Free from Obsessive-Compulsive Disorder. New Harbinger Publications; 2010.

Jang JH, Kim JH, Jung WH, Choi JS, Jung MH, Lee JM, Choi CH, Kang DH, Kwon JS. Functional connectivity in Fronto-subcortical circuitry during the resting state in obsessive-compulsive disorder. Neuroscience letters. 2010 474(3):158–62.

Jhung K, Ku J, Kim SJ, Lee H, Kim KR, An SK, Kim SI, Yoon KJ, Lee E. Distinct functional connectivity of limbic network in the washing type obsessive–compulsive disorder. Progress in Neuro-Psychopharmacology and Biological Psychiatry. 2014;53:149–55.

Kang DH, Jang JH, Han JY, Kim JH, Jung WH, Choi JS, Choi CH, Kwon JS. Neural correlates of altered response inhibition and dysfunctional connectivity at rest in obsessive–compulsive disorder. Progress in Neuro-Psychopharmacology and Biological Psychiatry. 2013;40:340–6.

Kho AN, Hayes MG, Rasmussen-Torvik L, Pacheco JA, Thompson WK, Armstrong LL, Denny JC, Peissig PL, Miller AW, Wei WQ, Bielinski SJ. Use of diverse electronic medical record systems to identify genetic risk for type 2 diabetes within a genome-wide association study. Journal of the American Medical Informatics Association. 2012;19(2):212–8.

Koek RJ, Schwartz HN, Scully S, Langevin JP, Spangler S, Korotinsky A, Jou K, Leuchter A. Treatment-refractory posttraumatic stress disorder (TRPTSD): A review and framework for the future. Progress in Neuro-Psychopharmacology and Biological Psychiatry. 2016;70:170–218.

Kurlan R. Tourette's syndrome and 'PANDAS' Will the relation bear out? Neurology. 1998;50(6):1530–4.

Lenane MC, Swedo SE, Leonard H, Pauls DL, Sceery W, Rapoport JL. Psychiatric disorders in first degree relatives of children and adolescents with obsessive compulsive disorder. Journal of the American Academy of Child & Adolescent Psychiatry. 1990;29(3):407–12.

Machalińska A, Kawa MP, Marlicz W, Machaliński B. Complement system activation and endothelial dysfunction in patients with age-related macular degeneration (AMD): Possible relationship between AMD and atherosclerosis. Acta Ophthalmologica. 2012;90(8):695–703.

Mahjani B, Klei L, Mattheisen M, Halvorsen MW, Reichenberg A, Roeder K, Pedersen NL, Boberg J, de Schipper E, Bulik CM, Landén M. The genetic architecture of obsessive-compulsive disorder: Contribution of liability to OCD from alleles across the frequency spectrum. American Journal of Psychiatry. 2022;179(3):216–25.

Maltby N, Tolin DF, Worhunsky P, O'Keefe TM, Kiehl KA. Dysfunctional action monitoring hyperactivates frontal–striatal circuits in obsessive–compulsive disorder: An event-related fMRI study. Neuroimage. 2005;24(2):495–503.

Marincowitz C, Lochner C, Stein DJ. The neurobiology of obsessive-compulsive personality disorder: A systematic review. CNS Spectrums. 2021:1–39.

Mataix-Cols D, Wooderson S, Lawrence N, Brammer MJ, Speckens A, Phillips ML. Distinct neural correlates of washing, checking, and hoarding symptom dimensions in obsessive-compulsive disorder. Archives of General Psychiatry. 2004;61(6):564–76.

McCarthy MI, Hirschhorn JN. Genome-wide association studies: Potential next steps on a genetic journey. Human Molecular Genetics. 2008 Oct 15;17(R2):R156–65.

Menzies L, Achard S, Chamberlain SR, Fineberg N, Chen CH, Del Campo N, Sahakian BJ, Robbins TW, Bullmore E. Neurocognitive endophenotypes of obsessive-compulsive disorder. Brain. 2007; 130(12):3223–36.

Moheb N, Charuworn K, Ashla MM, Desarzant R, Chavez D, Mendez MF. Repetitive behaviors in frontotemporal dementia: Compulsions or impulsions? The Journal of Neuropsychiatry and Clinical Neurosciences. 2019:132–6.

Morein-Zamir S, Voon V, Dodds CM, Sule A, van Niekerk J, Sahakian BJ, Robbins TW. Divergent subcortical activity for distinct executive functions: Stopping and shifting in obsessive compulsive disorder. Psychological Medicine. 2016;46(4):829–40.

Muiños-Gimeno M, Guidi M, Kagerbauer B, Martín-Santos R, Navinés R, Alonso P, Menchón JM, Gratacos M, Estivill X, Espinosa-Parrilla Y. Allele variants in functional microRNA target sites of the neurotrophin-3 receptor gene (NTRK3) as susceptibility factors for anxiety disorders. Human Mutation. 2009;30(7):1062–71.

Murayama K, Nakao T, Sanematsu H, Okada K, Yoshiura T, Tomita M, Masuda Y, Isomura K, Nakagawa A, Kanba S. Differential neural network of checking versus washing symptoms in obsessive-compulsive disorder. Progress in Neuro-Psychopharmacology and Biological Psychiatry. 2013;40:160–6.

Naj AC, Jun G, Reitz C, Kunkle BW, Perry W, Park YS, Beecham GW, Rajbhandary RA, Hamilton-Nelson KL, Wang LS, Kauwe JS. Effects of multiple genetic loci on age at onset in late-onset Alzheimer disease: A genome-wide association study. JAMA Neurology. 2014;71(11):1394–404.

Nestadt G, Lan T, Samuels J, Riddle M, Bienvenu III OJ, Liang KY, Hoehn-Saric R, Cullen B, Grados M, Beaty TH, Shugart YY. Complex segregation analysis provides compelling evidence for a major gene underlying obsessive-compulsive disorder and for heterogeneity by sex. The American Journal of Human Genetics. 2000;67(6):1611–6.

Nicki A. The abused mind: Feminist theory, psychiatric disability, and trauma. Hypatia. 2001;16(4):80–104.

Okada K, Nakao T, Sanematsu H, Murayama K, Honda S, Tomita M, Togao O, Yoshiura T, Kanba S. Biological heterogeneity of obsessive–compulsive disorder: A voxel-based morphometric study based on dimensional assessment. Psychiatry and Clinical Neurosciences. 2015 (7):411–21.

Paula-Pérez I. Differential diagnosis between obsessive compulsive disorder and restrictive and repetitive behavioural patterns, activities and interests in autism spectrum disorders. Revista de Psiquiatría y Salud Mental (English Edition). 2013;6(4):178–86.

Pauls DL. The genetics of obsessive compulsive disorder: A review of the evidence. American Journal of Medical Genetics Part C: Seminars in Medical Genetics. 2008 May 15;148(2):133–9. Wiley Subscription Services, Inc., A Wiley Company.

Pauls DL, Towbin KE, Leckman JF, Zahner GE, Cohen DJ. Gilles de la Tourette's syndrome and obsessive-compulsive disorder: Evidence supporting a genetic relationship. Archives of General Psychiatry. 1986;43(12):1180–2.

Pincus T, Morley S. Cognitive-processing bias in chronic pain: A review and integration. Psychological Bulletin. 2001;127(5):599.

Ping L, Su-Fang L, Hai-Ying H, Zhang-Ye D, Jia L, Zhi-Hua G, Hong-Fang X, Yu-Feng Z, Zhan-Jiang L. Abnormal spontaneous neural activity in obsessive-compulsive disorder: A resting-state functional magnetic resonance imaging study. PLoS One. 2013;8(6):e67262.

Pinto A, Greenberg BD, Grados MA, Bienvenu III OJ, Samuels JF, Murphy DL, Hasler G, Stout RL, Rauch SL, Shugart YY, Pauls DL. Further development of YBOCS dimensions in the OCD Collaborative Genetics study: Symptoms vs. categories. Psychiatry Research. 2008;160(1):83–93.

Posner J, Marsh R, Maia TV, Peterson BS, Gruber A, Simpson HB. Reduced functional connectivity within the limbic cortico-striato-thalamo-cortical loop in unmedicated adults with obsessive-compulsive disorder. Human Brain Mapping. 2014 Jun;35(6):2852–60.

Pritchard JK, Cox NJ. The allelic architecture of human disease genes: Common disease–common variant . . . or not? Human Molecular Genetics. 2002;11(20):2417–23.

Rasmussen SA, Eisen JL. The epidemiology and differential diagnosis of obsessive compulsive disorder. The Journal of Clinical Psychiatry. 1994;55:5–10.

Richards A, Kanady JC, Neylan TC. Sleep disturbance in PTSD and other anxiety-related disorders: An updated review of clinical features, physiological characteristics, and psychological and neurobiological mechanisms. Neuropsychopharmacology. 2020;45(1):55–73.

Rucci P, Gherardi S, Tansella M, Piccinelli M, Berardi D, Bisoffi G, Corsino MA, Pini S. Subthreshold psychiatric disorders in primary care: Prevalence and associated characteristics. Journal of Affective Disorders. 2003;76(1–3):171–81.

Ruzilawati AB, Islam MA, Muhamed SK, Ahmad I. Smoking Genes: A Case–Control Study of Dopamine Transporter Gene (SLC6A3) and Dopamine Receptor Genes (DRD1, DRD2 and DRD3) Polymorphisms and smoking behaviour in a malay male cohort. Biomolecules. 2020;10(12):1633.

Sakai Y, Narumoto J, Nishida S, Nakamae T, Yamada K, Nishimura T, Fukui K. Corticostriatal functional connectivity in non-medicated patients with obsessive-compulsive disorder. European Psychiatry. 2011:(7):463–9.

Samuels JF, Bienvenu III OJ, Pinto A, Fyer AJ, McCracken JT, Rauch SL, Murphy DL, Grados MA, Greenberg BD, Knowles JA, Piacentini J. Hoarding in obsessive–compulsive disorder: Results from the OCD Collaborative Genetics Study. Behaviour Research and Therapy. 2007;45(4):673–86.

Saxena S, Brody AL, Schwartz JM, Baxter LR. Neuroimaging and frontal-subcortical circuitry in obsessive-compulsive disorder. The British Journal of Psychiatry. 1998 Aug;173(S35):26–37.

Saxena S, Rauch SL. Functional neuroimaging and the neuroanatomy of obsessive-compulsive disorder. Psychiatric Clinics of North America. 2000 Sep 1;23(3):563–86.

Stein DJ, Kogan CS, Atmaca M, Fineberg NA, Fontenelle LF, Grant JE, Matsunaga H, Reddy YC, Simpson HB, Thomsen PH, Van Den Heuvel OA. The classification of obsessive–compulsive and related disorders in the ICD-11. Journal of Affective Disorders. 2016;190:663–74.

Sterneckert JL, Reinhardt P, Schöler HR. Investigating human disease using stem cell models. Nature Reviews Genetics. 2014 Sep;15(9):625–39.

Sutcliffe JS, Delahanty RJ, Prasad HC, McCauley JL, Han Q, Jiang L, Li C, Folstein SE, Blakely RD. Allelic heterogeneity at the serotonin transporter locus (SLC6A4) confers susceptibility to autism and rigid-compulsive behaviors. The American Journal of Human Genetics. 2005;77(2):265–79.

Tang R, Noh HJ, Wang D, Sigurdsson S, Swofford R, Perloski M, Duxbury M, Patterson EE, Albright J, Castelhano M, Auton A. Candidate genes and functional noncoding variants identified in a canine model of obsessive-compulsive disorder. Genome Biology. 2014;15(3):1–5.

Thiel A, Thiel J, Oddo S, Langnickel R, Brand M, Markowitsch HJ, Stirn A. Obsessive-compulsive disorder patients with washing symptoms show a specific brain network when confronted with aggressive, sexual, and disgusting stimuli. Neuropsychoanalysis. 2014;16(2):83–96.

Thorsen AL, van den Heuvel OA, Hansen B, Kvale G. Neuroimaging of psychotherapy for obsessive–compulsive disorder: A systematic review. Psychiatry Research: Neuroimaging. 2015;233(3):306–13.

Tian L, Meng C, Jiang Y, Tang Q, Wang S, Xie X, Fu X, Jin C, Zhang F, Wang J. Abnormal functional connectivity of brain network hubs associated with symptom severity in treatment-naive patients with obsessive–compulsive disorder: A resting-state functional MRI study. Progress in Neuro-Psychopharmacology and Biological Psychiatry. 2016;66:104–11.

van Velzen LS, Vriend C, de Wit SJ, van den Heuvel OA. Response inhibition and interference control in obsessive–compulsive spectrum disorders. Frontiers in Human Neuroscience. 2014 11;8:419.

Walitza S, Wendland JR, Gruenblatt E, Warnke A, Sontag TA, Tucha O, Lange KW. Genetics of early-onset obsessive–compulsive disorder. European Child & Adolescent Psychiatry. 2010;19(3):227–35.

Whiteside SP, Port JD, Abramowitz JS. A meta–analysis of functional neuroimaging in obsessive–compulsive disorder. Psychiatry Research: Neuroimaging. 2004;132(1):69–79.

Whitfield-Gabrieli S, Ford JM. Default mode network activity and connectivity in psychopathology. Annual Review of Clinical Psychology. 2012 29;8.

Whitton AE, Henry JD, Grisham JR. Cognitive and psychophysiological correlates of disgust in obsessive–compulsive disorder. British Journal of Clinical Psychology. 2015;54:16–33.

Zalsman G, Frisch A, Apter A, Weizman A. Genetics of suicidal behavior: Candidate association genetic approach. Israel Journal of Psychiatry. 2002;39(4):252.

Zhang T, Wang J, Yang Y, Wu Q, Li B, Chen L, Yue Q, Tang H, Yan C, Lui S, Huang X. Abnormal small-world architecture of top–down control networks in obsessive–compulsive disorder. Journal of Psychiatry and Neuroscience. 2011;36(1):23–31.

Zhou B, Zhu Z, Ransom BR, Tong X. Oligodendrocyte lineage cells and depression. Molecular Psychiatry. 2021;26(1):103–17.

Zhu Y, Fan Q, Zhang Z, Zhang H, Tong S, Li Y. Spontaneous neuronal activity in insula predicts symptom severity of unmedicated obsessive compulsive disorder adults. In 2015 37th Annual International Conference of the IEEE Engineering in Medicine and Biology Society (EMBC) 2015 (pp. 5445–8). IEEE.

Section 3

Treatment and Intervention of OCD

11 Dietary Nutrients and Their Role in Obsessive-Compulsive Disorder

Shubham Singh,[a] *Hanan Fahad Alharbi,*[b] *and Senthilkumar Rajagopal*[c,*]

[a] Department of Molecular Metabolism, Harvard T.H. Chan School of Public Health, Harvard University, Boston, MA, USA

[b] Maternity and Child Health Nursing Department, College of Nursing, Princess Nourah bint Abdulrahman University

[c] Maternity and Child Health Nursing Department, College of Nursing, Princess Nourah bint Abdulrahman University

11.1 INTRODUCTION

Living beings are brought to their existent forms and physiology by a complex set of biochemical reactions which are in turn tightly regulated. Perturbations in these processes lead to diseases which can be classified based on the tissue it damages and phenotypes exhibited, like neurological ones affect nervous system and show neural defects. Neurological disorders can be degenerative or developmental; degenerative ones like Alzheimer's are largely affected by the age of the animals. Neurodevelopmental disorders largely arise due to genetic reasons at the developmental stage of the animals and show mild to severe behavioral defects. One such mental disorder is obsessive-compulsive disorder syndrome, and as the name suggests, it's marked by obsession for some behavior the subject would want to do in an unnecessarily repetitive fashion.[1] This behavioral obsession could be for things as simple as handwashing, cleaning study tables, making sure lights are off or doors are closed.[1] These probably cause insecurity and anxiety in the subject, and the person would want to exhaustively make very sure that a few things are done, hence repeating the same exercise again and again. It affects one's quality of life negatively, turning severe as the subjects often find it difficult to settle with other peers. They waste a lot of time on thinking, checking, and performing activities of their compulsion. There are no established laboratory diagnostic tests for OCD; mostly it is detected based on the observation of the behavior and by clinical interview. In children, OCD is more difficult to detect as children often repeat behaviors, and their cognition is not developed enough to make informed decisions. OCD is moderately inheritable, suggesting the role of genetic factors behind the development of OCD.[2] OCDs are found to co-occur with other neurological disorders like Tourette's syndrome, tic disorder, schizophrenia, or Huntington's disease.[3–8]

*Corresponding author, Dr. R. Senthilkumar, PhD, Associate Professor, Dept of Biotechnology, School of Applied Sciences, REVA University, Bangalore-560064, KA, India. Email: senthilanal@yahoo.com

DOI: 10.1201/9781003387893-14

11.1.1 Molecular Biology and Pathophysiology

A series of imaging tools like diffusion tensor imaging, positron emission tomography (PET), single photon emission computed tomography (SPECT), and functional MRI (fMRI) have shown the loss of white matter in a few brain regions, elevated metabolism and hyper synaptic activity in the cerebral regions.[9, 10] OCD is strongly associated with serotonin metabolism and signaling in the brain, often OCD patients seem to have hypersensitive postsynaptic serotonin receptors as polymorphisms in serotonin transporters and receptors is strongly associated with OCD.[11] In addition to, serotonin mediated signaling in the brain, glutamate and dopamine biochemistry and signaling too seems to be dysregulated.[12, 13] Glutamate transporters Sapap3 and SLC1A1 have been implicated in the OCD.[13–17] A few other genes related to OCD pathophysiology include dopamine receptor D-sub-4-gene,[18] catechol-O-methyltransferse, and monoamine oxidase and genes at BDNF locus, SLITRK1, GABAR1, OLIG2, and myelin oligodendrocyte glycoprotein.[2, 19] A few brain areas which have been implicated for the OCD include frontal orbito-striatal area, caudate nucleus, and dorsolateral prefrontal cortex.[20] Brain region responsible for motor activities arising out of learned behaviors, like the basal ganglia, seem to be the most crucial part of the brain that is affected in patients with OCD.[21, 22] There are several reports which show that lesions in the basal ganglia can lead to OCD symptoms.[23, 24] Neural physiology of serotonin along with glutamate and dopamine system has been found to play a role in development of OCD.[25–27] A few reports strongly suggest that cortico-striato-thalamo-cortical neural circuit in the brain that controls self-control is defective in OCD subjects.[28–30]

11.2 IMMUNE SYSTEM

There are some reports that emphasize that streptococcal infection, for example, group A β-haemolytic *Streptococcus* bacteria (GABHS) mediated infection in basal ganglia may be responsible for childhood onset of OCD.[31] It is additionally supported by the cases of OCD which can be symptomatically treated with antibiotics if administered at early onset of the disorder. Antibodies developed by immune system to fight *Streptococcus* cross reacts with nerve tissue especially in basal ganglia, which follows OCD or OCD with tics.[32] *Streptococcus*-related OCD is called as pediatric autoimmune neuropsychiatric disorder associated with *Streptococcus* (PANDAS)[33] and is marked by increase in anti-Streptolysin O antibodies (used for diagnosis of PANDAS but not a robust method). OCD patients in PANDAS category have high amounts of anti-basal ganglia antibody and anti-phosphatidylethanolamine antibody,[34] marking the onset of OCD.

Monocytes and microglial physiology is also reported to be disrupted in the child subjects of nervous disorders.[35, 36] OCD patients were found to have increased number of total monocytes and $CD16^+$ monocytes (both non-classical and intermediate), and monocytes from OCD patients secrete more pro-inflammatory molecules like GM-CSF, IL-1β, IL-6, IL-8, TNF-α on immune challenge triggered on treatment with LPS.[37] Psychoactive drugs used to treat OCD helped manage the excessive cytokine secretion from monocytes.[37] This suggests that immune cells of OCD are prone to respond more fiercely on immune challenge. Though contrary to peripheral monocytes, peripheral blood leukocytes from OCD subjects produced less IL-6 compared to healthy controls.[38] Microglial abnormality in OCD patients is not directly reported, but it's indirectly inferenced from animal studies, to be specific from Hoxb8 KO mice, which display a few repetitive behaviors as Hoxb8 lineage cell are major microglia in the brain.[39–41]

11.2.1 Diacylglycerol Kinase δ

Diacylglycerol kinase δ animal models for OCD is well reviewed in detail.[42, 43] Brain-specific diacylglycerol kinase δ (DGKδ) knockout mice is one of the good models of OCD sensitive to inhibitors of serotonin transporters and recapitulate several OCD behaviors. Serotonin transporter (SERT) is a multi-span transmembrane protein localized at neuronal synapses in the brain and mediates influx/reuptake of 5-HT into the presynaptic neuron from the synaptic clefts.[44] 5-HT influxes into the presynaptic neuron is metabolized or recycled for further chemical synaptic transmission. Drugs targeting SERT come under the class of serotonin reuptake inhibitors, and their use in the pharmacological treatment of OCD is well established.[45] DGKδ levels negatively correlate with SERT levels, and SERT protein is upregulated in DGKδ KO mice brain,[46] suggesting that OCD symptoms in brain-specific DGKδ KO mice are due to upregulated SERT and increased serotonin reuptake. DGKδ2 interacts with SERT, melanoma antigen gene-D1 and Praja-1 E3 ubiquitin protein ligase and induces degradation of SERT in Praja-1-dependent ubiquitination proteasome system.[47] DGKδ KO mice also have reduced levels of enzyme responsible for 5-HT synthesis, that is, tryptophan hydroxylase-2 and increased levels of enzymes that degrades 5-HT, that is, monoamine oxidase-A, and as a consequence 5-HT levels were reduced in DGKδ KO mice brain.[47]

11.3 ESTABLISHED AND FRONTLINE TREATMENT REGIME FOR OCD

There are no direct effective treatments for OCD, but a few therapeutic interventions are available, which can improve the quality of life for the subjects. These therapies include psychotherapy, pharmacotherapy, and brain modulation. Psychotherapy is based on the verbal and comforting communication to alleviate the destructive/disturbing thought between the physiologist and the subject like cognitive-behavior therapy (CBT).[48] It largely improves the quality of life of the subject by allowing one to get rid of unhealthy ways of thinking and resulting behavior in consequence. Scientists and medical personnel are pushing hard to widen the scope and utility of CBT, and there are several online websites or app-based applications that provide CBT.[49–52]

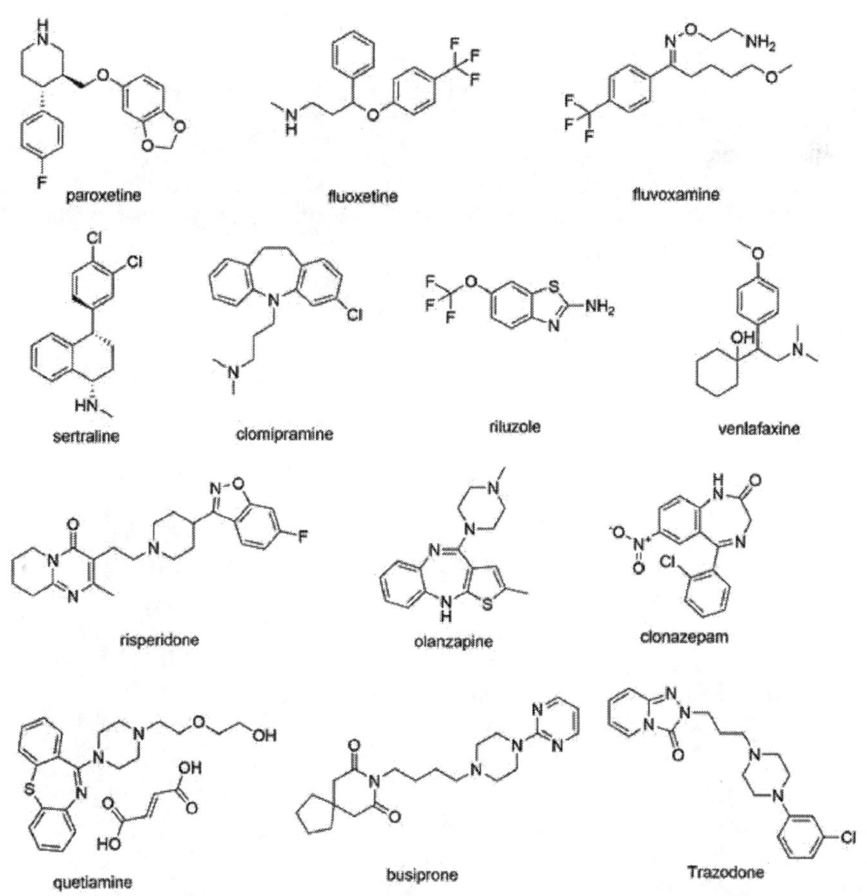

FIGURE 11.1 Structure of small molecules used to manage OCD symptoms (pharmacotherapy) to OCD subjects.

Small molecule drug which inhibit serotonin reuptake at neuronal synapses (FDA approved—paroxetine, fluoxetine, fluvoxamine, sertraline, etc.) and a few antidepressants like clomipramine are in use as first line of treatment to pharmacologically treat symptoms of OCD.[26, 27] Discontinuation of drugs cause significant relapse of OCD symptoms.[53] Other than drugs targeting serotonin pathway in brain, drugs targeted at glutamate and dopamine too helps in OCD. Glutamate antagonist like riluzole is found to be effective in children subjects of OCD.[54, 55] In cases where OCD subjects are refractory to these drugs, administration of antipsychotic drugs like risperidone, olanzapine, quetiapine, buspirone, clonazepam, trazodone, and venlafaxine along with serotonin reuptake inhibitors are found to be of some use to alleviate symptoms of OCD.[56, 57]

Last line of treatment to OCD involves brain modulation like electroconvulsive therapy,[58] transcranial magnetic stimulation[59] (used for depolarizing or hyperpolarizing neurons), and deep brain stimulation.[60] Transcranial magnetic stimulation of supplementary motor area and prefrontal cortex seem to be effective in alleviating OCD, while stimulation of dorsolateral prefrontal cortex is ineffective.[60]

In addition to all previously mentioned three modes of treatments, nutrients have been tested to improve the quality of OCD patients. We would discuss the routine dietary or clinical nutrients that can alleviate the symptoms of OCD and enhance the lifestyle of OCD subjects.

11.4 DIETARY NUTRIENTS IN THE INTERVENTION OF OCD

Diet is one of the most crucial components of well-being, and healthy diet also mirrors a healthy individual in most cases. Dietary input provides antioxidants, enzyme co-factors (vitamins, inorganic metals), and a large array of precursors or building units for the complex biomolecules. Lost homeostasis due to genetic or biochemical perturbations can be tried to bring back near normal by planning dietary intake cautiously though it's not possible in all the cases. Here, we try to discuss a few of the reports that investigate the possible link between dietary supplements and OCD. A few clinical evidences suggest that administration of nutrients like myo-inositol, N-acetyl cysteine, glycine, and eicosapentaenoic acid (EPA) alleviate the symptoms of OCD in humans. Let's discuss the reported nutrients one by one and their possible mechanisms through which they might be benefitting OCD patients on administration.

11.4.1 Role of Myo-Inositol

Myo-inositol is a glucose isomer, and it is crucial for the synthesis of phosphatidylinositol and its phosphorylated derivatives, which are very crucial and potent secondary signaling molecules acting below a series of neuronal receptors including adrenergic, cholinergic, serotonergic, metabotropic, glutamatergic receptors.[61] In addition to acting as precursor to inositol lipids, myo-inositol can act as feedback signal as it directly affects the levels of inositol monophosphatase—an enzyme that acts as phospholipase C to remove phosphate group on inositol of inositol monophosphate.[62] A great challenge in the therapy of neurological disorders is active compound delivery to the brain as blood brain barrier restricts several molecules from entering into the brain, and myo-inositol is no exception, but an acute myo-inositol administration (5 g/kg) can increase up to 17% of myo-inositol levels in the mice brain.[63] Chronic myo-inositol administration (1.2 g/kg for 10 days) can increase up to 5% of myo-inositol levels in the mice brain.[63] Myo-inositol administration for four to five weeks alleviates OCD symptoms and when used as monotherapy at dose of 12–18 g/day.[64, 65] It seems to act by modulating the serotonin reuptake (5-HT) and increases the 5-HT_2 receptor density at synapses directly or indirectly.[61] It prevents desensitization of 5-HT_2 receptor, and at higher doses it increases D2-recepor density in selective brain regions especially striatum.[66, 67] Myo-inositol is also found to be reduced in the brains of suicide subjects, but no parallel studies are available for OCD.[68]

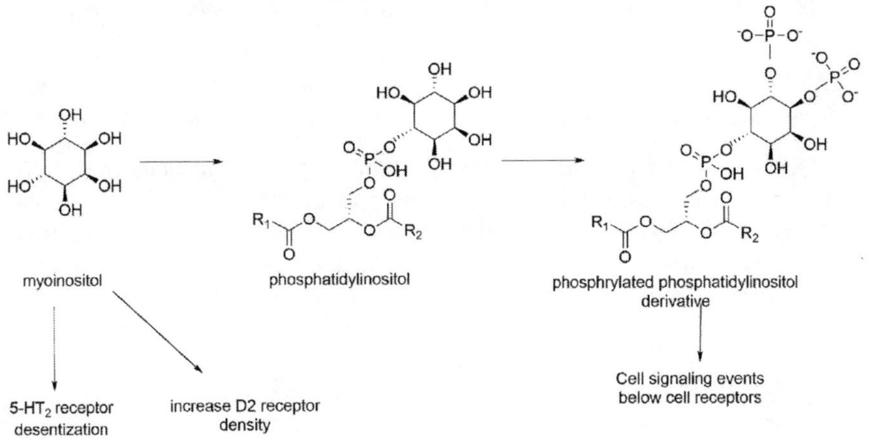

FIGURE 11.2 Myo-inositol participation in cell signaling and receptor modulation that might alleviate symptoms in OCD.

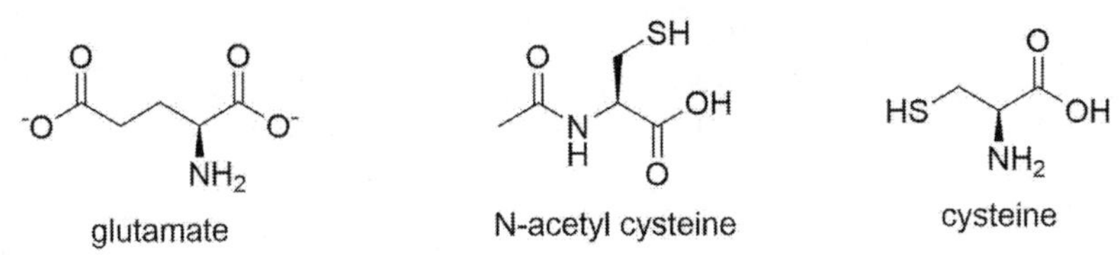

FIGURE 11.3 Structure of glutamate, N-acetyl cysteine, and cysteine.

11.4.2 Regulation of Glutamate, N-Acetyl Cysteine in OCD

Perturbed glutamate signaling in the brain has been implicated in OCD symptoms, and increased levels of glutamate in CSF and a few brain regions has been reported in OCD patients.[13] Glutamate excess causes excitotoxicity and oxidative stress,[69] and OCD subjects have perturbed antioxidant system, as will be discussed in the upcoming section. Its release at the synapses is inhibited by N-acetyl cysteine.[70–72] N-acetyl cysteine exhibits greater bioavailability, stability, solubility, and is resistant to oxidation than L-cysteine and makes it a better source for cysteine instead of parenteral cysteine.[73, 74] It's safe at high doses as high as 8,000 mg/day without significant adverse effects.[75] It is metabolized to L-cysteine in the liver, and it probably crosses blood brain barrier to reach the brain.[75] N-acetyl cysteine alleviated OCD symptoms by a seemingly dual mode of action—firstly by an inhibitory effect on synaptic glutamate release and secondly by attenuating inflammatory reactions and acting as antioxidants.[70] N-acetyl cysteine is also explored as augmentation therapy along with fluvoxamine and citalopram independently[76] and reported that it helps as a treatment for OCD. It shows several bioactive effects, including neurochemical, antioxidant, anti-inflammatory, and cyto-protective effects.[77]

11.4.3 Glycine Administration in the OCD

Glycine at higher doses ~60 g/day shows therapeutic benefits against OCD, but such high doses lead to side effects triggering nausea.[78] Glycine acts as agonist to the NMDA receptors, and it modulates cortical glutamatergic function.[79] Glycine is also reported to act as anti-inflammatory agent by modulating chloride channels in immune cells.[80] It needs further exploration in case of therapy to PANDAS.

11.4.4 Herbal Medicines in the OCD

A few herbal extracts like *Hypericum perforatum* (Hp), *Silybum marianum* (Sm), and *Echium amoenum* (Ea) are reported to help in the management of neurological disorders. A few clinical trials have examined the effects of these herbal extracts (if any) in the OCD patients. Hp extracts[81] as high dose as 600–1,800 mg or Sm extract[82] dose of 200 mg didn't improve the condition of the OCD patients. Ea extracts administration for four to six weeks produced a slight decrease in OCD symptoms with significant decrease in anxiety.[83] *Valeriana officinalis* L better known for its use as perfume and its extract have shown to be effective in treating severe anxiety, insomnia, and migraine.[84] It acts as a sedative and pain reliever. It was tested for its efficacy to manage OCD in a placebo-controlled manner and was found to be useful in the treatment of OCD, and its effects were very rapid.[84, 85] One side effect to Valeriana extract is reduced libido.[85]

A few sections of this chapter discuss the possible link between oxidative stress, inflammation, and OCD. Curcumin (a bioactive compound found in turmeric) is an established anti-inflammatory agent, and it helps in the management of a series of pathological conditions, like arthritis, common cold, cancer, stomach disorders, pain, and swelling.[86] Curcumin is never tested against OCD in humans, but quinpirole (dopamine agonist) induced OCD in animal rats was alleviated on curcumin administration for 35 days. Quinpirole induces compulsive checking and ritualistic behaviors, and these were reversed by curcumin up to the same degree as on paroxetine administration. Borage extracts showed some beneficial effects after four to six weeks of administration.[87]

11.4.5 Role of Essential free Fatty Acids and Lipids

Essential free fatty acid supplementation has been found to play a therapeutic role in neurological disorders like eicosapentaenoic acid in schizophrenia and docosahexaenoic acid along with eicosapentaenoic acid in bipolar disorder.[88, 89] Omega-3 eicosapentaenoic acid was tested at very a small scale for OCD but was not found beneficial for the OCD; it needs extensive exploration and clinical study.[90]

In one of the case studies, OCD had higher low density lipoprotein and triglyceride levels but lower high density lipoprotein than control subjects. Subjects of OCD with panic attacks had higher serum concentrations, but OCD subjects without panic attacks or alexithymic do not have any change in levels of serum lipids.[91, 92]

11.4.6 Atorvastatin and OCD

There is one report of association of nitric oxide (NO) signaling in OCD along with streptococcal infection triggered inflammation.[93] Atorvastatin (inhibitor of HMG-CoA reductase) belongs to

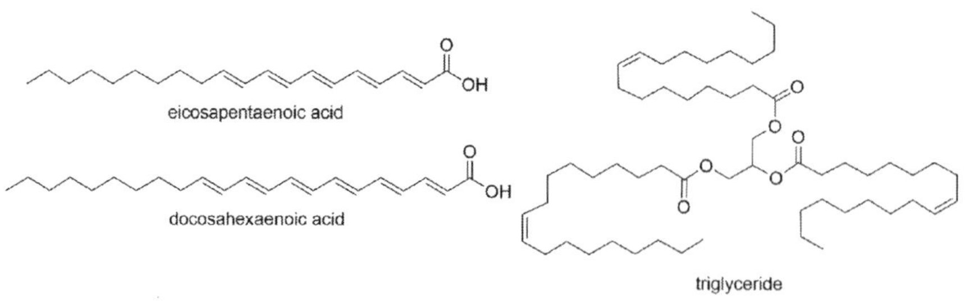

FIGURE 11.4A Structure of free fatty acid and triacylglyceride.

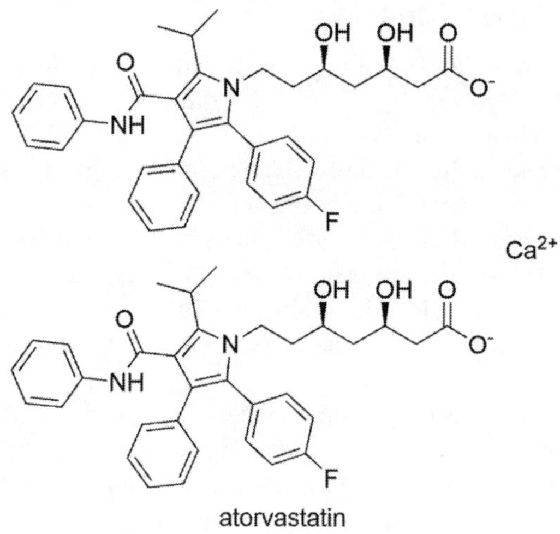

FIGURE 11.4 Structure of atorvastatin.

the category of statins used to treat dyslipidaemia (high cholesterol) and reduce bad cholesterol.[94] Atorvastatin is also found to modulate inflammation,[95] NO metabolism,[96] and increase GABA and glutathione in mice.[97] This led to the testing of atorvastatin as a therapeutic drug for OCD. Atorvastatin was tested in a small number of treatment-resistant OCD subjects using 26 subjects for 12-week-long double blind randomized study. Atorvastatin-treated group showed significant reduction in obsession subtotal score.[98]

11.4.7 Effect of Antioxidants in OCD

Inflammation, oxidative stress, and neurological disorders almost coexist except at very initial stages of the pathophysiological condition. Since inflammation-associated OCD is reported, there were few studies which tried to find the link between the cellular redox control enzymes and OCD. Superoxide dismutase activity was found to be slightly high, not statistically significant in OCD subjects without major depressive disorder (MDD), and were higher in OCD subjects with MDD compared to healthy controls.[99] Glutathione peroxidases were higher in both the groups of OCD subjects with or without MDD compared to healthy controls. Catalase activity was higher in OCD + MDD subjects but not in OCD-MDD subjects.[99] Malondialdehyde levels were higher in OCD + MCD and OCD-MDD compared to controls. These measurements were done in blood samples of enrolled subjects. Measurements in specific tissues (brain) or their parts (basal ganglia) might provide more significant findings in relevance to OCD. Lack of suitable model organisms to study OCD and lack of tools to measure activity of redox enzymes in live human subjects make these studies difficult to perform. In an another study, levels of MDA and SOD in erythrocytes were higher in OCD patients, while CAT, GSH-Px, and serum Se were lower than in the healthy controls.[100]

Free radical induced lipid peroxidation leads to the formation of thiobarbituric acid reacting substances (TBARS); hence it acts as a marker of the oxidative stress and lipid peroxidation. TBARS in serum of newly diagnosed OCD patients with no drug therapy to manage OCD was significantly higher in OCD subjects.[101] Serum TBARS correlate strongly and positively with the severity of OCD.[101] There were no changes in 8-F2-isoprostane, thioredoxin, and thioredoxin reductase levels don't change in child OCD subjects.[102]

11.5 ISOPRENOID SYSTEM AND OCD

Human hypothalamus secretes digoxin, which is a steroidal glycoside synthesized by isoprenoid pathway[103] and is normally extracted from digitalis plant. It is reported to regulate cation transport,[104] neurotransmitters transport across membranes at synapses. Serum levels of HMG CoA reductase and the concentration of digoxin and doichol (all from isoprenoid pathway) were decreased in OCD patients, while serum ubiquinone and magnesium and the activity of erythrocyte membrane Na+-K+ ATPase is increased. A+-K+ ATPase activity was attributed to by authors to the fact that hyperpolarizing neurotransmitters (dopamine, morphine, and noradrenaline) were increased, and depolarizing neuroactive molecules (serotonin, strychnine, nicotine, and quinolinic acid) were decreased.[103] There is a distinct metabolic state of isoprenoid pathways in the right and left hemisphere of the brain in OCD patients.[103]

Total serum glycosaminoglycans were found to be decreased in the OCD patients. A few other serum biomolecules like heparin sulfate, heparin, dermatan sulfate, chondroitin sulfates, and hyaluronic acid were decreased. Sugars like hexose, fucose, and sialic acid in the glycoproteins of the serum along with other molecules in serum like ganglioside, glycosyl-diglycerides, cerebrosides, sulfatides were decreased in the OCD patients.[103]

11.5.1 Mood Stabilizers in the Management of the OCD

Two mood stabilizers like divalproex (DVP) and lamotrigine (LMT) have been used in the management of bipolar disorder (acute mania and depressive episodes). Lamotrigine tested up to 100 mg/d in OCD subjects with significant improvement of irritability, depressive symptoms, and OCD but triggered hypomanic episode.[105] Administration of 1,000 mg/d divalproex for six weeks showed improvement in OCD symptoms, but OCD symptoms relapsed on termination of the drug administration.[105]

11.5.2 Effect of Tryptophan Metabolism and Dietary Intake

Tryptophan is essential branched amino acids, and mammals are dependent on dietary intake for it. It acts as a precursor of serotonin also called 5-hydroxytryptamine (5-HT). 5-HT plays an active role in a series of behavioral and emotional activities and their dysfunctional states like anxiety, depression, etc.[106] Dysfunctional serotonin uptake at synapses in the brain has been implicated in OCD. OCD patients were found to have reduced serum concentration of tryptophan, quinolinic acid, and serotonin and decreased levels of tyrosine dopamine and noradrenaline in serum.[103]

11.5.3 Effects of Vitamin D and B Complex

Vitamin D deficiency in serum measurements is found in several neurological disorders like schizophrenia,[107] autism,[108] and depression.[109] Vitamin D_3 (1, 25-dihydroxy-vitamin D_3) regulates rate

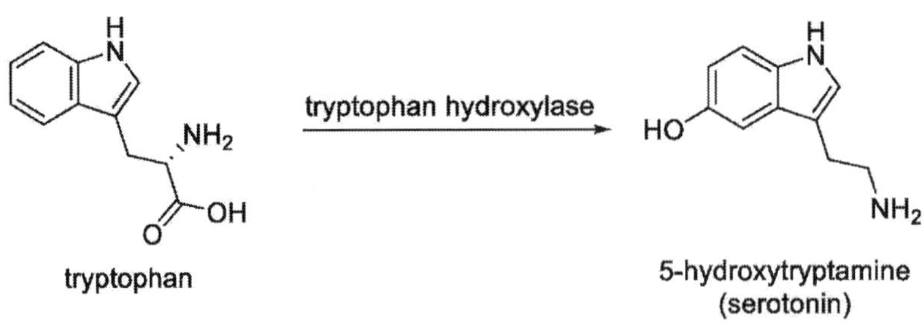

FIGURE 11.5 Synthesis of serotonin from tryptophan.

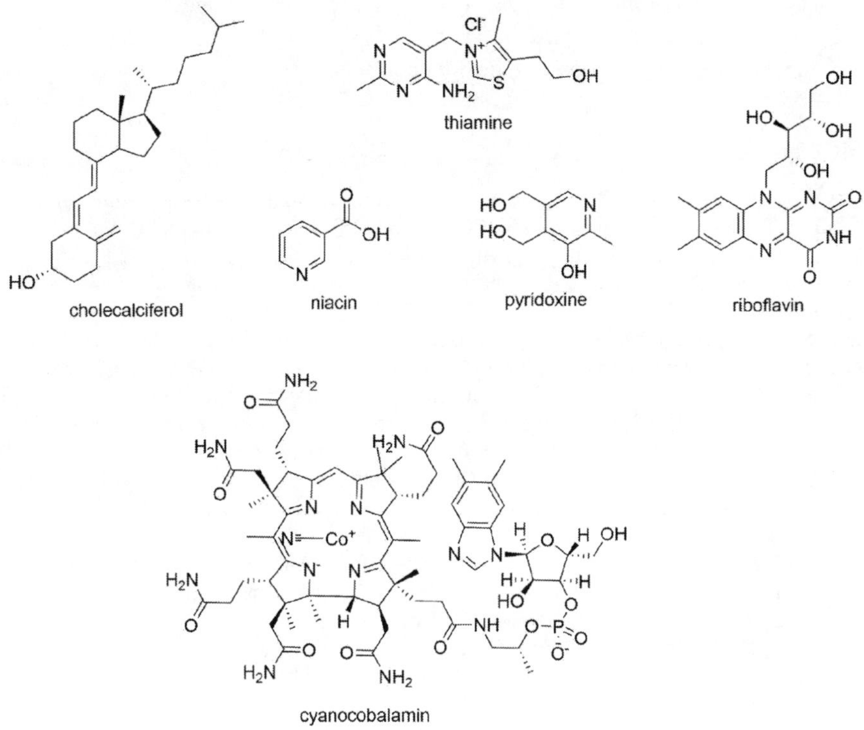

FIGURE 11.6 Structure of vitamin D and B types. Cholecalciferol (Vitamin D), thiamine (Vitamin B1), niacin (Vitamin B3), pyridoxine (Vitamin B6), riboflavin (Vitamin B2), cyanocobalamin (Vitamin B12).

limiting enzyme for serotonin synthesis—tyrosine hydrolase—and rate limiting enzyme for dopamine, epinephrine, and norepinephrine synthesis—tryptophan hydrolase.[110, 111] The deficiency of Vitamin D might be contributing to the OCD etiology by the serotonin and catecholamine metabolism. Vitamin D is also reported to act as antioxidant in the system and is reported to inhibit essential enzyme inducible nitric oxide synthase for nitric oxide synthesis.[112, 113] An increased level of nitric oxide is found in OCD patients, thus associating Vitamin D, redox system, and neurotransmitters.

Few case studies of OCD suggest Vitamin B_{12} deficiency[114–116] and homocysteine abundance[116–118] in subjects, and these fluctuations in Vitamin B_{12} and homocysteine correlate with the symptom severity of OCD.[117] Folate levels do not show any change in OCD subjects compared to the healthy controls.[118] Folate was also tried as a supplement with fluoxetine to manage OCD, but folate group didn't show any significant change in OCD symptoms.[119]

11.5.4 Role of Inorganic Trace Elements

Micronutrient cocktail along with solution of free amino acids has been reported to alleviate symptoms of OCD in patients in a few case studies and solution of free amino acids also helps to alleviate withdrawal symptoms.[120] It was reported that selenium in plasma circulation was low in OCD patients who correlated well with the plasma glutathione peroxidase activity.[100] Other metallic micronutrients like manganese and calcium were high, while zinc, iron, and magnesium were lower in the plasma of the OCD patients.[121] Zinc modulates both inhibitory and excitatory transmission such as GABA and NMDA.[122] Keeping this in mind, zinc was tried in an eight-week clinical trial along with fluoxetine (20 mg/day), and zinc group (440 mg/day as zinc sulfate) alleviated OCD symptoms.[123] Zinc is a potent antioxidant, and the beneficial effect of zinc could also be due to its antioxidant activity.[124, 125] An element-to-element correlative study suggested increase in Ca:Mg, Ca:Zn, Ca:Fe, Mg:Fe, Cu:Fe and decrease in Ca:Mn, Mg:Cu, Mg:Mn, Zn:Cu, Zn:Mn, Cu:Mn, and Fe:Mn ratios in OCD subjects compared to healthy controls.[121]

REFERENCES

1. Richter, P. M. A., and Ramos, R. T. (2018) Obsessive-compulsive disorder, Continuum 24, 828–844.
2. Prakash, J., Chaudhury, S., Chatterjee, K., Kumar, S. (2022) Do food and nutrition have therapeutic value for disorders of mood and conation? Industrail Psychiartiy Journal 31(1), 1–5.
3. Lombroso, P. J., and Scahill, L. (2008) Tourette syndrome and obsessive-compulsive disorder, Brain and Development 30, 231–237.
4. Pallanti, S., Grassi, G., Sarrecchia, E. D., Cantisani, A., and Pellegrini, M. (2011) Obsessive-compulsive disorder comorbidity: Clinical assessment and therapeutic implications, Frontiers in Psychiatry 2, 70.
5. Pallanti, S., Grassi, G., Sarrecchia, E. D., Cantisani, A., and Pellegrini, M. (2011) Obsessive-compulsive disorder comorbidity: Clinical assessment and therapeutic implications, Frontiers in Psychiatry 2, 70.
6. Conelea, C. A., Walther, M. R., Freeman, J. B., Garcia, A. M., Sapyta, J., Khanna, M., and Franklin, M. (2014) Tic-related obsessive-compulsive disorder (OCD): Phenomenology and treatment outcome in the Pediatric OCD Treatment Study II, Journal of the American Academy of Child and Adolescent Psychiatry 53, 1308–1316.
7. Sujita Kumar Kar, Ipsita Choudhury. (2016) Review article: An empirical review on oxidative stress markers and their relevance in obsessive-compulsive disorder, International Journal of Nutrition, Pharmacology, Neurological Diseases 6(4), 139–145.
8. Sharma, L. P., and Reddy, Y. C. J. (2019) Obsessive-compulsive disorder comorbid with schizophrenia and bipolar disorder, Indian journal of Psychiatry 61, S140-S148.
9. Anderson, K. E., Gehl, C. R., Marder, K. S., Beglinger, L. J., Paulsen, J. S., and Huntington's Study, G. (2010) Comorbidities of obsessive and compulsive symptoms in Huntington's disease, The Journal of Nervous and Mental Disease 198, 334–338.
10. Milad, M. R., and Rauch, S. L. (2012) Obsessive-compulsive disorder: Beyond segregated cortico-striatal pathways, Trends in Cognitive Sciences 16, 43–51.
11. Koch, K., Reess, T. J., Rus, O. G., Zimmer, C., and Zaudig, M. (2014) Diffusion tensor imaging (DTI) studies in patients with obsessive-compulsive disorder (OCD): A review, Journal of Psychiatric Research 54, 26–35.
12. Greenberg, B. D., Benjamin, J., Martin, J. D., Keuler, D., Huang, S. J., Altemus, M., and Murphy, D. L. (1998) Delayed obsessive-compulsive disorder symptom exacerbation after a single dose of a serotonin antagonist in fluoxetine-treated but not untreated patients, Psychopharmacology 140, 434–444.
13. Greist, J. H., Jefferson, J. W., Kobak, K. A., Katzelnick, D. J., and Serlin, R. C. (1995) Efficacy and tolerability of serotonin transport inhibitors in obsessive-compulsive disorder. A meta-analysis, Archives of General Psychiatry 52, 53–60.
14. Chakrabarty, K., Bhattacharyya, S., Christopher, R., and Khanna, S. (2005) Glutamatergic dysfunction in OCD, Neuropsychopharmacology: Official publication of the American College of Neuropsychopharmacology 30, 1735–1740.
15. Welch, J. M., Lu, J., Rodriguiz, R. M., Trotta, N. C., Peca, J., Ding, J. D., Feliciano, C., Chen, M., Adams, J. P., Luo, J., Dudek, S. M., Weinberg, R. J., Calakos, N., Wetsel, W. C., and Feng, G. (2007) Cortico-striatal synaptic defects and OCD-like behaviours in Sapap3-mutant mice, Nature 448, 894–900.
16. Arnold, P. D., Sicard, T., Burroughs, E., Richter, M. A., and Kennedy, J. L. (2006) Glutamate transporter gene SLC1A1 associated with obsessive-compulsive disorder, Archives of General Psychiatry 63, 769–776.
17. Dickel, D. E., Veenstra-VanderWeele, J., Cox, N. J., Wu, X., Fischer, D. J., Van Etten-Lee, M., Himle, J. A., Leventhal, B. L., Cook, E. H., Jr., and Hanna, G. L. (2006) Association testing of the positional and functional candidate gene SLC1A1/EAAC1 in early-onset obsessive-compulsive disorder, Archives of General Psychiatry 63, 778–785.
18. Stewart, S. E., Fagerness, J. A., Platko, J., Smoller, J. W., Scharf, J. M., Illmann, C., Jenike, E., Chabane, N., Leboyer, M., Delorme, R., Jenike, M. A., and Pauls, D. L. (2007) Association of the SLC1A1 glutamate transporter gene and obsessive-compulsive disorder, American Journal of Medical Genetics. Part B, Neuropsychiatric Genetics: The Official Publication of the International Society of Psychiatric Genetics 144B, 1027–1033.

19. Camarena, B., Loyzaga, C., Aguilar, A., Weissbecker, K., and Nicolini, H. (2007) Association study between the dopamine receptor D(4) gene and obsessive-compulsive disorder, European Neuropsychopharmacology: The Journal of the European College of Neuropsychopharmacology 17, 406–409.
20. Ozomaro, U., Cai, G., Kajiwara, Y., Yoon, S., Makarov, V., Delorme, R., Betancur, C., Ruhrmann, S., Falkai, P., Grabe, H. J., Maier, W., Wagner, M., Lennertz, L., Moessner, R., Murphy, D. L., Buxbaum, J. D., Zuchner, S., and Grice, D. E. (2013) Characterization of SLITRK1 variation in obsessive-compulsive disorder, PLoS One 8, e70376.
21. Whiteside, S. P., Port, J. D., and Abramowitz, J. S. (2004) A meta-analysis of functional neuroimaging in obsessive-compulsive disorder, Psychiatry Research 132, 69–79.
22. Pearlman, D. M., Vora, H. S., Marquis, B. G., Najjar, S., and Dudley, L. A. (2014) Anti-basal ganglia antibodies in primary obsessive-compulsive disorder: Systematic review and meta-analysis, The British Journal of Psychiatry: The Journal of Mental Science 205, 8–16.
23. Welter, M. L., Burbaud, P., Fernandez-Vidal, S., Bardinet, E., Coste, J., Piallat, B., Borg, M., Besnard, S., Sauleau, P., Devaux, B., Pidoux, B., Chaynes, P., Tezenas du Montcel, S., Bastian, A., Langbour, N., Teillant, A., Haynes, W., Yelnik, J., Karachi, C., Mallet, L., and French Stimulation dans Trouble Obsessionnel Compulsif Study, G. (2011) Basal ganglia dysfunction in OCD: Subthalamic neuronal activity correlates with symptoms severity and predicts high-frequency stimulation efficacy. Translational Psychiatry 1, e5.
24. Chacko, R. C., Corbin, M. A., and Harper, R. G. (2000) Acquired obsessive-compulsive disorder associated with basal ganglia lesions, The Journal of Neuropsychiatry and Clinical Neurosciences 12, 269–272.
25. Maia, T. V., Cooney, R. E., and Peterson, B. S. (2008) The neural bases of obsessive-compulsive disorder in children and adults, Development and Psychopathology 20, 1251–1283.
26. Nemoda, Z., Szekely, A., and Sasvari-Szekely, M. (2011) Psychopathological aspects of dopaminergic gene polymorphisms in adolescence and young adulthood, Neuroscience and Biobehavioral Reviews 35, 1665–1686.
27. Hesse, S., Muller, U., Lincke, T., Barthel, H., Villmann, T., Angermeyer, M. C., Sabri, O., and Stengler-Wenzke, K. (2005) Serotonin and dopamine transporter imaging in patients with obsessive-compulsive disorder, Psychiatry Research 140, 63–72.
28. Billett, E. A., Richter, M. A., King, N., Heils, A., Lesch, K. P., and Kennedy, J. L. (1997) Obsessive compulsive disorder, response to serotonin reuptake inhibitors and the serotonin transporter gene, Molecular Psychiatry 2, 403–406.
29. Zhang, H., Wang, B., Li, K., Wang, X., Li, X., Zhu, J., Zhao, Q., Yang, Y., Lv, L., Zhang, M., and Zhang, H. (2019) Altered functional connectivity between the cerebellum and the cortico-striato-thalamo-cortical circuit in obsessive-compulsive disorder, Frontiers in Psychiatry 10, 522.
30. Zhao, Q., Xu, T., Wang, Y., Chen, D., Liu, Q., Yang, Z., and Wang, Z. (2019) Limbic cortico-striato-thalamo-cortical functional connectivity in drug-naive patients of obsessive-compulsive disorder, Psychological Medicine, 1–13.
31. Gao, J., Zhou, Y., Yang, X., Luo, J., Meng, F., Zheng, D., and Li, Z. (2019) Abnormalities within and beyond the cortico-striato-thalamo-cortical circuitry in medication-free patients with OCD revealed by the fractional amplitude of low-frequency fluctuations and resting-state functional connectivity, Neuroscience Letters 712, 134449.
32. Macerollo, A., and Martino, D. (2013) Pediatric autoimmune neuropsychiatric disorders associated with streptococcal infections (PANDAS): An evolving concept, Tremor and other Hyperkinetic Movements 3.
33. Snider, L. A., and Swedo, S. E. (2004) PANDAS: Current status and directions for research, Molecular Psychiatry 9, 900–907.
34. da Rocha, F. F., Correa, H., and Teixeira, A. L. (2008) Obsessive-compulsive disorder and immunology: A review, Progress in Neuro-psychopharmacology & Biological Psychiatry 32, 1139–1146.
35. Williams, K. A., and Swedo, S. E. (2015) Post-infectious autoimmune disorders: Sydenham's chorea, PANDAS and beyond, Brain Research 1617, 144–154.
36. Frick, L., and Pittenger, C. (2016) Microglial Dysregulation in OCD, Tourette Syndrome, and PANDAS, Journal of Immunology Research 2016, 8606057.
37. Baufeld, C., O'Loughlin, E., Calcagno, N., Madore, C., and Butovsky, O. (2018) Differential contribution of microglia and monocytes in neurodegenerative diseases, Journal of Neural Transmission 125, 809–826.
38. Rodriguez, N., Morer, A., Gonzalez-Navarro, E. A., Serra-Pages, C., Boloc, D., Torres, T., Garcia-Cerro, S., Mas, S., Gasso, P., and Lazaro, L. (2017) Inflammatory dysregulation of monocytes in pediatric patients with obsessive-compulsive disorder, Journal of Neuroinflammation 14, 261.

39. Fluitman, S., Denys, D., Vulink, N., Schutters, S., Heijnen, C., and Westenberg, H. (2010) Lipopolysaccharide-induced cytokine production in obsessive-compulsive disorder and generalized social anxiety disorder, Psychiatry Research 178, 313–316.
40. Trankner, D., Boulet, A., Peden, E., Focht, R., Van Deren, D., and Capecchi, M. (2019) A Microglia Sublineage Protects from Sex-Linked Anxiety Symptoms and Obsessive Compulsion, Cell Reports 29, 791–799 e793.
41. Nagarajan, N., Jones, B. W., West, P. J., Marc, R. E., and Capecchi, M. R. (2018) Corticostriatal circuit defects in Hoxb8 mutant mice, Molecular Psychiatry 23, 1868–1877.
42. De, S., Van Deren, D., Peden, E., Hockin, M., Boulet, A., Titen, S., and Capecchi, M. R. (2018) Two distinct ontogenies confer heterogeneity to mouse brain microglia, Development 145.
43. Yadin, E., Friedman, E., and Bridger, W. H. (1991) Spontaneous alternation behavior: An animal model for obsessive-compulsive disorder? Pharmacology, Biochemistry, and Behavior 40, 311–315.
44. Joel, D. (2006) Current animal models of obsessive compulsive disorder: A critical review, Progress in Neuro-Psychopharmacology & Biological Psychiatry 30, 374–388.
45. Grunblatt, E., Marinova, Z., Roth, A., Gardini, E., Ball, J., Geissler, J., Wojdacz, T. K., Romanos, M., and Walitza, S. (2018) Combining genetic and epigenetic parameters of the serotonin transporter gene in obsessive-compulsive disorder, Journal of Psychiatric Research 96, 209–217.
46. Locher, C., Koechlin, H., Zion, S. R., Werner, C., Pine, D. S., Kirsch, I., Kessler, R. C., and Kossowsky, J. (2017) Efficacy and safety of selective serotonin reuptake inhibitors, serotonin-norepinephrine reuptake inhibitors, and placebo for common psychiatric disorders among children and adolescents: A systematic review and meta-analysis, JAMA Psychiatry 74, 1011–1020.
47. Lu, Q., Komenoi, S., Usuki, T., Takahashi, D., and Sakane, F. (2018) Abnormalities of the serotonergic system in diacylglycerol kinase delta-deficient mouse brain, Biochemical and Biophysical Research Communications 497, 1031–1037.
48. Lu, Q., Murakami, C., Hoshino, F., Murakami, Y., and Sakane, F. (2020) Diacylglycerol kinase delta destabilizes serotonin transporter protein through the ubiquitin-proteasome system, Biochimica et biophysica acta. Molecular and Cell Biology of Lipids 1865, 158608.
49. Chinnery PF. (2022) Traumatic brain injury advances since 2017: What has changed? The Lancet: Neurology 21 (11), 953–956.
50. Pascual-Vera, B., Roncero, M., Doron, G., and Belloch, A. (2018) Assisting relapse prevention in OCD using a novel mobile app-based intervention: A case report, Bulletin of the Menninger Clinic 82, 390–406.
51. Miller, K. E., Kuhn, E., Owen, J. E., Taylor, K., Yu, J. S., Weiss, B. J., Crowley, J. J., and Trockel, M. (2019) Clinician perceptions related to the use of the CBT-I coach mobile app, Behavioral Sleep Medicine 17, 481–491.
52. Sun, M., Tang, S., Chen, J., Li, Y., Bai, W., Plummer, V., Lam, L., Qin, C., and Cross, W. M. (2019) A study protocol of mobile phone app-based cognitive behaviour training for the prevention of postpartum depression among high-risk mothers, BMC Public Health 19, 710.
53. Mantani, A., Kato, T., Furukawa, T. A., Horikoshi, M., Imai, H., Hiroe, T., Chino, B., Funayama, T., Yonemoto, N., Zhou, Q., and Kawanishi, N. (2017) Smartphone cognitive behavioral therapy as an adjunct to pharmacotherapy for refractory depression: Randomized controlled trial, Journal of Medical Internet Research 19, e373.
54. Alina Arulsamy, Mohd Farooq Shaikh. (2016) Review Article: The impact of epilepsy on the manifestation of anxiety disorder, International Journal of Nutrition, Pharmacology, Neurological Diseases 6(1), 3–11.
55. Coric, V., Taskiran, S., Pittenger, C., Wasylink, S., Mathalon, D. H., Valentine, G., Saksa, J., Wu, Y. T., Gueorguieva, R., Sanacora, G., Malison, R. T., and Krystal, J. H. (2005) Riluzole augmentation in treatment-resistant obsessive-compulsive disorder: An open-label trial, Biological Psychiatry 58, 424–428.
56. de Boer, J. N., Vingerhoets, C., Hirdes, M., McAlonan, G. M., Amelsvoort, T. V., and Zinkstok, J. R. (2019) Efficacy and tolerability of Riluzole in psychiatric disorders: A systematic review and preliminary meta-analysis, Psychiatry Research 278, 294–302.
57. Biria, M., Huang, F. X., Worbe, Y., Fineberg, N. A., Robbins, T. W., and Fernandez-Egea, E. (2019) A cross sectional study of impact and clinical risk factors of antipsychotic-induced OCD, European Neuropsychopharmacology: The Journal of the European College of Neuropsychopharmacology 29, 905–913.
58. Del Casale, A., Sorice, S., Padovano, A., Simmaco, M., Ferracuti, S., Lamis, D. A., Rapinesi, C., Sani, G., Girardi, P., Kotzalidis, G. D., and Pompili, M. (2019) Psychopharmacological treatment of obsessive-compulsive disorder (OCD), Current Neuropharmacology 17, 710–736.
59. Hanisch, F., Friedemann, J., Piro, J., and Gutmann, P. (2009) Maintenance electroconvulsive therapy for comorbid pharmacotherapy-refractory obsessive-compulsive and schizoaffective disorder, European Journal of Medical Research 14, 367–368.

60. Gomes, P. V., Brasil-Neto, J. P., Allam, N., and Rodrigues de Souza, E. (2012) A randomized, double-blind trial of repetitive transcranial magnetic stimulation in obsessive-compulsive disorder with three-month follow-up, The Journal of Neuropsychiatry and Clinical Neurosciences 24, 437–443.
61. Rapinesi, C., Kotzalidis, G. D., Ferracuti, S., Sani, G., Girardi, P., and Del Casale, A. (2019) Brain stimulation in obsessive-compulsive disorder (OCD): A systematic review, Current Neuropharmacology 17, 787–807.
62. Harvey, B. H., Brink, C. B., Seedat, S., and Stein, D. J. (2002) Defining the neuromolecular action of myo-inositol: Application to obsessive-compulsive disorder, Progress in Neuro-Psychopharmacology & Biological Psychiatry 26, 21–32.
63. Murray, M., and Greenberg, M. L. (2000) Expression of yeast INM1 encoding inositol monophosphatase is regulated by inositol, carbon source and growth stage and is decreased by lithium and valproate, Molecular Microbiology 36, 651–661.
64. Pettegrew, J. W., Panchalingam, K., Levine, J., McClure, R. J., Gershon, S., and Yao, J. K. (2001) Chronic myo-inositol increases rat brain phosphatidylethanolamine plasmalogen, Biological Psychiatry 49, 444–453.
65. Fux, M., Levine, J., Aviv, A., and Belmaker, R. H. (1996) Inositol treatment of obsessive-compulsive disorder, The American Journal of Psychiatry 153, 1219–1221.
66. Leppink, E. W., Redden, S. A., and Grant, J. E. (2017) A double-blind, placebo-controlled study of inositol in trichotillomania, International Clinical Psychopharmacology 32, 107–114.
67. Rahman, S., and Neuman, R. S. (1993) Myo-inositol reduces serotonin (5-HT2) receptor induced homologous and heterologous desensitization, Brain Research 631, 349–351.
68. Harvey, B. H., Scheepers, A., Brand, L., and Stein, D. J. (2001) Chronic inositol increases striatal D(2) receptors but does not modify dexamphetamine-induced motor behavior. Relevance to obsessive-compulsive disorder, Pharmacology, Biochemistry, and Behavior 68, 245–253.
69. Shimon, H., Agam, G., Belmaker, R. H., Hyde, T. M., and Kleinman, J. E. (1997) Reduced frontal cortex inositol levels in postmortem brain of suicide victims and patients with bipolar disorder, The American Journal of Psychiatry 154, 1148–1150.
70. Burdo, J., Dargusch, R., and Schubert, D. (2006) Distribution of the cystine/glutamate antiporter system xc- in the brain, kidney, and duodenum, The Journal of Histochemistry and Cytochemistry: Official Journal of the Histochemistry Society 54, 549–557.
71. Oliver, G., Dean, O., Camfield, D., Blair-West, S., Ng, C., Berk, M., and Sarris, J. (2015) N-acetyl cysteine in the treatment of obsessive compulsive and related disorders: A systematic review, Clinical Psychopharmacology and Neuroscience: The Official Scientific Journal of the Korean College of Neuropsychopharmacology 13, 12–24.
72. Moran, M. M., McFarland, K., Melendez, R. I., Kalivas, P. W., and Seamans, J. K. (2005) Cystine/glutamate exchange regulates metabotropic glutamate receptor presynaptic inhibition of excitatory transmission and vulnerability to cocaine seeking, The Journal of Neuroscience: The Official Journal of the Society for Neuroscience 25, 6389–6393.
73. Baker, D. A., Shen, H., and Kalivas, P. W. (2002) Cystine/glutamate exchange serves as the source for extracellular glutamate: Modifications by repeated cocaine administration, Amino Acids 23, 161–162.
74. Berk, M., Malhi, G. S., Gray, L. J., and Dean, O. M. (2013) The promise of N-acetylcysteine in neuropsychiatry, Trends in Pharmacological Sciences 34, 167–177.
75. Bonanomi, L., and Gazzaniga, A. (1980) Toxicological, pharmacokinetic and metabolic studies on acetylcysteine, European Journal of Respiratory Diseases. Supplement 111, 45–51.
76. Atkuri, K. R., Mantovani, J. J., Herzenberg, L. A., and Herzenberg, L. A. (2007) N-Acetylcysteine—a safe antidote for cysteine/glutathione deficiency, Current Opinion in Pharmacology 7, 355–359.
77. Lafleur, D. L., Pittenger, C., Kelmendi, B., Gardner, T., Wasylink, S., Malison, R. T., Sanacora, G., Krystal, J. H., and Coric, V. (2006) N-acetylcysteine augmentation in serotonin reuptake inhibitor refractory obsessive-compulsive disorder, Psychopharmacology 184, 254–256.
78. Ng, F., Berk, M., Dean, O., and Bush, A. I. (2008) Oxidative stress in psychiatric disorders: Evidence base and therapeutic implications, The International Journal of Neuropsychopharmacology 11, 851–876.
79. Cleveland, W. L., DeLaPaz, R. L., Fawwaz, R. A., and Challop, R. S. (2009) High-dose glycine treatment of refractory obsessive-compulsive disorder and body dysmorphic disorder in a 5-year period, Neural Plasticity 2009, 768398.
80. Singer, H. S., Morris, C., and Grados, M. (2010) Glutamatergic modulatory therapy for Tourette syndrome, Medical Hypotheses 74, 862–867.
81. Stoffels, B., Turler, A., Schmidt, J., Nazir, A., Tsukamoto, T., Moore, B. A., Schnurr, C., Kalff, J. C., and Bauer, A. J. (2011) Anti-inflammatory role of glycine in reducing rodent postoperative inflammatory ileus, Neurogastroenterology and Motility: The Official Journal of the European Gastrointestinal Motility Society 23, 76–87, e78.

82. Kobak, K. A., Taylor, L. V., Bystritsky, A., Kohlenberg, C. J., Greist, J. H., Tucker, P., Warner, G., Futterer, R., and Vapnik, T. (2005) St John's wort versus placebo in obsessive-compulsive disorder: Results from a double-blind study, International Clinical Psychopharmacology 20, 299–304.
83. Sayyah, M., Boostani, H., Pakseresht, S., and Malayeri, A. (2010) Comparison of Silybum marianum (L.) Gaertn. with fluoxetine in the treatment of obsessive-compulsive disorder, Progress in Neuro-Psychopharmacology & Biological Psychiatry 34, 362–365.
84. Sayyah, M., Boostani, H., Pakseresht, S., and Malaieri, A. (2009) Efficacy of aqueous extract of Echium amoenum in treatment of obsessive-compulsive disorder, Progress in Neuro-Psychopharmacology & Biological Psychiatry 33, 1513–1516.
85. Andreatini, R., Sartori, V. A., Seabra, M. L., and Leite, J. R. (2002) Effect of valepotriates (valerian extract) in generalized anxiety disorder: A randomized placebo-controlled pilot study, Phytotherapy Research: PTR 16, 650–654.
86. Pakseresht, S., Boostani, H., and Sayyah, M. (2011) Extract of valerian root (Valeriana officinalis L.) vs. placebo in treatment of obsessive-compulsive disorder: A randomized double-blind study, Journal of Complementary & Integrative Medicine 8.
87. Hewlings, S. J., and Kalman, D. S. (2017) Curcumin: A review of its' effects on human health, Foods 6.
88. Chimakurthy, J., and Murthy, T. E. (2010) Effect of curcumin on quinpirole induced compulsive checking: An approach to determine the predictive and construct validity of the model, North American journal of medical sciences 2, 81–86.
89. Fusar-Poli, P., and Berger, G. (2012) Eicosapentaenoic acid interventions in schizophrenia: Meta-analysis of randomized, placebo-controlled studies, Journal of Clinical Psychopharmacology 32, 179–185.
90. Osher, Y., Bersudsky, Y., and Belmaker, R. H. (2005) Omega-3 eicosapentaenoic acid in bipolar depression: Report of a small open-label study, The Journal of Clinical Psychiatry 66, 726–729.
91. Fux, M., Benjamin, J., and Nemets, B. (2004) A placebo-controlled cross-over trial of adjunctive EPA in OCD, Journal of Psychiatric Research 38, 323–325.
92. Agargun, M. Y., Dulger, H., Inci, R., Kara, H., Ozer, O. A., Sekeroglu, M. R., and Besiroglu, L. (2004) Serum lipid concentrations in obsessive-compulsive disorder patients with and without panic attacks, Canadian Journal of Psychiatry. Revue canadienne de psychiatrie 49, 776–778.
93. De Berardis, D., Serroni, N., Marini, S., Rapini, G., Carano, A., Valchera, A., Iasevoli, F., Mazza, M., Signorelli, M., Aguglia, E., Perna, G., Martinotti, G., Varasano, P. A., Pressanti, G. L., and Di Giannantonio, M. (2014) Alexithymia, suicidal ideation, and serum lipid levels among drug-naive outpatients with obsessive-compulsive disorder, Revista brasileira de psiquiatria 36, 125–130.
94. Umathe, S. N., Bhutada, P. S., Jain, N. S., Mundhada, Y. R., Borkar, S. S., and Dhumal, B. (2009) Role of nitric oxide in obsessive-compulsive behavior and its involvement in the anti-compulsive effect of paroxetine in mice, Nitric Oxide: Biology and Chemistry 21, 140–147.
95. Nawrocki, J. W., Weiss, S. R., Davidson, M. H., Sprecher, D. L., Schwartz, S. L., Lupien, P. J., Jones, P. H., Haber, H. E., and Black, D. M. (1995) Reduction of LDL cholesterol by 25% to 60% in patients with primary hypercholesterolemia by atorvastatin, a new HMG-CoA reductase inhibitor, Arteriosclerosis, Thrombosis, and Vascular Biology 15, 678–682.
96. Marz, W., and Koenig, W. (2003) HMG-CoA reductase inhibition: Anti-inflammatory effects beyond lipid lowering? Journal of Cardiovascular Risk 10, 169–179.
97. Seker, F. B., Kilic, U., Caglayan, B., Ethemoglu, M. S., Caglayan, A. B., Ekimci, N., Demirci, S., Dogan, A., Oztezcan, S., Sahin, F., Yilmaz, B., and Kilic, E. (2015) HMG-CoA reductase inhibitor rosuvastatin improves abnormal brain electrical activity via mechanisms involving eNOS, Neuroscience 284, 349–359.
98. Sehar, N., Agarwal, N. B., Vohora, D., and Raisuddin, S. (2015) Atorvastatin prevents development of kindling by modulating hippocampal levels of dopamine, glutamate, and GABA in mice, Epilepsy & Behavior: E&B 42, 48–53.
99. Rahim, F., and Sayyah, M. (2018) Effects of atorvastatin on treatment-resistant obsessive-compulsive disorder: A double-blind randomized trial, Psychiatria Polska 52, 719–729.
100. Kuloglu, M., Atmaca, M., Tezcan, E., Gecici, O., Tunckol, H., and Ustundag, B. (2002) Antioxidant enzyme activities and malondialdehyde levels in patients with obsessive-compulsive disorder, Neuropsychobiology 46, 27–32.
101. Ozdemir, E., Cetinkaya, S., Ersan, S., Kucukosman, S., and Ersan, E. E. (2009) Serum selenium and plasma malondialdehyde levels and antioxidant enzyme activities in patients with obsessive-compulsive disorder, Progress in Neuro-Psychopharmacology & Biological Psychiatry 33, 62–65.
102. Chakraborty, S., Singh, O. P., Dasgupta, A., Mandal, N., and Nath Das, H. (2009) Correlation between lipid peroxidation-induced TBARS level and disease severity in obsessive-compulsive disorder, Progress in Neuro-Psychopharmacology & Biological Psychiatry 33, 363–366.

103. Bilgic, A., Colak Sivri, R., and Kilinc, I. (2018) 8-F2-isoprostane, thioredoxin and thioredoxin reductase levels in children with obsessive-compulsive disorder, Nordic Journal of Psychiatry 72, 484–488.
104. Kurup, R. K., and Kurup, P. A. (2002) Hypothalamic digoxin deficiency in obsessive compulsive disorder and la Tourette's syndrome, The International Journal of Neuroscience 112, 797–816.
105. Hougen, T. J., Lloyd, B. L., and Smith, T. W. (1979) Effects of inotropic and arrhythmogenic digoxin doses and of digoxin-specific antibody on myocardial monovalent cation transport in the dog, Circulation Research 44, 23–31.
106. Bisol, L. W., and Lara, D. R. (2009) Improvement of obsessive-compulsive disorder with divalproex and lamotrigine in two patients with bipolar II disorder, Pharmacopsychiatry 42, 37–39.
107. Marazziti, D. (2017) Understanding the role of serotonin in psychiatric diseases, F1000Research 6, 180.
108. Chiang, M., Natarajan, R., and Fan, X. (2016) Vitamin D in schizophrenia: A clinical review, Evidence-Based Mental Health 19, 6–9.
109. Wang, T., Shan, L., Du, L., Feng, J., Xu, Z., Staal, W. G., and Jia, F. (2016) Serum concentration of 25-hydroxyvitamin D in autism spectrum disorder: A systematic review and meta-analysis, European Child & Adolescent Psychiatry 25, 341–350.
110. Parker, G. B., Brotchie, H., and Graham, R. K. (2017) Vitamin D and depression, Journal of Affective Disorders 208, 56–61.
111. Kaneko, I., Sabir, M. S., Dussik, C. M., Whitfield, G. K., Karrys, A., Hsieh, J. C., Haussler, M. R., Meyer, M. B., Pike, J. W., and Jurutka, P. W. (2015) 1,25-Dihydroxyvitamin D regulates expression of the tryptophan hydroxylase 2 and leptin genes: Implication for behavioral influences of vitamin D, FASEB Journal: Official Publication of the Federation of American Societies for Experimental Biology 29, 4023–4035.
112. Cui, X., Pertile, R., Liu, P., and Eyles, D. W. (2015) Vitamin D regulates tyrosine hydroxylase expression: N-cadherin a possible mediator, Neuroscience 304, 90–100.
113. Garcion, E., Sindji, L., Montero-Menei, C., Andre, C., Brachet, P., and Darcy, F. (1998) Expression of inducible nitric oxide synthase during rat brain inflammation: Regulation by 1,25-dihydroxyvitamin D3, Glia 22, 282–294.
114. Behl, A., Swami, G., Sircar, S. S., Bhatia, M. S., and Banerjee, B. D. (2010) Relationship of possible stress-related biochemical markers to oxidative/antioxidative status in obsessive-compulsive disorder, Neuropsychobiology 61, 210–214.
115. Valizadeh, M., and Valizadeh, N. (2011) Obsessive compulsive disorder as early manifestation of B12 deficiency, Indian Journal of Psychological Medicine 33, 203–204.
116. Sharma, V., and Biswas, D. (2012) Cobalamin deficiency presenting as obsessive compulsive disorder: Case report, General Hospital Psychiatry 34, 578 e577–578.
117. Hermesh, H., Weizman, A., Shahar, A., and Munitz, H. (1988) Vitamin B12 and folic acid serum levels in obsessive compulsive disorder, Acta Psychiatrica Scandinavica 78, 8–10.
118. Atmaca, M., Tezcan, E., Kuloglu, M., Kirtas, O., and Ustundag, B. (2005) Serum folate and homocysteine levels in patients with obsessive-compulsive disorder, Psychiatry and Clinical Neurosciences 59, 616–620.
119. Esnafoglu, E., and Yaman, E. (2017) Vitamin B12, folic acid, homocysteine and vitamin D levels in children and adolescents with obsessive compulsive disorder, Psychiatry Research 254, 232–237.
120. Tural, U., Corapcioglu, A., Bosgelmez, S., Koroglu, G., Unver, H., Duman, C., and Onder, E. (2019) Double blind controlled study of adding folic acid to fluoxetine in the treatment of OCD, Psychiatria Danubina 31, 69–77.
121. Rodway, M., Vance, A., Watters, A., Lee, H., Bos, E., and Kaplan, B. J. (2012) Efficacy and cost of micronutrient treatment of childhood psychosis, BMJ Case Reports. Nov 9;2012:bcr2012007213.
122. Shohag, H., Ullah, A., Qusar, S., Rahman, M., and Hasnat, A. (2012) Alterations of serum zinc, copper, manganese, iron, calcium, and magnesium concentrations and the complexity of interelement relations in patients with obsessive-compulsive disorder, Biological Trace Element Research 148, 275–280.
123. Peters, S., Koh, J., and Choi, D. W. (1987) Zinc selectively blocks the action of N-methyl-D-aspartate on cortical neurons, Science 236, 589–593.
124. Sayyah, M., Olapour, A., Saeedabad, Y., Yazdan Parast, R., and Malayeri, A. (2012) Evaluation of oral zinc sulfate effect on obsessive-compulsive disorder: A randomized placebo-controlled clinical trial, Nutrition 28, 892–895.
125. Takeda, A. (2000) Movement of zinc and its functional significance in the brain, Brain Research. Brain Research Reviews 34, 137–148.

12 Dietary Management of Obsessive-Compulsive Disorder—Self Care

Krishnavignesh Lakshmanan[a,*] *and Sudha Devi Manickam*[b]

[a] Head, Academics and Research, Chennai Fertility Centre and Research Institute, Chennai-600030, Tamil Nadu, India

[b] Assistant Professor, Department of Biochemistry, Biotechnology and Bioinformatics, School of Biosciences, Avinashilingam Institute for Home Science and Higher Education for Women, Coimbatore - 641 043, Tamil Nadu, India

12.1 INTRODUCTION

Disease burden from psychiatric disorders is a significant public health issue worldwide and should be addressed proactively. The development of psychiatric disorders involves both genetic and environmental factors (Matsuoka & Hamazaki, 2016). *Diagnostic and Statistical Manual of Mental Disorders* observed four out of the ten leading causes of disability in the US and other developed countries are mental disorders. Major depression, bipolar disorder, schizophrenia, and OCD are among the most common mental disorders (Lakhan & Vieira, 2008).

OCD is an illness that can cause marked distress and disability. It often goes unrecognized and is undertreated (Fenske & Schwenk, 2009). It is a neuropsychiatric disorder that is characterized by obsessions and compulsions (Maher & Theodore, 2012). Clinically, OCD is characterized by persistent and recurrent thoughts, or images (obsessions), and/or repetitive behaviors or mental acts (compulsions), which cause significant impairment in the daily functioning of the individuals affected. It often begins early in life, it waxes and wanes over time, and remission is rare (Zghoul & Blier, 2003).

Common obsessions include fear of contamination, fear of harming self or others, intrusive violent images, recurrent forbidden or perverse sexual thoughts, a need to save items of perceived value, concern with sacrilege or morality, a need for symmetry, and excessive concern about the appearance. Examples of compulsions performed in response to obsessions include excessive

*Corresponding author, Dr. Krishnavignesh Lakshmanan, Head, Academics and Research, Chennai Fertility Centre and Research Institute, 79/129, Nelson Manickam Road, Aminjikarai, Chennai-600030, Tamil Nadu, India. E. mail: krishnavignesh.l@gmail.com; Mob.: +91 8428278218

 DOI: 10.1201/9781003387893-15

handwashing, ritualized bathing or grooming, checking behaviors, mental rituals, need to repeat activities, rereading the text, hoarding behaviors, and superstitious behaviors. Although the intrusive thoughts seen in OCD are often described as being ego-dystonic, many individuals struggle to determine whether or not their time-consuming behaviors are excessive (Pittenger *et al.*, 2005).

12.2 OBSESSIONS AND COMPULSIONS

Many people who have OCD know that their thoughts and habits don't make sense. They don't do them because they enjoy them but because they can't quit. And if they stop, they feel so bad that they start again (Baup *et al.*, 2008).

Obsessive thoughts can include

- Worries about yourself or other people getting hurt.
- Constant awareness of blinking, breathing, or other body sensations.
- The suspicion that a partner is unfaithful, with no reason to believe it.

Compulsive habits can include

- Doing tasks in a specific order every time or a certain "good" number of times.
- Needing to count things, like steps or bottles.
- Fear of touching doorknobs, using public toilets, or shaking hands.

OCD was once thought to be extremely rare, but recent epidemiological studies have shown it to be the fourth most common psychiatric disorder (after substance abuse, specific phobias, and major depression). OCD is often a chronic disorder that produces significant morbidity when not properly diagnosed and treated (Pittenger *et al.*, 2005). If left untreated, OCD can worsen to the point that the sufferer develops physical problems, becomes unable to function, or experiences suicidal thoughts. About 1% of OCD sufferers die by suicide. The mainstay of treatment includes cognitive-behavioral therapy and medication management (Pittenger *et al.*, 2005). As current treatments are estimated to address only one-third of the disease burden of psychological disorders, there is a need for new approaches to prevent or to delay its progression.

A poor diet can be detrimental to mental health. Recent findings showed that diet may influence the onset of mood disorders and specifically depression. Anxiety disorders, which, along with depression, are also classified as "common mental disorders" because of their prevalence across the globe, with approximately one in five people experiencing one of these conditions for any given year (Firth *et al.*, 2019). The human brain uses a substantial portion of the body's total energy and nutrient intake. Nutrients strongly influence both brain structure and function, and nutrition also affects neurodevelopment and neurotrophic function. It has been recognized in recent years that diet and nutrition may be an important factor contributing to psychiatric morbidity, and the prevention or treatment of psychiatric disorders could be conducted by addressing diet and nutrition (Matsuoka & Hamazaki, 2016). The role of nutrition in health and disease has been appreciated from time immemorial. Around 400 BC, Hippocrates wrote, "Let food be thy medicine and medicine be thy food." Now, in the twenty-first century, we are bombarded by claims in the media of "superfoods," wondrous nutritional supplements, and special diets that promise to cure or prevent disease, improve health, and restore functioning. Much of the focus has been on neurological disease, brain health, and psychological functioning (behavior, cognition, and emotion) (Brandt, 2019). This chapter gives a detailed insight into the prevalence, causes, comorbidities, and management of OCD through the natural and nutritional way.

12.3 PREVALENCE OF OCD

OCD is the fourth most common psychiatric disorder. It is estimated 2.5% of adults and 1% of children meet the *Diagnostic and Statistical Manual of Mental Disorders* (*DSM-IV*) classification for OCD. It usually first appears late in adolescence or early adulthood. OCD incidence is higher in females than in males. The age of onset is usually earlier in males than in females, 6–15 years compared to 20–29 years, respectively (Colodny & Hoffman, 1998).

Recent studies on the epidemiology of OCD estimate 50 million patients suffer from OCD worldwide, thus making it a global problem. It is a highly debilitating condition with a lifetime prevalence of 2% to 3%, more than twice that of schizophrenia, and is characterized by intrusive, troubling thoughts and/or repetitive, compulsive behaviors or mental rituals. The economic and social burden of OCD is difficult to quantify, but published data estimate $8 billion in the United States annually (Dell'Osso *et al.*, 2005). In India, the lifetime prevalence of OCD was found to be 0.6%. This rate is considerably lower compared to the 2–3% rate reported in the European and North American studies (Reddy *et al.*, 2010).

According to the *Diagnostic and Statistical Manual of Mental Disorders*, four out of the ten leading causes of disability in the US and other developed countries are mental disorders. Major depression, bipolar disorder, schizophrenia, OCD are among the most common mental disorders that currently plague numerous countries and have varying incidence rates from 26% in America to 4% in China (Lakhan & Vieira, 2008). In North America itself it affects ~5.2–10.4 million people. Most commonly diagnosed in childhood and adolescence, it can result in years of struggle and has been listed among the top ten causes of disability by the World Health Organization (Macphail *et al.*, 2018).

12.4 CAUSES OF OCD

Structural, neurochemical, genetic, and immunological factors play a role in the etiology and pathophysiology of OCD (Karcı & Celik, 2020). The change in membrane permeability and neuron activity in serotoninergic pathways and increased inflammatory activity in the brain are thought to be a connection between a deficiency of essential fatty acids and the development of compulsive symptoms (Colodny & Hoffman, 1998). Neuroimaging studies show the involvement of the dorsolateral prefrontal cortex, basal ganglia, and thalamus. Because of the response to selective serotonin reuptake inhibitors (SSRIs), it is hypothesized that the serotonin system is heavily involved in the neurochemistry of OCD (Fenske& Schwenk, 2009). Growing evidence shows that dysfunction of the limbic basal ganglia (BG) network is implicated in repetitive behaviors, such as OCD and Tourette's syndrome (TS), in humans (Baup *et al.*, 2008). Increased levels of nitric oxide was observed in OCD. Vitamin D has antioxidant effects and inhibits an essential enzyme (inducible nitric oxide synthase) for nitric oxide synthesis. Thus, vitamin D deficiency may play a role in OCD by causing the deterioration of neuroprotection (Karcı & Celik, 2020). Notably, essential vitamins, minerals, and omega-3 fatty acids are often deficient in the general American population and other developed countries and are exceptionally deficient in patients suffering from mental disorders (Lakhan & Vieira, 2008). These findings have triggered the attention of the researchers to look for a cure through dietary supplementations.

12.5 INDICATIONS FOR OCD

Clinically there are several factors to be considered to understand and diagnose OCD. Elevated glutamate, glutamine, and gamma-aminobutyric acid (GABA) levels in the caudate nucleus and increased glutamate levels in the cerebrospinal fluid have been found in patients with OCD (Karcı & Celik, 2020). A change in CNS availability of inositol may produce altered brain signaling and eventually lead to the development of neurological disorders (Colodny & Hoffman, 1998). Diagnosis should focus on all the cited causes and indications; based on that, the treatment should be planned.

12.6 COMORBIDITIES ASSOCIATED WITH OCD

Psychiatric morbidity results from distress and the time devoted to compulsions, as well as compulsion-induced dysfunction (Maher & Theodore, 2012). OCD usually begins in childhood or adolescence. It is estimated that OCD is accompanied by at least one additional psychiatric condition in 32–90% of the patients. As a group, anxiety disorders are the most frequent comorbid conditions in OCD (Diniz *et al.*, 2012). Other comorbid disorders such as mood, psychotic and somatoform disorders are common in patients with OCD (Karcı & Celik, 2020).

An anxiety disorder severely affects the sufferer's QOL, and this may be particularly true of those with OCD. Obsession severity was found to significantly predict patient QOL. Given the importance of these symptoms, treatments that directly target obsessions, and secondary depression symptoms in OCD are warranted (Masellis *et al.*, 2003). Treatment-resistant OCD represents an ongoing challenge for clinicians and researchers, and over the last decade, different forms of treatment have been investigated, including pharmacologic augmentation strategies (Dell'Osso *et al.*, 2005).

12.7 MANAGEMENT OF OCD

Different types of treatments available to treat OCD are psychological treatments, pharmacotherapy, complementary and alternative medicine (Fenske & Schwenk, 2009). Standard treatments include psychopharmacologic approaches using serotonin reuptake inhibitors, such as fluoxetine, and cognitive-behavioral therapy, which promotes a kind of learning through exposure to the feared or unpleasant stimulus and prevention of the compulsive response. Limited response to both treatments is common, and various psychopharmacologic agents, including the atypical antipsychotics, have been tested for their abilities to augment serotonin reuptake inhibitors (Maher & Theodore, 2012).

Besides pharmacologic, behavioral, and neurosurgical approaches, different brain stimulation methods transcranial magnetic stimulation, deep brain stimulation, and electroconvulsive therapy have been investigated in treatment-resistant patients with OCD. However, available data about the use of these techniques in OCD treatment are quite limited in terms of sample size and study design, given the difficulty in conducting standard blinded trials for these procedures. In addition, none of the mentioned treatments have received Food and Drug Administration approval for the treatment of OCD. Treatment-resistant OCD represents an ongoing challenge for clinicians and researchers, and over the last decade, different forms of treatment have been investigated, including pharmacologic augmentation strategies (Dell'Osso *et al.*, 2005). Although the treatment of OCD is remarkably advanced compared to 30 years ago, there are a number of areas where improvements can be made (Reddy *et al.*, 2010).

Anxiety is an important aspect of OCD with bases on its psychopathological presentation, and so far, most interventions that have been proved effective for treating OCD are similar to the ones developed for other anxiety disorders, such as social anxiety disorder (SAD), generalized anxiety disorder (GAD), panic, and post-traumatic stress disorder (PTSD) (Diniz *et al.*, 2012). Currently, the preferred treatment for anxiety disorders is cognitive behavior therapy and pharmacologic agents. Beta-blockers or benzodiazepines are used for time-limited and predictable anxiety disorders, whereas selective serotonin reuptake inhibitors (SSRIs), selective serotonin-norepinephrine reuptake inhibitors, tricyclic antidepressants, buspirone (Buspar), or monoamine oxidase inhibitors are preferred for chronic or recurrent anxiety disorders. Use of complementary and alternative medicine has increased over the past decade. A variety of studies have suggested that this use is greater in persons with symptoms or diagnoses of anxiety and depression (Saeed *et al.*, 2007).

Serotonin 5-HT_{2A} and 5-HT_{2C} receptors have been implicated in the pathophysiology of OCD and in the mechanism mediating the anti-compulsive effects of serotonin reuptake inhibitors (Flaisher-Grinberg *et al.*, 2008). Standard treatments include psychopharmacologic approaches using serotonin reuptake inhibitors, such as fluoxetine, and cognitive-behavioral therapy, which promotes a kind of learning through exposure to the feared or unpleasant stimulus and prevention of the compulsive response. Limited response to both treatments is common, and various psychopharmacologic agents, including the atypical antipsychotics, have been tested for their abilities to augment serotonin reuptake inhibitors (Maher & Theodore, 2012).

12.8 DIETARY MANAGEMENT OF OCD

OCD pathogenesis has been linked to high levels of oxidative stress, cortisol, inflammation, gut microbial dysfunction, and genetic abnormalities associated to mitochondrial dysfunction. Many of these hypothesized pathways may be influenced by dietary consumption or quality. It has been demonstrated that plant-based foods' vitamins and polyphenols, as well as extra virgin olive oil and omega-3 from fish, have anti-inflammatory properties. The release of pro-inflammatory cytokines, on the other hand, is associated with calorie-dense diets, hydrogenated fats, and added sweets. The gut microbiota may be modulated by foods high in fiber, prebiotics, and probiotics, while Mediterranean and other plant-rich diets have been associated to enhanced microbial diversity. Preclinical research suggests that high-fat diets may be linked to aberrant mitochondrial biogenesis, which is linked to increased free radical generation, in terms of mitochondrial dysfunction (Nguyen *et al.*, 2021).

Research has shown nutritional deficiencies that correlate with some mental disorders. The most common nutritional deficiencies seen in mental disorder patients are of omega-3 fatty acids, B vitamins, minerals like zinc, magnesium, iron, and amino acids that are precursors to neurotransmitters (Saeed *et al.*, 2007). Dietary management includes supplementation with neurotransmitters, hormonal precursors, vitamins and micronutrients, and herbal supplements through well-designed nutritional management, which is described next.

12.8.1 Neurotransmitter and Hormonal Precursors

Saeed et al. (2007) found that the amino acids (l-phenylalanine/phenylalanine [norepinephrine precursor], l-arginine, l-lysine, l-glutamine, l-leucine); melatonin; pregnenolone; phytoestrogens (soy or Mexican yam); tyrosine (norepinephrine precursor); SAMe (S-adenosyl-l-methionine) are the vital precursors associated with OCD.

12.8.2 Role of Vitamins and Micronutrients

Vitamin B_{12} and folate are thought to be effective in OCD treatment due to their associations with neurotransmitters. Depending on their antioxidant effect, zinc and selenium can be used in augmentation therapy for OCD. Studies have shown that nutrients that increase serotonin levels will reduce the symptoms of OCD. Tryptophan, an amino acid, is a precursor to serotonin, and tryptophan supplements (which are better than 5-hydroxytryptophan) will increase serotonin levels and treat OCD (Karcı & Celik, 2020).

Inositol has been reported to be effective in treating central nervous system disorders such as depression, Alzheimer's disease, panic disorder, and obsessive-compulsive disorder. It is known that a change in the CNS concentration of inositol may lead to modified brain cell signaling pathways and possibly to the development of a psychiatric disorder. Recent evidence indicates inositol has psychoactive effects by interacting with the second messenger system and ultimately regulating the cytosolic concentration of calcium. The signaling by calcium is known to mediate an array of cellular functions (secretion, contraction, and conduction). Due to the role calcium plays, its regulation intracellularly is known to be a complex phenomenon involving a number of active and passive transport systems. Inositol is now established as a significant mediator of calcium mobilization in the endoplasmic reticulum. Modifying this mobilization of calcium may be effective in treating some CNS disorders like Alzheimer's disease, depression, panic disorder, OCD, and as an analgesic for pain control (Colodny & Hoffman, 1998).

12.8.3 Herbal Supplements

The herbal supplements found effective in the management of OCD are ashwagandha (*Withania somnifera*); Bach flower essences; bacopa; Berocca; borage juice (starflower); bugleweed (*Lycopus virginicus*); catnip; chamomile; damiana; fennel; feverfew; ginkgo; ginseng; golden root (*Rhodiola rosea*); gotu kola; hops; kanna; lemon balm; lemongrass leaves; licorice; meadowsweet; motherwort; mullein (*Verbascum sinuatum*); mulungu; noni (*Morinda citrifolia*); peppermint; pine bark extract; reishi (*Ganoderma lucidum*); Relora (magnolia/phellodendron); schisandra; scullcup (skullcap); verbena (blue vervain) (Saeed *et al.*, 2007).

Ashwagandha is one of the best natural treatments for OCD as it intervenes in several mechanisms. It balances the brain chemistry, improves resilience to stress, and stimulates brain-derived neurotrophic factor (BDNF). This Ayurvedic herb is beneficial in reducing compulsive urges, which increase during the manifestation of anxiety and/or stress (Lakhan & Vieira, 2008). St. John's wort, milk thistle, valerian root, curcumin, and borage also showed positive signs in treating OCD (Karcı & Celik, 2020).

12.8.4 Nutritional Management of OCD

Nutrients strongly influence both brain structure and function, and nutrition also affects neurodevelopment and neurotrophic function. It has been recognized in recent years that diet and nutrition may be an important factor contributing to psychiatric morbidity and that prevention or treatment of psychiatric disorders could be conducted by addressing diet and nutrition (Matsuoka & Hamazaki, 2016).

The mechanisms through which these dietary changes can benefit mental health have yet to be fully established. However, the diet may act via several pathways that are implicated in mental health. These include pathways related to inflammation, oxidative stress, and mitochondrial dysfunction, which are disrupted in people with mental disorders. Gut microbiota dysbiosis has also been implicated because of emerging research demonstrating the involvement of the microbiome in the modulation of the immune function, stress response, neurotransmission, and neurogenesis. A healthy diet typically contains a wide variety of bioactive compounds that can beneficially interact with these pathways. For example, fruits and vegetables contain, in addition to beneficial vitamins, minerals, and fiber, a high concentration of various polyphenols that seem to be associated with reduced rates of depression in limited observational studies, potentially because of their neuroprotective, anti-inflammatory, and prebiotic properties. In addition, minerals (e.g., zinc, magnesium), fatty acids (e.g., omega 3 fatty acids), vitamins (e.g., B vitamins), and fiber (e.g., resistant starch) as well as other bioactive components (e.g., probiotics), which are typically abundant in healthy dietary patterns, may also be protective from mental illness. Diet combined with exercise has shown to be effective (Firth *et al.*, 2019).

Nutrition can affect the brain, throughout the life cycle, with profound implications for mental health and degenerative disease. Many aspects of nutrition, from entire diets to specific nutrients, affect brain structure and function (Dauncey, 2009). Although in the cases of certain nutrients, further research needs to be done to determine the best-recommended doses of most nutritional supplements, psychiatrists can recommend doses of dietary supplements based on previous and current efficacious studies and then adjust the doses based on the results obtained (Lakhan & Vieira, 2008). Saeed *et al.* (2007) suggested nutritional supplements should contain adrenal extracts; carbohydrate-rich diet; garum armoricum (great bluefish); ginger; l-theanine (green tea); macrobiotic diet; milk peptides; oats; perilla oil (perilla frutescens); vitamins B3, B6, B12, C, and D for treating patients with OCD.

For a healthy life and healthy brain, the diet should include antioxidant- and vitamins-rich foods like fruits, nuts, dry fruits, vegetables, spinach, sprouts, fishes, eggs, oats, milk, curd, cereals, and pulses in their day-to-day life. Additional to the diet, keeping the body and mind healthy through exercise, performing yoga, meditation, and breathing exercises will do wonders in maintaining good physical and mental health, and thus providing synergistic self-care management towards OCD and related disorders.

12.9 CONCLUSION

OCD is an illness that can cause marked distress and disability that severely affects one's QOL. Despite the fact that the treatment of OCD is amazingly cutting-edge contrasted with 30 years back, there are various territories where enhancements can be made. Since most of the antidepressants and other prescribed drugs results in severe side effects, it is better and safer to formulate a nutritionally rich diet that may address and manage OCD. As dietary practices and lifestyle habits can be changed by individuals at any time, new integrated approaches to mental health from the viewpoint of diet and nutrition, that is, nutritional psychiatry could be applicable to a larger population. By reducing the consumption of unhealthy food associated with increased risk for depression, such as processed meats, refined carbohydrates, and other inflammatory foods that may negatively affect the neural pathways which are associated with OCD and other psychological disorders and intake of beneficial nutrients, we are able to manage a positive impact on mental well-being. The diet/nutrient supplements which increase serotonin levels will reduce the symptoms of OCD. Thus the regular diet can be formulated through herbal and nutritional food supplements with appropriate levels of previously discussed vitamins, micronutrients, amino acids (in particular neurotransmitters and hormonal precursors) combined with exercise will help in preventing and managing OCD.

REFERENCES

Baup, N., Grabli, D., Karachi, C., Mounayar, S., François, C., Yelnik, J., . . . Tremblay, L. (2008). High-frequency stimulation of the anterior subthalamic nucleus reduces stereotyped behaviors in primates. *Journal of Neuroscience*, *28*(35), 8785–8788.

Brandt, J. (2019). Diet in brain health and neurological disorders: Risk factors and treatments. *Brain Sciences*, *9*(9), 234.

Colodny, L., & Hoffman, R. L. (1998). Inositol-clinical applications for exogenous use. *Alternative Medicine Review*, *3*, 432–447.

Dauncey, M. J. (2009). New insights into nutrition and cognitive neuroscience: Symposium on 'Early nutrition and later disease: Current concepts, research and implications'. *Proceedings of the Nutrition Society*, *68*(4), 408–415.

Dell'Osso, B., Altamura, A. C., Allen, A., & Hollander, E. (2005). Brain stimulation techniques in the treatment of obsessive-compulsive disorder: Current and future directions. *CNS Spectrums*, *10*(12), 966–979.

Diniz, J. B., Miguel, E. C., Oliveira, A. R. D., Reimer, A. E., Brandão, M. L., Mathis, M. A. D., . . . Hoexter, M. Q. (2012). Outlining new frontiers for the comprehension of obsessive-compulsive disorder: A review of its relationship with fear and anxiety. *Brazilian Journal of Psychiatry*, *34*, 81–91.

Fenske, J. N., & Schwenk, T. L. (2009). Obsessive compulsive disorder: Diagnosis and management. *American Family Physician*, 80(3):239–245.

Firth, J., Marx, W., Dash, S., Carney, R., Teasdale, S. B., Solmi, M., . . . Sarris, J. (2019). The effects of dietary improvement on symptoms of depression and anxiety: A meta-analysis of randomized controlled trials. *Psychosomatic medicine*, *81*(3), 265.

Flaisher-Grinberg, S., Klavir, O., & Joel, D. (2008). The role of 5-HT2A and 5-HT2C receptors in the signal attenuation rat model of obsessive–compulsive disorder. *International Journal of Neuropsychopharmacology*, *11*(6), 811–825.

Karcı, C. K., & Celik, G. G. (2020). Nutritional and herbal supplements in the treatment of obsessive compulsive disorder. *General Psychiatry*, *33*(2).

Lakhan, S. E., & Vieira, K. F. (2008). Nutritional therapies for mental disorders. *Nutrition Journal*, *7*(1), 2.

Macphail, E. C., Reimer, R., & Arnold, P. D. (2018). A314 a characterization of nutrition status and gut microbiota in obsessive-compulsive disorder (ocd) in youth. *Journal of the Canadian Association of Gastroenterology*, *1*(suppl_2), 451–452.

Maher, A. R., & Theodore, G. (2012). Summary of the comparative effectiveness review on off-label use of atypical antipsychotics. *Journal of Managed Care Pharmacy*, *18*(5 Supp B), 1–20.

Masellis, M., Rector, N. A., & Richter, M. A. (2003). Quality of life in OCD: Differential impact of obsessions, compulsions, and depression comorbidity. *The Canadian Journal of Psychiatry*, *48*(2), 72–77.

Matsuoka, Y., & Hamazaki, K. (2016). Considering mental health from the viewpoint of diet: The role and possibilities of nutritional psychiatry. *Seishin shinkeigaku zasshi= Psychiatria et neurologia Japonica*, *118*(12), 880–894.

Nguyen, T., Cribb, L., Ng, C., Byrne, G., Castle, D., Brakoulias, V., . . . Sarris, J. (2021). Dietary quality and nutrient intake in adults with obsessive–compulsive disorder. *BJPsych Open*, 7(6), E218. doi:10.1192/bjo.2021.1039.

Pittenger, C., Kelmendi, B., Bloch, M., Krystal, J. H., & Coric, V. (2005). Clinical treatment of obsessive compulsive disorder. *Psychiatry (Edgmont (Pa.: Township))*, *2*(11), 34–43.

Reddy, Y. J., Rao, N. P., & Khanna, S. (2010). An overview of Indian research in obsessive compulsive disorder. *Indian Journal of Psychiatry*, *52*(Suppl1), S200.

Saeed, S. A., Bloch, R. M., & Antonacci, D. J. (2007). Herbal and dietary supplements for treatment of anxiety disorders. *American Family Physician*, *76*(4), 549–556.

Zghoul, T., & Blier, P. (2003). Enhancing action of LSD on neuronal responsiveness to serotonin in a brain structure involved in obsessive–compulsive disorder. *International Journal of Neuropsychopharmacology*, *6*(1), 13–21.

13 Obsessive-Compulsive Disorder and Treatment with Effective Nature Supplements

Thilagar Gobinath[a] *and Samuthirapandian Ravichandran,*[a,b,*]
[a] Centre of Advanced Study in Marine Biology, Annamalai University, Parangipettai, TamilNadu, India
[b] Government Thirumagal Mills College, Gudiyatham, TN, India

13.1 INTRODUCTION

Obsessive-compulsive disorder (OCD) is a weakening mental disorder with a lifetime prevalence of 2 to 3% in the global population, and it is estimated to be the tenth leading cause of disability around the world (Pittenger *et al.*, 2005). OCD is characterized by repetitive, unwanted thoughts that make nervousness (obsessions) and stereotyped behaviors (compulsions) (Stein *et al.*, 2016). OCD typically begins in teenage years but sometimes may start in adulthood or childhood, with comorbid conditions like anxiety, mood, attention deficit hyperactivity disorder (AD/HD), oppositional-defiant, somatoform disorders, and substance use disorders (Boileau, 2011). OCD affects the quality of patient's life by significantly interfering with the individual's normal levels of functioning (Huppert *et al.*, 2009).

Genetic, environmental, neurological, behavioral, autoimmune response plays a significant role in the development and pathophysiology of OCD (Nestadt *et al.*, 2010). An imbalance or alterations in certain neurotransmitters is thought to be involved in OCD. Imbalance in serotonin and dysfunction in the availability of serotonin transporter (5-HTT) has been implicated in OCD indications (Nicolini, 2010).

In addition, emerging evidence from neurochemical imaging suggested that glutamate and gamma-aminobutyric acid (GABA) abnormalities in dysfunction of cortico-striato-thalamo-cortical (CSTC) may contribute to the pathophysiology of OCD (Sheshachala and Narayanaswamy, 2019). Also, the glutamatergic ionotropic N-methyl D-aspartate (NMDA) receptor has been associated with OCD. NMDA receptors are made up of one obligatory (NR1) and two additional (NR2, NR3) subunits. Mutations in genes coding for subunit NR2 have been associated with OCD risk and the presence of some specific symptoms (contamination obsessions and cleaning) (Sheshachala and Narayanaswamy, 2019).

The recommended treatments for OCD are cognitive behavioral therapy (CBT), selective serotonin reuptake inhibitors (SSRI), a combination of SSRI and CBT or pharmacotherapy with selective serotonin reuptake inhibitors (SSRI) like clomipramine, fluoxetine, fluvoxamine, paroxetine hydrochloride, sertraline, citalopram, and escitalopram (Kellner, 2010). But still, limited effects on the patients and side effects of present treatments have led to the search for the alternatives strategies. It is well-known that various nutritional deficiencies can be detected in patients with psychiatric disorders. Therefore, bioactive compounds and nutritional supplements are thought to be effective in treatment. In this paper, the use of nutritional and herbal bioactive compounds in the treatment of OCD will be reviewed briefly.

*Corresponding author.

 DOI: 10.1201/9781003387893-16

13.2 METHODS

This review is focused on herbal and nutritional treatments for OCD. Therefore, PubMed (Medline) and Google Scholar databases were searched using the terms "Obsessive-Compulsive Disorder" or "OCD" in combination with "herbal" or "plant" or "nutrition" or "nutraceuticals" or "supplement" or "minerals" or "vitamin."

13.3 RESULTS

13.3.1 Trace Elements

13.3.1.1 Magnesium

Magnesium is a positively charged ion and the fourth most abundant mineral in the human body, which is involved in several molecular functions of the body and has been linked to OCD (Botturi *et al.*, 2020). A study done by Shohag et al. (2012) and his colleagues shows that low concentration of magnesium levels contributes to and worsens many psychiatric problems that include OCD. Patients with OCD have significantly lower levels of magnesium in their blood compared to people without OCD. This study shows that imbalance in the magnesium relations with OCD compared with controls and therefore suggest a disturbance in elemental homeostasis.

13.3.1.2 Zinc

Zinc is an essential mineral for sound mental health. Zinc is an antioxidant mineral that is significantly required for a molecular process like gene expression, protein synthesis, and enzymatic reactions (Roohani *et al.*, 2013). In addition, zinc presents in the central nervous system, cerebral cortex, pineal gland, and hippocampus and modulates inhibitory and excitatory transmissions such as GABA and NMDA (Takeda, 2000). But unfortunately, most people with psychiatric problems are deficient in zinc concentration (Rafalo *et al.*, 2016). It is estimated that approximately 2 billion peoples are zinc deficient, which includes OCD patients. In one study Shohag et al. (2012) and his colleagues have found that patients with OCD tend to have much lower levels of zinc compared to people shorn of OCD. Another study showed that the zinc supplementation had greater improvements like reduced obsessions and compulsive behavior without any side effects (Kuygun Karcl and Gül Celik, 2020).

13.3.1.3 Iron

Iron is an important trace element in our body which is present in almost every single living cell. Iron helps in carrying blood oxygen from one organ to another organ (Abbaspour, Hurrell and Kelishadi, 2014). In addition, research shows that iron is an important cofactor in the synthesis of serotonin (Kim and Wessling-Resnick, 2014). Iron deficiency can increase the risk of developing psychiatric disorders. In another study, 48 OCD patients were compared with 48 healthy control, and researchers found that serum iron levels were significantly lower in patients with mild and moderate OCD (Shohag *et al.*, 2012). Unlike other minerals, intake of iron without any proper guidance may lead to health problems, since too much iron causes health issues and increases anxiety condition (Richardson *et al.*, 2015).

13.3.1.4 Selenium

Selenium is an essential trace element that plays a role in the antioxidant properties of glutathione peroxidase. Ozdemir et al. (2009) found that serum selenium level was lower in patients with OCD ($n = 28$). They reported a positive correlation between plasma glutathione peroxidase activity and selenium concentration. In a pilot, randomized, double-blind, placebo-controlled study, patients with OCD ($n = 30$) who had taken an SSRI for at least three months were divided into two groups.

While one group took 400 mg selenium capsules plus SSRI, the other group took placebo plus SSRI. At the end of the six weeks, it was reported that there was a significant difference in Y-BOCS scores between the two groups (Sayyah, Andishmand and Ganji, 2018).

13.3.2 N-Acetyl Cysteine

N-acetyl cysteine (NAC) is a modified form of the amino acid cysteine. NAC is a better option for cysteine compared to the parental cysteine due to its less reactivity, less toxicity, more tough to oxidation and water-soluble ability. Current environmental conditions rapidly deplete our source of cysteine and glutathione. Oral administration of NAC can be converted into cysteine in the liver and later used for glutathione synthesis (Mokhtari *et al.*, 2017). Increasing the levels of cysteine in our body can combat against oxidative stress in our brain. Cysteine can cross the blood-brain barrier (BBB). So intake of NAC inhibit the synaptic glutamate release by cysteine-glutamate exchange, which can help in treating several psychiatric disorders, including OCD (Tardiolo, Bramanti and Mazzon, 2018). Afshar et al. have conducted a double-blind placebo-controlled trial of NAC in 48 patients with OCD. The NAC-treated group (600 mg/d–2,400 mg/d) showed more significant changes in Y-BOCS score ($P = 0.003$) over time compared to the placebo group, but patients' comorbidities have not been included in the investigation (Afshar *et al.*, 2012). Lafleur et al. have conducted a case study of NAC (600 mg/d–3,000 mg/d) with the augmentation of fluvoxamine (300 mg). Over time NAC treatment showed a significant improvement in OCD symptoms, and Y-BOCS score decreased dramatically with time (Lafleur *et al.*, 2006). Sarris et al. have conducted a randomized, double-blind, placebo-controlled clinical trial of NAC augmentation in patients with OCD who have been taking psychotropic drugs or adjunctive therapies. The NAC-treated group had more reduction in Y-BOCS score compared with the placebo group (Sarris *et al.*, 2015). Costa et al. have conducted a double-blind, placebo-controlled trial of NAC in patients with OCD. The NAC-treated group 3 g/d showed significant reduction in Y-BOCS scores. The NAC treated group had a reduction of 4.3 points compared with 3.0 points for the placebo group. This clinical trial did not demonstrate significant benefits of NAC in reducing OCD symptoms in patients with treatment-resistant OCD, but secondary analysis suggested that NAC might have some beneficial effects in reducing anxiety symptoms in treatment-resistant OCD patients (Costa *et al.*, 2017). Paydary et al. conducted a randomized, double-blind, placebo-controlled trial to evaluate the possible effects of NAC as an augmentation therapy compared to fluvoxamine monotherapy in treatment of patients with established moderate-to-severe OCD. There was no significant difference in Y-BOCS scores between two groups; however NAC-treated patients accomplished remission more than the control group (Paydary *et al.*, 2016). Ghanizadeh et al. performed a randomized, double-blind, placebo-controlled study that searched NAC as an immunological agent to SSRI for treating children and adolescents with OCD. After a while, Y-BOCS scores of NAC group significantly decreased from 21.0(8.2) to 11.3(5.7) during this study (Ghanizadeh *et al.*, 2017).

13.3.3 Inositol

Inositol is a naturally occurring molecule found in nearly all plants and animals. It plays a key role in various biological processes. Inositol is an isomer of glucose and has been used in the treatment of psychiatric disorders for a long time (Thomas, Mills and Potter, 2016). Inositol has been found in fruits, grains, nuts, and beans. Humans have the highest concentration of inositol where it plays an important role in neurotransmitters, including serotonin. It used to be considered a B vitamin, but currently inositol is no longer considered as an essential nutrient since inositol is synthesized de novo from glucose-6-phosphate in the brain.

A sustained supply of inositol is essential for the synthesis of membrane phospholipids and intracellular secondary messenger cycle. It is suggested that MI may modulate the reuptake of serotonin and increase 5-HT_2 receptor density. Since supplementation with inositol can still help reduce symptoms, OCD may be due to these mechanisms (Kuygun Karcl and Gül Celik, 2020). In one study an obsessive-compulsive disorder patient took 18 g of inositol or placebo daily for six weeks. At the end of the trial, the patient who took inositol had a significantly lower Y-BOCS score. In contrast, Fux, Benjamin, and Belmaker suggested that there were significant placebo effects but no inositol effects. In the open trial, Seedat and Stein used inositol as augmentation for SSRIs treatment. Existing clinical studies suggested that inositol can effectively treat OCD, and also researchers warned patients need to take high doses (12 to 18 grams daily) if patient wants to experience the anxiety-reducing benefits of inositol (Seedat and Stein, 1999).

13.3.4 Glycine

Glycine is an amino acid commonly found in protein-rich diets, fish, meat, collagen, and gelatin. Glycine is one of the three amino acids that our body uses to make glutathione. In addition, glycine is associated with cortical glutamatergic function (Gould and Pazdro, 2019). Glycine has significant potential to reduce the symptoms of OCD symptoms due to its agonist effects on NMDA receptors (Pittenger, 2015). Glycine investigated in a single placebo-controlled study of patients with refractory OCD, supplementing with 60 g of glycine daily has been shown to reduce OCD symptoms, and unfortunately the daily large dose of glycine has significant side effects of unpleasant odor and taste and frequently causes nausea. While these clinical conditions make glycine a difficult gene to use in treatment, however, this case study suggested that modulating the coagonist site may have therapeutic potential. Brain glycine concentration is regulated reuptake by the transporters GLY-T1 and GLY-T2. Inhibition of these transporters can increase the glycine concentration. Sarcosine, or N-methylglycine, a naturally occurring low-affinity GLY-T inhibitor is available as a dietary supplement (Harsing Jr. *et al.*, 2006). A single controlled study has investigated the effects of sarcosine in OCD. Improvement was seen in three sets of patients, either alone or as augmentation of stable SSRI therapy (Wu *et al.*, 2011). In addition, one placebo-controlled, double-blind study on glycine augmentation in patients with OCD showed promise, but unfortunately dropout rates of patients in this study were significant due to the side effects of glycine. However, evaluable patients showed significant improvements in Y-BOCS scores (Greenberg *et al.*, 2009).

13.3.5 Curcumin

Curcuma longa or turmeric. Utilization of turmeric began in India back almost 4,000 years and now has become related with numerous societies in Asia, Japan, and Middle East nations. As of late it has gotten progressively far-reaching in the West when the remainder of the world perceived its therapeutic qualities (Mishra and Palanivelu, 2008; Prasad and Aggarwal, 2011). Because of its brilliant yellow color, turmeric is also known as "Indian saffron." The agent responsible for its coloring is called curcuminoids. Curcuminoids consists of curcumin demethoxycurcumin, 5'-methoxycurcumin, and dihydrocurcumin, which are found to be potent antioxidants (Henrotin, Priem and Mobasheri, 2013). The disease-preventive properties of turmeric are mainly due to the presence of curcumin and two related compounds, dimethoxy curcumin and bisdemethoxycurcumin. These bioactive compounds have been accounted for to have various pharmacological properties, including as cancer prevention agent, mitigating liver issues, anorexia, ailment, diabetic injuries, runny nose, hack, and sinusitis. Turmeric can enhance humoral and cell-mediated immunity (Kumar, 2020). The potential pharmacological effects of curcumin in mood disorders are associated with decreasing the levels of inflammatory interleukin 1β and tumor necrosis factor α; it also increases plasma brain-derived neurotrophic factor levels, decreases salivary cortisol concentrations, monoaminergic imbalances, hypothalamus-pituitary-adrenal axis disturbances, oxidative stress, and mitochondrial disturbances (Aggarwal and Harikumar, 2009; Prasad and Aggarwal, 2011). No human clinical studies have been completed inspecting the effects of curcumin for OCD patients. In an OCD animal model, rats were injected with the dopamine agonist quinpirole, and compulsive checking was induced. Oral administration of curcumin for 35 uninterrupted days significantly reduced compulsive checking and ritualistic behaviors and reversed quinpirole-induced deteriorations in spatial learning. Curcumin also reversed quinpirole-induced increases in dopamine in the brain and increased serotonin concentrations (Chimakurthy and Murthy, 2010).

13.3.6 Caffeine

Coffee is one of the most popular and consumed brewed drink worldwide. Coffee contains caffeine, the most widely used psychoactive substance along with several other biologically active substances (Cappelletti *et al.*, 2014). Research studies have reported that caffeine has potential health benefits on a number of diseases, including type 2 diabetes mellitus, Parkinson's disease, liver disease, stroke risk, Alzheimer's disease, some cancers and depression (Bidel and Tuomilehto, 2013). However, coffee is a surprising natural remedy for OCD treatment. Caffeine is a well-known antagonist of adenosine receptors and a phosphodiesterase 3 (PDE3) inhibitor (Hall *et al.*, 2015). As a result of this molecular target, caffeine usually increases anxiety and stress, but interestingly, two recent research studies suggested that caffeine can improve OCD symptoms. These clinical trials have proved that caffeine significantly reduced the severity of OCD symptoms in the people with treatment-resistant OCD condition. In a double-blind study of a patient (n = 62) with treatment-resistant OCD, Shams et al. showed that there was significant improvement with OCD symptoms due to the 300 mg/d treatment (Shams *et al.*, 2019). In another double-blind study of dextroamphetamine versus caffeine augmentation with patients with treatment-resistant OCD (n = 24), Koran, Aboujaoude, and Gamel showed that there was a significant improvement in OCD symptoms due to the caffeine augmentation of dextroamphetamine (Koran, Aboujaoude and Gamel, 2009). Based on these findings, caffeine can reduce the symptoms of treatment-resistance OCD symptoms and serve as an adjunct therapy in the obsessive-compulsive disorder.

13.3.7 Probiotics

The microbiome, a very crucial microbial ecological system, contains approximately 100 trillion microorganisms and functions to establish the intestinal lining and supports in its maintenance. Microbiome is influenced by several factors such as genetics, sex, diet, and mental health (Conlon and Bird, 2015). As you probably already know, the health of your gut and its trillion bacteria significantly influence your mental health and modify emotional behavior (Foster, 2013). So not surprisingly, probiotics have been investigated as another possible natural remedy for the treatment of OCD. Although serotonin is synthesized in the brain, 90% of serotonin supply is found in the digestive tract and blood platelets, which makes the digestive tract a crucial target in OCD treatment. Probiotic supplements can enhance the microbiome in your gut, and this will help in creating a better environment for the synthesis of serotonin (Wallace and Milev, 2017). A systematic review of 38 studies concluded that probiotics can improve psychiatric-disorder-related behaviors, including anxiety and OCD (Wang *et al.*, 2016). Not only with OCD but most of the psychiatric-related diseases, and effects of probiotics were studied using *Bifidobacterium longum, Bifidobacterium breve,* and *Lactobacillus rhamnosus* strains at the dose of 10^9 and 10^{10} colony-forming units. Researchers believe these probiotics can help with OCD.

13.3.8 Saffron

Saffron is a spice derived from the plant *Crocus sativus*, which is grown in Spain in the west to Kashmir in the east. Due to the potential bioactive compounds in saffron, which has antioxidants, it is considered anti-cancer, alleviates premenstrual syndrome, helps in weight loss, reduces chances of heart attack, reduces blood sugar, reduces age-related macular degeneration, improves memory in patients with Alzheimer's, and modulates mood and depression (Melnyk, Wang and

Marcone, 2010). Also safranal and crocetin, two crucial bioactive compounds within saffron, has been found to stimulate GABA receptors and increase serotonin levels (Mokhtari-Zaer, Khazdair and Boskabady, 2015). Although there are no human studies that examine the benefits of saffron in OCD treatment, in an OCD animal model, and compared with antidepressant 1-(3-chlorophenyl) piperazine, after the intraperitoneal injection (i.p.) 30 and 50 mg/kg, the study revealed that crocins' effects on animal model of OCD cannot be attributed to changes in locomotor activity. Also, this study suggested that crocins might play a big role in compulsive behavior and support a functional interaction between crocins and the serotonergic system (Georgiadou, Tarantilis and Pitsikas, 2012).

13.3.9 Ashwagandha

Withania somnifera, also called ashwagandha, is Indian ginseng, poison gooseberry, or winter cherry. It is an evergreen shrub in the solanaceae or nightshade family indigenous in India, the Middle East, and North Africa. The active secondary metabolites constituents of *W. somnifera* include alkaloids like isopelletierine, anaferine, cuseohygrine, anahygrine, steroidal lactones (withanolides, withaferins), and saponins. Itoindosides and acylsterylglucosides in ashwagandha are anti-stress agents (Singh *et al.*, 2011). We have utilized ashwagandha for this reason since olden times, particularly in the conventional arrangement of medication in India (Ayurveda) for its wide scope of medical advantages. In society cures the world over, *W. somnifera* is utilized to treat a wide arrangement of sicknesses, including diabetes and cancer, relieves arthritis, high blood pressure, asthma, rheumatism, stress, and has antioxidant, anticonvulsant, and antibacterial properties (Singh *et al.*, 2015). In a randomized double-blind placebo-controlled trial of ashwagandha in OCD treatment (n = 30), they found that ashwagandha was more effective in OCD treatment compared with placebo. No side effects with ashwagandha extract was an important advantage of this double-blind placebo-controlled trial (Jahanbakhsh *et al.*, 2016). Apart from human studies, the examination of benefits of ashwagandha in OCD treatment was conducted in an OCD animal model. Mice were injected with serotonergic neurotoxin. After the i.p. administration, marble burying behavior activity were significantly reduced without affecting locomotor activity. These effects were comparable with standard fluoxetine, ritanserin, and parachlorophenylalanine (Kaurav *et al.*, 2012). Animal model studies also suggested that ashwagandha has anti-anxiety effects, reduces OCD-like behavior, and improves stress tolerance in rats (Archana and Namasivayam, 1998; Bhattacharya *et al.*, 2000; Bhattacharya and Muruganandam, 2003; Kokare *et al.*, 2010).

13.3.10 Milk thistle

Silybum marianum, also called milk thistle, Mary thistle, or holy thistle. It is a prickly plant from the daisy family indigenous in the Mediterranean basin and is now widespread throughout the world. The active and secondary metabolite compounds of milk thistle include silydianin, silychristin, and silibinin (Bijak, 2017). Silymarin is considered as a most biologically active compound according to the in vitro studies. Silymarin in *S. marianum* is recognized as a potential antidepressant. We have utilized milk thistle for this reason since olden times, particularly in the herbal medication in the West and Chinese traditional medicinal practice for its wide scope of medical advantages. Milk thistle is utilized to treat a wide array of sickness including snake bites, liver disease, infertility, sexual dysfunction, psoriasis, and remembrance complications. In one double-blind and randomized trial of milk thistle extract in OCD patients (n = 35), they found that milk thistle extract (600 mg/d) does not have any significant difference in OCD treatment compared with fluoxetine (30 mg/day). This study justified that the milk thistle has similar effects with SSRI antidepressant fluoxetine in OCD symptoms and its positive effect (Sayyah *et al.*, 2010). Osuchowski *et al.* (2004) investigated the effects of silymarin on the regional brain neurotransmitters in BALB/c mice model. Study results revealed that silymarin had a brain serotonergic, dopaminergic, and noradrenergic effects.

13.3.11 Borage

Echium amoenum (borage) belongs to the family Boraginaceae and is a large annual plant found mainly in Europe and in the northern parts of Iran. Borage and its oil have an extensive history of utilization as a traditional medication all through the world. Borage holds a prime place of respect in Iranian traditional medicine (Abolhassani, 2010). Borage contains loads of bioactive compounds that have a number of pharmacological effects, antioxidant, antibacterial, antiviral, anti-inflammatory, and immunomodulatory and neuroprotective. In traditional treatment in Iran, borage is utilized as emulcent, anti-inflammatory and analgesic, anxiolytic, and sedative (Rabbani, Sajjadi and Khalili, 2011). Essential oil from borage is very high in gamma linoleic acid (GLA), which is an essential fatty acid that must be obtained from diet. In one double blind, parallel-group trial study, borage extract (500 mg/day) for 24 OCD patients has been compared with placebo (n = 20). While no significant difference was observed between the groups in the first week, a significant difference occurred at week 4 and week 6. In addition, anxiety symptoms were also significantly reduced with borage extract treatment (Sayyah *et al.*, 2009).

13.3.12 Valerian

Valerian or *Valeriana officinalis* belongs to the family Caprifoliaceae. Throughout the world valerian is mostly used for sleeping disorders, especially for people who cannot sleep (insomnia) (Yuan *et al.*, 2004). Valerian contains more than 150 bioactive compounds, majorly pyridine alkaloids, and some organic acids and terpenes (valepotriates, esterified iridoid-monoterpenes) (Patočka and Jakl, 2010). In a double-blind randomized trial of patients with OCD, Pakseresht, Boostani, and Sayyah showed that there were reduced OCD symptoms in 25% of patients due to the 765 mg/d (Pakseresht, Boostani and Sayyah, 2011). Several research studies have suggested that valerian reduces OCD symptoms by following molecular mechanisms such as partially activating serotonin receptors, maintaining serotonin levels, reducing stress hormone levels, binding to GABA receptors in the amygdala, a brain region associated with fear and anxiety, increasing GABA levels, inhibiting the breakdown of GABA in the brain (Houghton, 1999; Marder *et al.*, 2003; Yuan *et al.*, 2004; Fernández *et al.*, 2004; Dietz *et al.*, 2005; Khom *et al.*, 2007; Hattesohl *et al.*, 2008; Briley, 2010; Murphy *et al.*, 2010; Nam *et al.*, 2013; Jung *et al.*, 2014, 2015).

13.3.13 St. John's Wort

Hypericum perforatum, popularly known as St. John's wort, is the best-known herbal antidepressant. For several years Europeans used St. John's wort to treat mood and anxiety disorders. Cui and Zheng's (2016) investigation found that the potential bioactive compound from St. John's wort is known as hypericum, which appears to have a great influence on the brain serotonin system (Cui and Zheng, 2016). In a double-blind study of a patient (n = 60) with OCD treated with St. John's wort (600–1800 mg/day), Kobak et al. showed that there was no significant improvement with OCD symptoms while compared with placebo (Kobak *et al.*, 2005). Another study with 12 OCD patients found that St. John's wort produced significant reduction in OCD symptoms similar to SSRI. Another randomized-controlled trial investigated the effects of St. John's wort in reducing OCD symptoms. In that study OCD patients (n = 60) who took St. John's Wort did not have any significant effects than those who took placebo (Baek, Nierenberg and Kinrys, 2014). This investigation proves the effectiveness of St. John's wort as an OCD treatment.

13.3.14 Vitamin D

Vitamin D has several health benefits. Reasonable levels of the vitamin D help your body to maintain normal levels of calcium and phosphorus. In addition, vitamin D helps in the absorption of calcium, which keeps your bone strong. Vitamin D plays a crucial role in normal brain development, neurotransmission, neuroprotection, proliferation, and differentiation (Khazai, Judd and Tangpricha, 2008). Several research studies have supported that vitamin D deficiency is associated with several mental disorders such as autism, major depressive disorder, schizophrenia, and OCD. The link between vitamin D and OCD is not well characterized as like other diseases. However, people with OCD frequently have low levels of vitamin D. The pathophysiology of OCD is complex. There are several possible relationships between vitamin D and OCD. One of these possible mechanisms involved between vitamin D active from vitamin D3 (1,25-dihydroxy-vitamin D3) and tyrosine/tryptophan hydroxylase. Tyrosine hydroxylase is the rate-limiting enzyme in dopamine, epinephrine, and norepinephrine synthesis. Instead, tryptophan hydroxylase is the rate-limiting enzyme in serotonin synthesis (Kuygun Karcl and Gül Celik, 2020). Concentration of the tyrosine/tryptophan hydroxylase is directly influenced by vitamin D3; therefore, vitamin D deficiency may lead to OCD etiology. Another possible relationship between vitamin D and OCD is neuroprotective effects of vitamin D. Some research studies have shown that higher concentration of nitric oxide was present in OCD patients. Vitamin D has potential antioxidant activity that can inhibit the essential enzyme for nitric oxide synthesis. Vitamin D deficiency may play crucial role in OCD by causing neurodegeneration. Yazici, Percinel Yazici, and Ustundag have compared vitamin D levels in children and adolescents (7–15 years age) with OCD. Researchers found that where was a difference in vitamin D levels but not statistically different than healthy controls. Thus, they do not recommend the relation between low concentration of serum vitamin D and OCD (Yazici, Percinel Yazici and Ustundag, 2018). On the other hand, Esnafoğlu and Yaman have compared vitamin D, vitamin B12, homocysteine, and folic acid levels between OCD patients (n = 52) and healthy controls (n = 30). Researchers found that there was a negative correlation between disease severity of OCD and vitamin D concentration. However, this study proves vitamin D deficiency can play a crucial role in the pathophysiology of OCD (Esnafoğlu and Yaman, 2017). Another study on OCD patients with streptococcal infections (n = 60) were compared with 60 healthy controls. Çelık et al. found that serum 25-hydroxyvitamin D (25(OH) D) (active form of vitamin D) were not significantly differentiated. However, vitamin D deficiency was more frequent in the OCD patients compared to the control group (Çelık *et al.*, 2016).

13.3.15 Vitamin B12

Vitamin B12 is one of the important vitamins affecting several systems of your body. Hematologic, neurologic, gastrointestinal, and mental disorders will arise in case of deficiency (Ankar and Bhimji, 2018). The psychiatric symptoms may not be concurrent with symptoms arising from other systems and even may precede them. Vitamin B12 symptoms include agitation, irritability, negativism, confusion, disorientation, amnesia, impaired concentration and attention, and insomnia (Türksoy *et al.*, 2014). Mental disorders that may be diagnosed with vitamin B12 deficiency include depression, bipolar disorder, panic disorder, psychosis, phobias, OCD, and dementia (Cornish and Mehl-Madrona, 2008).

In adult patients, the clinical picture may especially involve affective or psychotic symptoms. The significance of vitamin B12, folate, and homocysteine in the metabolism of carbon transfer (methylation) needed for the development of serotonin, other monoamine neurotransmitters, and catecholamines can be explained by these observations (Kennedy, 2016). In one study, Hermesh et al. have compared 30 patients with OCD, 30 patients with chronic schizophrenia, and 30 healthy controls. Twenty percent of the OCD patients who had low levels of vitamin B12 were observed at non-significant levels (Hermesh *et al.*, 1988). However, the same study concluded that vitamin B12 deficiency could be a consequence of OCD rather than the cause of OCD. Atmaca et al. found lower folate levels and higher homocysteine levels in patients with OCD (n = 23) while comparing to those without (n = 23), and those symptoms were significantly correlated with the severity symptoms of OCD (Atmaca *et al.*, 2005). That study concluded that OCD patients might have folate deficiency, higher homocysteine levels, and probable impaired metylation and monoamine metabolism. Türksoy et al. investigated effects of Vitamin B12, folate, and homocysteine levels in patients with OCD (n = 35) against healthy controls (n = 22). It was found that decreased level of vitamin B12 and increased homocysteine levels was identified in some patients (Türksoy *et al.*, 2014). In addition, increased homocysteine levels were positively correlated in Yale-Brown compulsion and Yale-Brown total scales (Türksoy *et al.*, 2014). Likewise, Esnafoğlu and Yaman found lower levels of vitamin B12 and vitamin D and higher levels of homocysteine in the patient group compared to control group. However, folate levels don't seems to change in both groups (Esnafoğlu and Yaman, 2017). Interestingly, a double-blind study was conducted with 36 OCD patients. OCD patients were randomly assigned to two different treatment groups, folic acid + fluoxetine or placebo + fluoxetine. The study concluded that there was no significant difference in YBOCS score between both groups (Tural *et al.*, 2019).

13.4 CONCLUSION

There is only a limited number of studies that investigate the herbal and nutritional supplements for the treatment of OCD. Depending on their antioxidant effects, trace elements such as magnesium, zinc, iron, and selenium can be used alone or as an augmentation therapy for OCD. Some clinical studies have reported that NAC showed some significant improvement in OCD treatment. Investigations on the glycine in OCD treatment are quite disappointing, maintaining that treatment with glycine is very difficult due to its adverse effects. The existing clinical studies data suggest inositol may potentially be effective as monotherapy for OCD. Some natural nutraceuticals may provide positive effects with pharmaceuticals. Curcumin, saffron, ashwagandha, caffeine, St. John's wort, valerian, borage, milk thistle are the herbal supplements that have been investigated in OCD treatment. Despite the few studies on herbal supplements for OCD treatment, double-blind randomized controlled studies with standardized preparations of these herbals are needed. Although previous studies showed that some herbal and nutritional supplements might be potentially effective in OCD treatment, further double-blind randomized controlled longitudinal studies with structured measurement procedures in larger samples are needed. OCD and vitamin D association has been studied only in children and adolescents. Reduced vitamin D levels in OCD patients' data were limited, and correlation between vitamin D deficiency and OCD symptoms were also negative. Similar to vitamin D, vitamin B12 and folate are thought to be effective in OCD treatment due to their associations with neurotransmitters. However, low levels of vitamin B12 and folate are recorded in OCD patients, but these levels can be affected by diet.

CONFLICT OF INTEREST

The authors declare that they have no known competing financial interests or personal relationships that could have appeared to influence the work reported in this paper.

FUNDING

This research received no specific grant from any funding agency in the public, commercial, or not-for-profit sectors.

REFERENCES

Abbaspour, N., Hurrell, R. and Kelishadi, R. (2014) 'Review on iron and its importance for human health', *Journal of Research in Medical Sciences*. Isfahan University of Medical Sciences (IUMS), pp. 164–174. Available at: /pmc/articles/PMC3999603/?report=abstract (Accessed: 3 October 2020).

Abolhassani, M. (2010) 'Antiviral activity of borage (Echium amoenum)', *Archives of Medical Science*, 6(3), pp. 366–369. doi: 10.5114/aoms.2010.14256.

Afshar, H. *et al.* (2012) 'N-acetylcysteine add-on treatment in refractory obsessive-compulsive disorder: A randomized, double-blind, placebo-controlled trial', *Journal of Clinical Psychopharmacology*, 32(6), pp. 797–803. doi: 10.1097/JCP.0b013e318272677d.

Aggarwal, B. B. and Harikumar, K. B. (2009) 'Potential therapeutic effects of curcumin, the anti-inflammatory agent, against neurodegenerative, cardiovascular, pulmonary, metabolic, autoimmune and neoplastic diseases', *International Journal of Biochemistry and Cell Biology*. NIH Public Access, pp. 40–59. doi: 10.1016/j.biocel.2008.06.010.

Ankar, A. and Bhimji, S. S. (2018) *Vitamin B12 Deficiency (Cobalamin), StatPearls*. StatPearls Publishing. Available at: www.ncbi.nlm.nih.gov/pubmed/28722952 (Accessed: 4 October 2020).

Archana, R. and Namasivayam, A. (1998) 'Antistressor effect of Withania somnifera', *Journal of Ethnopharmacology*, 64(1), pp. 91–93. doi: 10.1016/S0378-8741(98)00107-X.

Atmaca, M. *et al.* (2005) 'Serum folate and homocysteine levels in patients with obsessive-compulsive disorder', *Psychiatry and Clinical Neurosciences*, 59(5), pp. 616–620. doi: 10.1111/j.1440-1819.2005.01425.x.

Baek, J. H., Nierenberg, A. A. and Kinrys, G. (2014) 'Clinical applications of herbal medicines for anxiety and insomnia; Targeting patients with bipolar disorder', *Australian and New Zealand Journal of Psychiatry*. SAGE Publications Ltd, pp. 705–715. doi: 10.1177/0004867414539198.

Bhattacharya, S. K. and Muruganandam, A. V. (2003) 'Adaptogenic activity of Withania somnifera: An experimental study using a rat model of chronic stress', *Pharmacology Biochemistry and Behavior*, 75(3), pp. 547–555. doi: 10.1016/S0091-3057(03)00110-2.

Bhattacharya, S. K. *et al.* (2000) 'Anxiolytic-antidepressant activity of Withania somnifera glycowithanolides: An experimental study', *Phytomedicine*, 7(6), pp. 463–469. doi: 10.1016/S0944-7113(00)80030-6.

Bidel, S. and Tuomilehto, J. (2013) 'The emerging health benefits of coffee with an emphasis on type 2 diabetes and cardiovascular disease', *European Endocrinology*, 9(2), pp. 99–106. doi: 10.17925/ee.2013.09.02.99.

Bijak, M. (2017) 'Silybin, a major bioactive component of milk thistle (Silybum marianum L. Gaernt.)—chemistry, bioavailability, and metabolism', *Molecules*. MDPI AG. doi: 10.3390/molecules22111942.

Boileau, B. (2011) 'A review of obsessive-compulsive disorder in children and adolescents', *Dialogues in Clinical Neuroscience*. Les Laboratoires Servier, pp. 401–411. Available at: www.dialogues-cns.org (Accessed: 3 October 2020).

Botturi, A. *et al.* (2020) 'The role and the effect of magnesium in mental disorders: A systematic review', *Nutrients*. MDPI AG, pp. 1–21. doi: 10.3390/nu12061661.

Briley, M. (2010) 'Improvement of social adaptation in depression with serotonin and norepinephrine reuptake inhibitors', *Neuropsychiatric Disease and Treatment*, 6, p. 647. doi: 10.2147/ndt.s13171.

Cappelletti, S. *et al.* (2014) 'Caffeine: Cognitive and physical performance enhancer or psychoactive drug?' *Current Neuropharmacology*, 13(1), pp. 71–88. doi: 10.2174/1570159x13666141210215655.

Çelık, G. *et al.* (2016) 'Vitamin D deficiency in obsessive–compulsive disorder patients with pediatric autoimmune neuropsychiatric disorders associated with streptococcal infections: A case control study', *Noropsikiyatri Arsivi*, 53(1), pp. 31–34. doi: 10.5152/npa.2015.8763.

Chimakurthy, J. and Murthy, T. E. (2010) 'Effect of curcumin on quinpirole induced compulsive checking: An approach to determine the predictive and construct validity of the model.', *North American Journal of Medical Sciences*, 2(2), pp. 81–816. doi: 10.4297/najms.2010.281.

Conlon, M. A. and Bird, A. R. (2015) 'The impact of diet and lifestyle on gut microbiota and human health', *Nutrients*. MDPI AG, pp. 17–44. doi: 10.3390/nu7010017.

Cornish, S. and Mehl-Madrona, L. (2008) 'The role of vitamins and minerals in psychiatry', *Integrative Medicine Insights*, 2008(3), pp. 33–42. doi: 10.4137/117863370800300003.

Costa, D. L. C. *et al.* (2017) 'Randomized, double-blind, placebo-controlled trial of <em>N</em>-acetylcysteine augmentation for treatment-resistant obsessive-compulsive disorder', *The Journal of Clinical Psychiatry*, 78(7), pp. e766–e773. doi: 10.4088/JCP.16m11101.

Cui, Y. H. and Zheng, Y. (2016) 'A meta-analysis on the efficacy and safety of St John's wort extract in depression therapy in comparison with selective serotonin reuptake inhibitors in adults', *Neuropsychiatric Disease and Treatment*, 12, pp. 1715–1723. doi: 10.2147/NDT.S106752.

Dietz, B. M. *et al.* (2005) 'Valerian extract and valerenic acid are partial agonists of the 5-HT 5a receptor in vitro', *Molecular Brain Research*, 138(2), pp. 191–197. doi: 10.1016/j.molbrainres.2005.04.009.

Esnafoğlu, E. and Yaman, E. (2017) 'Vitamin B12, folic acid, homocysteine and vitamin D levels in children and adolescents with obsessive compulsive disorder', *Psychiatry Research*, 254, pp. 232–237. doi: 10.1016/j.psychres.2017.04.032.

Fernández, S. *et al.* (2004) 'Sedative and sleep-enhancing properties of linarin, a flavonoid-isolated from Valeriana officinalis', *Pharmacology Biochemistry and Behavior*, 77(2), pp. 399–404. doi: 10.1016/j.pbb.2003.12.003.

Foster, J. A. (2013) 'Gut feelings: Bacteria and the brain', *Cerebrum: The Dana Forum on Brain Science*, p. 9. Available at: www.ncbi.nlm.nih.gov/pubmed/24116266 (Accessed: 4 October 2020).

Georgiadou, G., Tarantilis, P. A. and Pitsikas, N. (2012) 'Effects of the active constituents of Crocus Sativus L., crocins, in an animal model of obsessive-compulsive disorder', *Neuroscience Letters*, 528(1), pp. 27–30. doi: 10.1016/j.neulet.2012.08.081.

Ghanizadeh, A. *et al.* (2017) 'Efficacy of N-acetylcysteine augmentation on obsessive compulsive disorder: A multicenter randomized double blind placebo controlled clinical trial', *Iranian Journal of Psychiatry*, 12(2), pp. 134–141. Available at: /pmc/articles/PMC5483239/?report=abstract (Accessed: 30 September 2020).

Gould, R. L. and Pazdro, R. (2019) 'Impact of supplementary amino acids, micronutrients, and overall diet on glutathione homeostasis', *Nutrients*. MDPI AG. doi: 10.3390/nu11051056.

Greenberg, W. M. *et al.* (2009) 'Adjunctive glycine in the treatment of obsessive-compulsive disorder in adults', *Journal of Psychiatric Research*, 43(6), pp. 664–670. doi: 10.1016/j.jpsychires.2008.10.007.

Hall, S. *et al.* (2015) 'A review of the bioactivity of coffee, caffeine and key coffee constituents on inflammatory responses linked to depression', *Food Research International*. Elsevier Ltd, pp. 626–636. doi: 10.1016/j.foodres.2015.07.027.

Harsing Jr., L. *et al.* (2006) 'Glycine transporter type-1 and its Inhibitors', *Current Medicinal Chemistry*, 13(9), pp. 1017–1044. doi: 10.2174/092986706776360932.

Hattesohl, M. *et al.* (2008) 'Extracts of Valeriana officinalis L. s.l. show anxiolytic and antidepressant effects but neither sedative nor myorelaxant properties', *Phytomedicine*, 15(1–2), pp. 2–15. doi: 10.1016/j.phymed.2007.11.027.

Henrotin, Y., Priem, F. and Mobasheri, A. (2013) 'Curcumin: A new paradigm and therapeutic opportunity for the treatment of osteoarthritis: Curcumin for osteoarthritis management', *SpringerPlus*. SpringerOpen, pp. 1–9. doi: 10.1186/2193-1801-2-56.

Hermesh, H. *et al.* (1988) 'Vitamin B12 and folic acid serum levels in obsessive compulsive disorder', *Acta Psychiatrica Scandinavica*, 78(1), pp. 8–10. doi: 10.1111/j.1600-0447.1988.tb06294.x.

Houghton, P. J. (1999) 'The scientific basis for the reputed activity of Valerian', *Journal of Pharmacy and Pharmacology*, 51(5), pp. 505–512. doi: 10.1211/0022357991772772.

Huppert, J. D. *et al.* (2009) 'Quality of life and functional impairment in obsessive-compulsive disorder: A comparison of patients with and without comorbidity, patients in remission, and healthy controls', *Depression and Anxiety*, 26(1), pp. 39–45. doi: 10.1002/da.20506.

Jahanbakhsh, S. P. *et al.* (2016) 'Evaluation of the efficacy of Withania somnifera (Ashwagandha) root extract in patients with obsessive-compulsive disorder: A randomized double-blind placebo-controlled trial', *Complementary Therapies in Medicine*, 27, pp. 25–29. doi: 10.1016/j.ctim.2016.03.018.

Jung, H. Y. *et al.* (2014) 'Valeriana officinalis root extract suppresses physical stress by electric shock and psychological stress by nociceptive stimulation-evoked responses by decreasing the ratio of monoamine neurotransmitters to their metabolites', *BMC Complementary and Alternative Medicine*, 14(1). doi: 10.1186/1472-6882-14-476.

Jung, H. Y. *et al.* (2015) 'Valerenic acid protects against physical and psychological stress by reducing the turnover of serotonin and norepinephrine in mouse hippocampus-amygdala region', *Journal of Medicinal Food*, 18(12), pp. 1333–1339. doi: 10.1089/jmf.2014.3412.

Kaurav, B. P. S. *et al.* (2012) 'Influence of Withania somnifera on obsessive compulsive disorder in mice', *Asian Pacific Journal of Tropical Medicine*, 5(5), pp. 380–384. doi: 10.1016/S1995-7645(12)60063-7.

Kellner, M. (2010) 'Tratamiento farmacológico del trastorno obsesivo-compulsivo', *Dialogues in Clinical Neuroscience*, 12(2), pp. 187–197. Available at: Www.dialogues-cns.org (Accessed: 3 October 2020).

Kennedy, D. O. (2016) 'B vitamins and the brain: Mechanisms, dose and efficacy—A review', *Nutrients*. MDPI AG, p. 68. doi: 10.3390/nu8020068.

Khazai, N., Judd, S. E. and Tangpricha, V. (2008) 'Calcium and vitamin D: Skeletal and extraskeletal health', *Current Rheumatology Reports*. NIH Public Access, pp. 110–117. doi: 10.1007/s11926-008-0020-y.

Khom, S. *et al.* (2007) 'Valerenic acid potentiates and inhibits GABAA receptors: Molecular mechanism and subunit specificity', *Neuropharmacology*, 53(1), pp. 178–187. doi: 10.1016/j.neuropharm.2007.04.018.

Kim, J. and Wessling-Resnick, M. (2014) 'Iron and mechanisms of emotional behavior', *Journal of Nutritional Biochemistry*. Elsevier Inc., pp. 1101–1107. doi: 10.1016/j.jnutbio.2014.07.003.

Kobak, K. A. *et al.* (2005) 'St John's wort versus placebo in obsessive-compulsive disorder: Results from a double-blind study', *International Clinical Psychopharmacology*, 20(6), pp. 299–304. doi: 10.1097/00004850-200511000-00003.

Kokare, D. M. *et al.* (2010) 'Involvement of α-MSH in the social isolation induced anxiety- and depression-like behaviors in rat', *Neuropharmacology*, 58(7), pp. 1009–1018. doi: 10.1016/j.neuropharm.2010.01.006.

Koran, L. M., Aboujaoude, E. and Gamel, N. N. (2009) 'Double-blind study of dextroamphetamine versus caffeine augmentation for treatment-resistant obsessive-compulsive disorder', *The Journal of Clinical Psychiatry*, 70(11), pp. 1530–1535. doi: 10.4088/JCP.08m04605.

Kumar, V. (2020) 'Seven spices of India-from kitchen to clinic', *Journal of Ethnic Foods*. BioMed Central Ltd., pp. 1–16. doi: 10.1186/s42779-020-00058-0.

Kuygun Karcl, C. and Gül Celik, G. (2020) 'Nutritional and herbal supplements in the treatment of obsessive compulsive disorder', *General Psychiatry*. BMJ Publishing Group, p. 100159. doi: 10.1136/gpsych-2019-100159.

Lafleur, D. L. *et al.* (2006) 'N-acetylcysteine augmentation in serotonin reuptake inhibitor refractory obsessive-compulsive disorder', *Psychopharmacology*, 184(2). doi: 10.1007/S00213-005-0246-6.

Marder, M. *et al.* (2003) '6-Methylapigenin and hesperidin: New valeriana flavonoids with activity on the CNS', *Pharmacology Biochemistry and Behavior*, 75(3), pp. 537–545. doi: 10.1016/S0091-3057(03)00121-7.

Melnyk, J. P., Wang, S. and Marcone, M. F. (2010) 'Chemical and biological properties of the world's most expensive spice: Saffron', *Food Research International*, pp. 1981–1989. doi: 10.1016/j.foodres.2010.07.033.

Mishra, S. and Palanivelu, K. (2008) 'The effect of curcumin (turmeric) on Alzheimer's disease: An overview', *Annals of Indian Academy of Neurology*. Medknow Publications and Media Pvt. Ltd, pp. 13–19. doi: 10.4103/0972-2327.40220.

Mokhtari, V. *et al.* (2017) 'A review on various uses of N-acetyl cysteine', *Cell Journal*. Royan Institute (ACECR), pp. 11–17. doi: 10.22074/cellj.2016.4872.

Mokhtari-Zaer, A., Khazdair, M. R. and Boskabady, M. H. (2015) 'Smooth muscle relaxant activity of Crocus sativus (saffron) and its constituents: Possible mechanisms', *Avicenna Journal of Phytomedicine*, 5(5), pp. 365–75. Available at: www.ncbi.nlm.nih.gov/pubmed/26468456 (Accessed: 1 October 2020).

Murphy, K. *et al.* (2010) 'Valeriana officinalis root extracts have potent anxiolytic effects in laboratory rats', *Phytomedicine*, 17(8–9), pp. 674–678. doi: 10.1016/j.phymed.2009.10.020.

Nam, S. M. *et al.* (2013) 'Valeriana officinalis extract and its main component, valerenic acid, ameliorate d-galactose-induced reductions in memory, cell proliferation, and neuroblast differentiation by reducing corticosterone levels and lipid peroxidation', *Experimental Gerontology*, 48(11), pp. 1369–1377. doi: 10.1016/j.exger.2013.09.002.

Nestadt, G., Grados, M. and Samuels, J. F. (2010) 'Genetics of obsessive-compulsive disorder', *Psychiatric Clinics of North America*. NIH Public Access, pp. 141–158. doi: 10.1016/j.psc.2009.11.001.

Nicolini, H. (2010) 'Serotonin transporter gene polymorphisms & obsessive-compulsive disorder', *Indian Journal of Medical Research*. Wolters Kluwer—Medknow Publications, pp. 663–665. Available at: www.ncbi.nlm.nih.gov/pmc/articles/PMC3102452/ (Accessed: 3 October 2020).

Osuchowski, M. F. *et al.* (2004) 'Alterations in regional brain neurotransmitters by silymarin, a natural antioxidant flavonoid mixture, in BALB/c mice', *Pharmaceutical Biology*, 42(4–5), pp. 384–389. doi: 10.1080/13880200490519712.

Ozdemir, E. *et al.* (2009) 'Serum selenium and plasma malondialdehyde levels and antioxidant enzyme activities in patients with obsessive-compulsive disorder', *Progress in Neuro-Psychopharmacology and Biological Psychiatry*, 33(1), pp. 62–65. doi: 10.1016/j.pnpbp.2008.10.004.

Pakseresht, S., Boostani, H. and Sayyah, M. (2011) 'Extract of Valerian root (Valeriana Officinalis L.) vs. Placebo in treatment of obsessive-compulsive disorder: A randomized double-blind study', *Journal of Complementary and Integrative Medicine*, 8(1). doi: 10.2202/1553-3840.1465.

Patočka, J. and Jakl, J. (2010) 'Biomedically relevant chemical constituents of Valeriana officinalis', *Journal of Applied Biomedicine*. University of South Bohemia, pp. 11–18. doi: 10.2478/v10136-009-0002-z.

Paydary, K. *et al.* (2016) 'N-acetylcysteine augmentation therapy for moderate-to-severe obsessive-compulsive disorder: Randomized, double-blind, placebo-controlled trial', *Journal of Clinical Pharmacy and Therapeutics*, 41(2), pp. 214–219. doi: 10.1111/jcpt.12370.

Pittenger, C. (2015) 'Glutamate modulators in the treatment of obsessive-compulsive disorder', *Psychiatric Annals*, 45(6), pp. 308–315. doi: 10.3928/00485713-20150602-06.

Pittenger, C. *et al.* (2005) 'Clinical treatment of obsessive compulsive disorder', *Psychiatry (Edgmont (Pa. : Township))*, 2(11), pp. 34–43. Available at: www.ncbi.nlm.nih.gov/pubmed/21120095 (Accessed: 3 October 2020).

Prasad, S. and Aggarwal, B. B. (2011) 'Turmeric, the golden spice: From traditional medicine to modern medicine', in *Herbal Medicine: Biomolecular and Clinical Aspects: Second Edition*. CRC Press, pp. 263–288. Available at: www.ncbi.nlm.nih.gov/books/NBK92752/ (Accessed: 4 October 2020).

Rabbani, M., Sajjadi, S. E. and Khalili, S. (2011) 'A Lack of tolerance to the anxiolytic action of Echium amoenum', *Research in Pharmaceutical Sciences*, 6(2), pp. 101–106. Available at: www.ncbi.nlm.nih.gov/pubmed/22224093 (Accessed: 4 October 2020).

Rafalo, A. *et al.* (2016) 'Zinc Deficiency and Depression', in *Nutritional Deficiency*. InTech. doi: 10.5772/63210.

Richardson, A. C. *et al.* (2015) 'Higher body iron is associated with greater depression symptoms among young adult men but not women: Observational data from the daily life study', *Nutrients*, 7(8), pp. 6055–6072. doi: 10.3390/nu7085270.

Roohani, N. *et al.* (2013) 'Zinc and its importance for human health: An integrative review', *Journal of Research in Medical Sciences*. Wolters Kluwer—Medknow Publications, pp. 144–157. doi: 10.1016/j.foodpol.2013.06.008.

Sarris, J. *et al.* (2015) 'N-Acetyl Cysteine (NAC) in the treatment of obsessive-compulsive disorder: A 16-week, double-blind, randomised, placebo-controlled study', *CNS Drugs*, 29(9), pp. 801–809. doi: 10.1007/s40263-015-0272-9.

Sayyah, M. *et al.* (2009) 'Efficacy of aqueous extract of Echium amoenum in treatment of obsessive-compulsive disorder', *Progress in Neuro-Psychopharmacology and Biological Psychiatry*, 33(8), pp. 1513–1516. doi: 10.1016/j.pnpbp.2009.08.021.

Sayyah, M. *et al.* (2010) 'Comparison of Silybum marianum (L.) Gaertn. with fluoxetine in the treatment of obsessive-compulsive disorder', *Progress in Neuro-Psychopharmacology and Biological Psychiatry*, 34(2), pp. 362–365. doi: 10.1016/j.pnpbp.2009.12.016.

Sayyah, M., Andishmand, M. and Ganji, R. (2018) 'Effect of selenium as an adjunctive therapy in patients with treatment-resistant obsessive-compulsive disorder: A pilot randomized double blind placebo-controlled clinical trial', *Archives of Psychiatry and Psychotherapy*, 4, pp. 57–65. doi: 10.12740/APP/99584.

Seedat, S. and Stein, D. J. (1999) 'Inositol augmentation of serotonin reuptake inhibitors in treatment-refractory obsessive—compulsive disorder: An open trial', *International Clinical Psychopharmacology*, 14(6), pp. 353–356. doi: 10.1097/00004850-199911000-00005.

Shams, J. *et al.* (2019) 'Using caffeine on the patients as therapeutic option against treatment-resistant obsessive-compulsive disorder', *Journal of Family Medicine and Primary Care*, 8(5), p. 1741. doi: 10.4103/jfmpc.jfmpc_93_19.

Sheshachala, K. and Narayanaswamy, J. (2019) 'Glutamatergic augmentation strategies in obsessive-compulsive disorder', *Indian Journal of Psychiatry*. Wolters Kluwer Medknow Publications, pp. S58–S65. doi: 10.4103/psychiatry.IndianJPsychiatry_520_18.

Shohag, H. *et al.* (2012) 'Alterations of serum zinc, copper, manganese, iron, calcium, and magnesium concentrations and the complexity of interelement relations in patients with obsessive-compulsive disorder', *Biological Trace Element Research*, 148(3), pp. 275–280. doi: 10.1007/s12011-012-9371-3.

Singh, N. *et al.* (2011) 'An overview on Ashwagandha: A Rasayana (Rejuvenator) of Ayurveda', *African Journal of Traditional, Complementary and Alternative Medicines*, 8(5 SUPPL.), pp. 208–213. doi: 10.4314/ajtcam.v8i5S.9.

Singh, P. *et al.* (2015) 'Biotechnological interventions in Withania somnifera (L.) Dunal', *Biotechnology and Genetic Engineering Reviews*, 31(1–2), pp. 1–20. doi: 10.1080/02648725.2015.1020467.

Stein, D. J. *et al.* (2016) 'The classification of obsessive–compulsive and related disorders in the ICD-11', *Journal of Affective Disorders*, 190, pp. 663–674. doi: 10.1016/j.jad.2015.10.061.

Takeda, A. (2000) 'Movement of zinc and its functional significance in the brain', *Brain Research Reviews*, pp. 137–148. doi: 10.1016/S0165-0173(00)00044-8.

Tardiolo, G., Bramanti, P. and Mazzon, E. (2018) 'Overview on the effects of N-acetylcysteine in neurodegenerative diseases', *Molecules*. MDPI AG. doi: 10.3390/molecules23123305.

Thomas, M. P., Mills, S. J. and Potter, B. V. L. (2016) 'The "other" inositols and their phosphates: Synthesis, biology, and medicine (with recent advances in myo-inositol chemistry)', *Angewandte Chemie—International Edition*, 55(5), pp. 1614–1650. doi: 10.1002/anie.201502227.

Tural, Ü. *et al.* (2019) 'Double blind controlled study of adding folic acid to fluoxetine in the treatment of OCD', *Psychiatria Danubina*, 31(1), pp. 69–77. doi: 10.24869/psyd.2019.69.

Türksoy, N. *et al.* (2014) 'Vitamin B12, folate, and homocysteine levels in patients with obsessive–compulsive disorder', *Neuropsychiatric Disease and Treatment*, 10, pp. 1671–1675. doi: 10.2147/NDT.S67668.

Wallace, C. J. K. and Milev, R. (2017) 'The effects of probiotics on depressive symptoms in humans: A systematic review', *Annals of General Psychiatry*. BioMed Central Ltd., p. 14. doi: 10.1186/s12991-017-0138-2.

Wang, H. *et al.* (2016) 'Effect of probiotics on central nervous system functions in animals and humans: A systematic review', *Journal of Neurogastroenterology and Motility*, pp. 589–605. doi: 10.5056/jnm16018.

Wu, P. L. *et al.* (2011) 'Sarcosine therapy for obsessive compulsive disorder: A prospective, open-label study', *Journal of Clinical Psychopharmacology*, 31(3), pp. 369–374. doi: 10.1097/JCP.0b013e3182189878.

Yazici, K. U., Percinel Yazici, I. and Ustundag, B. (2018) 'Vitamin D levels in children and adolescents with obsessive compulsive disorder', *Nordic Journal of Psychiatry*, 72(3), pp. 173–178. doi: 10.1080/08039488.2017.1406985.

Yuan, C. S. *et al.* (2004) 'The gamma-aminobutyric acidergic effects of valerian and valerenic acid on rat brainstem neuronal activity', *Anesthesia and Analgesia*, 98(2), pp. 353–358. doi: 10.1213/01.ANE.0000096189.70405.A5.

14 Nutritional and Dietary Role in Neuropsychiatric Disorders

Depression and Obsessive-Compulsive Disorder (OCD)

Supta Sarkar,[a,*] *and Subhra Sarkar*[b]

[a] Department of Foods and Nutrition, College of Community Science, Professor Jayashankar Telangana State Agricultural University, Hyderabad, Telangana, India, https://orcid.org/0000-0002-7210-6520

[b] Department of Psychiatric Nursing, Lokopriya Gopinath Bordoloi Regional Institute of Mental Health, Tezpur, Assam, India, https://orcid.org/0000-0002-0392-4642

14.1 INTRODUCTION

Neuropsychiatric disorders are organic brain disorders that lead to the manifestation of psychiatric symptoms. These involve interactions between brain lesions, location, dysfunction, and psychopathological exposition (Lyketsos, 2006). Some of the neuropsychiatric disorders include major depressive disorder, obsessive-compulsive disorder (OCD), anxiety disorder, schizophrenia, autism disorder, bipolar disorder, and attention deficit hyperactivity disorder (Lotan et al., 2014; Lewin et al., 2006). Neuropsychiatric disorders substantially burden the healthcare system, economy, and well-being of affected patients and their dependents worldwide. According to the World Health Organization's (WHO) report (2000) on Global Burden of Disease estimates, neuropsychiatric disorders account for 13% of disability-adjusted life years (DALYs), and 33% of the years lived with disability (YLD) are attributed to these disorders. As per the guidelines given by National Institute for Health and Care Excellence (NICE, 2009) on psychiatric treatment, successful treatment and management of neuropsychiatric disorder may be done by a few of the modes like pharmacotherapy, somatic therapy, and psychosocial interventions. However, the role of diet in the prevention and management of neuropsychiatric disorders such as depression and OCD is an interesting and essential field that has only emerged in the past few years.

*Corresponding author: Supta Sarkar, Assistant Professor, Department of Foods and Nutrition, College of Community Science, Professor Jayashankar Telangana State Agricultural University, Hyderabad, Telangana, India. Mobile No. +91–9148224553, https://orcid.org/0000-0002-7210-6520

DOI: 10.1201/9781003387893-17

In this chapter, we will discuss two common neuropsychiatric disorders: Depression and OCD and the role of diet and nutrition in their prevention and management. This chapter will give a glimpse into the relationships between particular nutrients and foods, diet patterns, including consumption of dietary fats, vitamins, minerals, and antioxidants, to provide a potential preventive and therapeutic approach for the adjuvant treatment of neuropsychiatric disorders.

14.2 DEPRESSION

Depression primarily is a disorder of mood which may affect various aspects of an individual's life, resulting in reduced productivity, poor psychosocial status, and decreased quality of life and well-being (Lai et al., 2014). The *Diagnostic and Statistical Manual of Mental Disorders, Fifth Edition* (*DSM-5*) classifies one of the principal forms of depression as Major Depressive Disorder (MDD), which is characterized by symptoms like sleep disturbances, suicidality, or altered appetite; energy, concentration, or interest deficit; agitation; or feelings of worthlessness, hopelessness, or regret (American Psychiatric Association, 2013). Due to depression's effects on a person's biological and other day-to-day activities, it is also becoming a leading cause of disability across the world, affecting 322 million lives worldwide (WHO, 2017). A recent study by Sagar et al., 2020 indicated that in 2017, 197.3 million people in India had mental disorders, including 45.7 million with depressive disorders, which contributed the most to the total mental disorders DALYs (33.8%). Depression was forecasted to be the world's second-largest contributor to the global disease burden by 2020 (Reddy, 2010), less than ischemic heart disease, and the leading cause of disease burden by 2030 (WHO, 2008).

14.2.1 Causes of Depression

Depression is a disorder of multiple origins. It may be caused due to psychosocial or biological factors. The well-known psychosocial causes of depression include psychological stress, personality with low self-esteem, sedentary behavior, advancing age, and environmental factors, like persistent exposure to violence, neglect, abuse, or poverty, which can play a crucial role in the development of depression (Huang et al., 2019; Bakian et al., 2020; Kheirouri and Alizadeh, 2019; Yang et al., 2015; Ellingson et al., 2018; Richardson et al., 2012). However, biological factors also play a critical role in the causation of this neuropsychiatric disorder. These biological factors broadly are neurobiological alterations in the brain involving neurochemical changes, inflammation, oxidation, neuroplasticity, and genetics (Su et al., 2003; Huang et al., 2019). Depression is linked to insufficiency of monoamine neurotransmitters, that is, serotonin, norepinephrine, and dopamine, in definite structures of the central nervous system (Ruhe et al., 2007; Moret and Briley, 2011; El Mansari et al., 2010), which may be due to either reduced presynaptic release of these neurotransmitters or aberrant signal transductions. Inflammatory responses also play an important role in depression. Proinflammatory cytokines, which are involved in inflammation, are frequently observed at an increased level in the serum or plasma of depressed patients. Major depression may occur secondary to a systemic inflammatory disorder in which proinflammatory cytokines may interfere with neurotransmitter metabolism, reduce plasma tryptophan level, alter neurotransmitter messenger ribonucleic acid (RNA), and inhibit brain-derived-neurotrophic-factor (BDNF) expression. BDNF is a peptide critical for axonal growth, neuronal survival, and synaptic plasticity and function. This peptide, which could be neuroprotective and stimulate neurogenesis, has been reported to be reduced among patients with depression (Sánchez-Villegas et al., 2011). Downregulation of BDNF not only decreases the neurogenesis signaling pathways, neuronal survival, or neuronal plasticity contributing to depression but also decreases further during chronic depression, causing a vicious circle and deteriorating depression over time (Li et al., 2017; Huang et al., 2019). Studies suggest that inflammatory cytokines, like interleukin-1β (IL-1β), IL-6, soluble IL-2 receptors, and tumor necrosis factor-alpha (TNF-α), which are the components of the immune system, play a crucial role in inflammatory responses and are found to be elevated in depressive patients, suggesting an association of immune system with depression (Reichenberg et al., 2001; Yang et al., 2015). It suggests that the innate immune system plays a vital role in developing the major depressive disorder and highlights the importance of anti-inflammatory therapy for depression mitigation.

14.2.2 Role of Nutrition in Depression

Emerging data indicate that antidepressants may cause several adverse effects, including nausea, increased appetite, weight gain, fatigue, drowsiness, sexual problems, sleep disturbances, gastrointestinal and nervous systems side effects, and may promote suicide (Huang et al., 2019; Gunnell and Ashby, 2004). Therefore, people are now thinking of using alternative approaches for which the consumption of a healthy diet is one of the most promising alternatives. In addition, there are several studies which strongly support an association between different nutrients and depression. In this section of the chapter, we will discuss the role of different nutrients which function in depression at large.

14.2.2.1 Antioxidants

Major depressive disorder is caused by oxidative stress and is frequently correlated with impaired antioxidant defenses, increased serum malondialdehyde (MDA), and reduced plasma ascorbic acid and vitamin E rates (Huang et al., 2019). MDA is a natural product of lipid peroxidation in oxidative stress. Increased MDA might be resultant of higher oxidation. Studies also show that reduced levels of antioxidants increase the risk of depression (Khanzode et al., 2003). Therefore, in several studies in recent years, the potential role of nutrients in depressive symptoms has been elucidated involving nutrients, especially antioxidant nutrients such as ascorbic acid, polyunsaturated fatty acids, folic acid, vitamin B12, vitamin D, and polyphenols (Huang et al., 2019; Kheirouri and Alizadeh, 2019), suggesting that antioxidants contribute in protecting individuals from depression.

14.2.2.2 Dietary Fat

There are different types of fatty acids that are mainly classified into three groups: Saturated fatty acids (SFA), monounsaturated fatty acids (MUFA), and polyunsaturated fatty acids (PUFA). Essential PUFAs are major components of the membranes of the neuronal cell (Maidment, 2000), which play an essential role in neuronal function. Linoleic acid (LA) and alpha-linolenic acid (ALA) are essential long-chain PUFAs that are synthesized endogenously to create the ω-6 PUFAs like arachidonic acid (AA) and docosapentaenoic acid (DPA), and the ω-3 PUFAs like eicosapentaenoic acid (EPA) and docosahexaenoic acid (DHA). ω-6 and ω-3 fatty acids are PUFA, which cannot be formed in the body and, therefore, must be consumed through the diet. The ω-6 PUFA are precursors for the proinflammatory series of eicosanoids, whereas ω-3-derived metabolites are precursors for the anti-inflammatory series of eicosanoids. Hence, the fatty acid composition of the diet, which has a high ratio of ω-6:ω-3 PUFA, is associated with increased rates of inflammatory diseases and also, thereby, increased incidence of depression (Mizunoya et al., 2013; Grosso et al., 2016; Jacka et al., 2004).

A study by Sánchez-Villegas et al. (2011) found a detrimental relationship between the intake of trans fatty acids (TFA) and the risk of depression. In contrast, negative relations were found for MUFA and PUFA with depression. A beneficial effect of MUFA intake in depression has been hypothesized because MUFA intake may improve the binding of serotonin to its receptors, whereas the beneficial effect of ω-3 PUFA is hypothesized on their ability to positively modulate the immune and inflammatory response. Besides the anti-inflammatory properties of ω-3 PUFA, it also has antioxidative properties; adjusting the capacity of the serotonin system, proinflammatory cytokines; increasing levels of BDNF, and declining ability of inflammatory markers (C-reactive protein), which can play a central role in counteracting the biochemical processes occurring in depression. PUFA composition defects in neuronal cell membranes may affect receptor activity, reuptake of neurotransmitters, and signal transmission and may, therefore, lead to psychiatric disorders. The essential ω-3 PUFA, particularly DHA, is the predominant structural fatty acid in the gray matter of the brain. There is compelling evidence that increased dietary EPA and DHA intake improves the provision of preformed DHA to the brain. This preformed DHA is a more potent source of DHA than equivalent precursors of ω-3 PUFA like ALA, which is found in vegetable sources (Huang et al., 2019; Jacka et al., 2004; Hoffmann et al., 2019).

14.2.2.3 Vitamins

The associations of several vitamins with depression have been examined over time in a different population-based study. The vitamins which are mostly reported to have a role in depression are vitamin B complex (thiamine [B1], pyridoxine [B6], folic acid or folate [B9], cyanocobalamine [B12]), vitamin C, and vitamin D.

14.2.2.3.1 Vitamin B Complex

Folic acid and vitamin B12 are involved in forming monoamine neurotransmitters and other essential methylation reactions in the brain (Gilbody et al., 2007). Studies have shown that low plasma levels of vitamin B1, folic acid, and vitamin B12 are correlated with an increased risk of depression (Singh et al., 2017; Beydoun et al., 2010; Sánchez-Villegas et al., 2009). Because of the strong association of folic acid, vitamin B12 and depression, supplements of folic acid have long been used to treat depression (Huang et al., 2019), which also indicated that diets rich in folic acid and vitamin B12 might substantially reduce the risk of depression.

14.2.2.3.2 Vitamin C

A study by Hashimoto et al. (2017) found substantially elevated levels of vitamin C in the cerebrospinal fluid of older people with depression but no cognitive impairment, which was interpreted as a result of higher oxidative stress in the central nervous system. Payne et al. (2012) analyzed the dietary patterns of depressed people, which included lower fruit and vegetable intakes, contributing to lower levels of vitamin C. The study revealed natural vitamin C intake to be more beneficial for depression.

14.2.2.3.3 Vitamin D

Vitamin D and its calcitriol metabolite could protect our brain from dopamine and serotonin depletion and thus contribute to brain health. It also has a neuroprotective role in hippocampal cell survival (Berk et al., 2007). The role of vitamin D in brain development is poorly understood; however animal studies revealed that maternal vitamin D depletion could cause an abnormal brain shape in the offspring, an increased number of mitotic cells, reduced nerve growth factor, reduced glial cell line-derived neurotrophic factor, and reduced low-affinity neurotrophin receptor, which is suggestive of an association between vitamin D and brain development (Eyles et al., 2003). Furthermore, many studies indicated reduced vitamin D serum levels in individuals suffering from depressive disorder (Hoffman et al., 2019). Also, there are several supplementary trials that have suggested significant improvement in depressive symptoms with supplementation of vitamin D from 50,000 IU/week to 20,000 IU/week (Penckofer et al., 2017; Stokes et al., 2016). However, there are few studies which could not show any significant effect of supplementation of vitamin D on depression (Menon et al., 2020). Therefore, more dose-dependent trials are required to be conducted for further validation.

14.2.2.4 Minerals

Minerals play a crucial role in the management of depression.

14.2.2.4.1 Zinc

This trace element may act as an antidepressant by modulating the activity of different neurotransmitter systems, such as enhancing the serotonergic system role, preventing the function of the N-methyl-D-aspartate (NMDA) receptor complex in the central nervous system, or increasing the BDNF levels. These systems play critical roles in the pathoetiology of depression. Decreased serum zinc concentrations have been reported in major depressive disorder, which corresponded to the severity of depressive symptoms, whereas its supplementation in humans significantly decreased several markers of inflammation, such as C-reactive protein, IL-6, and TNF-α. The upregulation of these inflammatory markers was associated with symptoms of depression (Li et al., 2017; Yary and Aazami, 2012). Zinc deficiency has also been associated with the overactivation of the hypothalamic-pituitary-adrenal (HPA) system, which is believed to be the central stress response system, and the dysfunction of this HPA system plays a role in depression (Yary and Aazami, 2012).

14.2.2.4.2 Magnesium

Magnesium plays a vital role in the pathophysiology of depression through several mechanisms. It regulates the HPA system; regulates the synthesis of adrenocorticotropic hormone, gamma-aminobutyric acid (GABA), and glutamatergic neurotransmitters; and activates several transduction pathways such as protein kinase C; modulates glutamatergic transmission by blocking NMDA receptors; increases neurotransmission by endogenous serotonin receptors (Sartori et al., 2012; Yary et al., 2013). Immune system mechanisms also mediate the relationship between magnesium and depression (Jacka et al., 2009). Enhanced inflammatory factors, TNF-α and magnesium deficiencies have been correlated with depression (Yary et al., 2013). One of the important inflammatory markers is C-reactive protein, which may be lowered by adequate dietary intake of magnesium (Bo et al., 2006). Chronic administration of magnesium may also increase the secretion of BDNF expression (Slutsky et al., 2010).

14.2.2.4.3 Selenium

Selenium, known to be a potent antioxidant, is an essential micronutrient which may be toxic in large doses (Rayman, 2000). Since depression is an inflammatory disorder accompanied by an accumulation of highly reactive oxygen species, the antioxidative and anti-inflammatory properties of selenium may play a significant role in its management. Selenium also has an important modulatory impact on dopamine, and dopamine plays a role in depression pathophysiology and other psychiatric disorders (Pasco et al., 2012), which indicate selenium may be a modifiable factor in the primary prevention and management of depression. Pasco et al., 2012 suggested that lower dietary selenium intake is associated with an increased risk of depression.

14.2.2.4.4 Iron

Iron, an essential trace element, plays a role in neurotransmitter syntheses, such as dopamine and serotonin; improves neurobehavioral functions by modulating neurotransmission pathways and neural activities; and prevents NMDA toxicity and reduces NMDA levels in all brain tissues (Li et al., 2017).

14.2.2.5 Polyphenol

Plant polyphenols, due to their antioxidant property, has a role as an antidepressant. Consumption of a higher amount of green tea (≥4 cups/day), which is known to be rich in polyphenols, was associated with a lower prevalence of depressive symptoms (Pham et al., 2014). Studies also reported a significant inverse association of dietary intake of phenolic acid, flavanones, and anthocyanins with depression (Godos et al., 2018).

14.2.2.6 Amino Acid

The amino acids tryptophan, tyrosine, phenylalanine, and methionine are often helpful in treating many mood disorders, including depression (McLean et al., 2004; Agnoli et al., 1976; Aan het Rot et al., 2006). Tryptophan is a precursor of serotonin and can also restore serotonin levels leading to diminished depression (Agnoli et al., 1976). Dietary supplements containing tyrosine and its precursor, phenylalanine, lead to arousal and alertness, while methionine combines with adenosine triphosphate (ATP) to produce S-adenosylmethionine (SAM), which supports the production or synthesis of neurotransmitters in the brain (De Leo, 1987).

14.2.2.7 Dietary Creatine

Creatine, an endogenous nitrogenous organic acid, has an important role in the homeostasis of brain energy. It is actively being researched for its antidepressant effects, which may be another promising nutraceutical candidate for depression. Creatine supplementation has generally shown improvement and acceleration in antidepressant response (Bakian et al., 2020).

14.2.3 Role of Diet in Depression

Nutrients are always consumed in combination in the form of diet. Thus, understanding the effect of the complexities of several nutrients in a diet on depression is more important to better understand the disease to facilitate preventive measures. In this section of the chapter, we will discuss the role of different dietary patterns in the prevention or management of depression.

14.2.3.1 Western Diet

The Western dietary pattern, often referred to as "unhealthy diet" and mainly characterized by high consumption of refined grains, red meat, processed meat, high-fat dairy products, butter, fast food, potatoes and high-fat gravy, sweets, carbonated drinks, and low intakes of fruits and vegetables (Lai et al., 2014; Li et al., 2017). A vast number of studies identified a positive association between Western dietary patterns and the increased risk of depression (Li et al., 2017; Kim et al., 2015). This association might be due to several underlying plausible mechanisms (Li et al., 2017). Firstly, the high consumption of processed food may cause inflammation and inflammatory diseases like cardiovascular diseases, which are known to be involved in the pathogenesis of depression. Secondly, a high intake of refined grains, processed meat, and foods with high sugar and high fat is associated with higher levels of low-grade inflammation (C-reactive protein) and subsequent brain atrophy, which are correlated with depression. Thirdly, the deleterious effect of the Western-style dietary pattern could also come from sugar-rich sweets and soft drinks because of the alteration of endorphin levels and oxidative stress.

14.2.3.2 Healthy Diet

In general, the healthy dietary pattern is characterized by high intakes of whole grains, vegetables and fruits, olive oil, fish, soy, poultry, and low-fat dairy and antioxidants (e.g., vitamin C, vitamin E, flavonoids, and carotenoids) (Li et al., 2017; Lai et al., 2014). A number of studies indicated that a healthy dietary pattern is associated with a decreased risk of depression (Li et al., 2017; Lai et al., 2014). The plausible underlying mechanisms, as indicated by Li et al. (2017), for the association of a healthy diet with depression might be due to the high consumption of vegetables and fruits, which may have beneficial protective effects against depression as they contain high amounts of antioxidants (e.g., vitamin C, vitamin E, and other carotenoids compounds). Antioxidants have the potential to reduce the level of oxidative stress, which induces neuronal damage, specifically hippocampal neurons, that are thought to be a major contributor in depression. Vegetables and fruits rich in folate may also play a protective role since studies have shown that folate deficiencies resulted in increased homocysteine concentrations and reduced availability of S-adenosylmethionine, which play a critical role in the pathophysiology of depression. High fish consumption in the healthy dietary pattern may also confer a protective effect as it contains a high content of long-chain ω-3 PUFA, which have anti-inflammatory properties and may contribute to brain functioning and serotonin neurotransmission. Studies demonstrated that adding ω-3 PUFA may improve the condition of depression (Grosso et al., 2016; Jacka et al., 2004).

14.2.3.3 Mediterranean Diet

The Mediterranean diet is a typical diet of the Mediterranean region, mainly characterized by a high intake of vegetables, fruits, cereals, legumes, bread, and potatoes; high intake of MUFA,

PUFA; moderately high intake of fish; low intake of saturated fats; low intake of red meat and poultry; low-to-moderate intake of milk products; and regular but moderate intake of alcohol. It is assumed that adherence to the Mediterranean dietary pattern decreases inflammatory, vascular, and metabolic processes associated with the risk of depression (Sánchez-Villegas et al., 2009). The abundant use of olive oil, rich in MUFA, can boost serotonin binding to its receptors (Logan 2005). Vegetables and fruits containing folate, vitamins B12 and B6 are involved in 1-carbon metabolism, which acts in multiple methylation reactions, such as those involving serotonin and other monoamine neurotransmitters (Bottiglieri, 1996). Such nutrient combinations in a Mediterranean diet can have a beneficial therapeutic impact on pathophysiology and depression pathoetiology.

14.2.4 General Dietary and Lifestyle Recommendation for Depression

The role and importance of different nutrients and dietary patterns have been explained previously. Based on all the information and evidence, in this section, we will discuss a few general dietary and lifestyle recommendations which may be implemented for the prevention and management of depression. However, it must be noted that it is always advised to take the guidance of a physician and dietitian before making any changes to better understand the complexities of the condition and its relation with other comorbidities, if any.

a. Higher added sugars, sugary foods, sweetened carbonated beverages, and lower-quality carbohydrate consumption were associated with depression risk (Sanchez-Villegas et al., 2018). Thus, it is advised to reduce sugar consumption to minimize the risk as they can modulate serotonin release, plasma concentrations of tryptophan, and large neutral amino acids (Markus, 2007).
b. High consumption of fast-food, processed food, and bakery food is associated with a higher risk of depression (Mikolajczyk et al., 2009; Sanchez-Villegas et al., 2018). Therefore, limit the intake of these foods as they are high in trans fats, saturated fat, refined carbohydrates, and added sugars.
c. Diets higher in fruit, vegetables, wholegrain cereals, pulses, nuts, and seeds are associated with a reduced risk for depression. Hence, increased consumption of these foods, which should form the majority of the diet as they are dense in nutrients, high in fiber, and low in saturated and trans fats.
d. Include high MUFA and ω-3 PUFA containing foods in the diet. Olive oil containing high content of MUFA and other specific antioxidants is often associated with a reduced risk of depression. Also, higher fish consumption, one of the primary sources of ω-3 PUFA, is associated with reduced depression risk. Other foods rich in ω-3 fatty acids are chia seeds, flax seeds, soybean, walnut, etc.
e. Higher adherence to a healthy diet like the Mediterranean diet has been associated with a reduced probability and/or risk for depression. Thus, following such dietary patterns, mainly including vegetables, fruits, fish, whole grains, olive oil, and avoiding processed foods, red meat, and saturated and trans fats (hydrogenated fat, palm oil) is recommended.
f. Mediterranean diet includes regular but low to moderate consumption of alcohol, which was reported to be associated with better health and cognitive performance (Stockwell et al., 2013). However, it must be noted that the co-occurrence of depression and alcohol use disorders is often associated (McHugh and Weiss, 2019). Thus, avoid excessive consumption of alcohol.
g. Physical activity can help reduce the risk of depression and improve cognitive function and overall psychological well-being (Boutcher, 2003; Miles, 2007). Therefore, include regular exercise in the daily regime, which might be highly beneficial.

14.3 OBSESSIVE-COMPULSIVE DISORDER (OCD)

According to American Psychiatric Association (2000), obsessions are recurring and persistent thoughts, urges, or images that are perceived as disturbing and inappropriate and cause significant anxiety or discomfort, while compulsions are repeated behaviors such as handwashing, ordering, evaluating, verifying; or mental activities such as constant praying, counting, silently repeating words that a person feels compelled to conduct in response to an addiction, or in response to an obsession which must be applied rigidly. Epidemiological studies have demonstrated that OCD usually begins in childhood or adolescence (Karcı and Celik, 2020). The literature on the occurrence of OCD in India is scanty. A recent study by Jaisoorya et al. (2015) among Indian adolescents found the point prevalence of OCD to be 0.8%. However, multiple studies indicate that the prevalence of OCD is much lesser than the subthreshold OCD, indicating a large number of people may suffer from obsessive-compulsive symptoms though not a full-blown OCD (Jaisoorya et al., 2017; Adam, 2012). The comorbidity of OCD has been frequently seen with a number of other neuropsychiatric disorders. OCD has also demonstrated depressive symptoms and comorbid depressive disorder, while obsessions and compulsions in patients with depressive disorder may be found. It also commonly co-occurs with disorders like anxiety, mood, attention deficit hyperactivity disorder, epilepsy, and psychosis (Bhattacharya et al., 2005; Pallanti et al., 2011; Karcı and Celik, 2020).

14.3.1 Causes of Obsessive-Compulsive Disorder (OCD)

The contemporary knowledge indicates that OCD has a multifactorial etiopathogenesis. Neurobiological, that is, neurochemical, structural, oxidative; genetic; immunological and psychosocial factors play an important role in the causation of OCD. Changes in certain neurotransmitters trigger OCD symptoms. Decreased levels of serotonin and the dysregulation of serotonin transporter lead to the development of OCD symptoms (Goddard et al., 2008; Barr et al., 1993). In addition to the serotonin transporter system, glutamate, glutamine, homocarnosine, and GABA have been found to be higher in OCD patients (Bolton et al., 2001; Rosenberg et al., 2000; Pittenger et al., 2006). Furthermore, evidence indicates that OCD is more closely related to the antioxidant system, increased oxidative processes leading to inflammation and dysfunction in neurotransmitters due to increased free-radical burden (Madhura, 2015; Behl et al., 2010). The primary free radicals in human tissues are superoxide, hydroxyl, hydrogen peroxide, singlet oxygen, and nitric oxide. Free radicals cause both direct damage, via lipid peroxidation, and indirect damage, via MDA. MDA is the major reactive aldehyde and the end product of the peroxidation of PUFA (Ersan et al., 2006). It is used as a measure of damage to tissues and has been found to be elevated in patients with OCD (Behl et al., 2010). Few other signs of such tissue damage are alterations in the level of substances with antioxidant effects, such as vitamins C and E, and melatonin, and the activities of enzymes such as superoxide dismutase (SOD), catalase, glutathione peroxidase (Kurutas, 2015). It has been reported that basal ganglia are particularly vulnerable to damage by free radicals due to the large concentration of catecholamines in this region of the brain. Apart from the neurobiological factors, evidence indicates the familial transmission of OCD. The morbid risk is higher among first-degree relatives of patients with OCD (Pauls et al., 1995). Furthermore, the age at the onset of obsessive-compulsive symptoms in the patient with OCD was strongly related to familiality (Nestadt et al., 2000). Also, alterations to the immune functioning have been associated with OCD (Arnold and Richter, 2001). In addition, psychosocial experiences, stress, and traumatic events and environmental factors like perinatal injuries, infections, and toxic pathogens may also play a role in the causation of this disorder (Grisham et al., 2008).

14.3.2 Role of Nutrients in Obsessive-Compulsive Disorder (OCD)

Although research in OCD has been done for years, still the role of nutrition, diet, or the impact of nutritional status on OCD is underexplored. There are multiple methods for OCD treatment, including pharmacotherapy with mega doses of antidepressants like selective serotonin reuptake inhibitors (SSRI), clomipramine, morphine, deep brain stimulation, and functional neurosurgery (Fontenelle et al., 2007). Most of these methods are invasive or unnatural, with chances of reoccurrence and also exerting severe adverse effects on the patients. Thus, finding an alternative solution like nutrients, diet, or nutritional supplement is very crucial. Also, with the increasing interest in natural therapies, nutritional and dietary approaches have become a more practical and promising intervention for managing OCD. In this section of the chapter, we will discuss the role of different nutrients and nutraceuticals in OCD.

14.3.2.1 Antioxidant

Oxidative stress is a result of an imbalance between free radicals and antioxidants. Free radicals and oxidative stress are known to have a pathophysiological role in OCD. Thus, antioxidants play a significant role in maintaining the balance of the system. One of the potentially devastating effects of free radicals is membrane lipid peroxidation (Behl et al., 2010). It has been reported that the brain and nervous system may be especially prone to free-radical-induced damage since the brain is a highly oxygenated organ (requiring one-fifth of the body's oxygen (Ozdemir et al., 2009) and is low in catalase activity (Paul et al., 1989), which makes the brain and neurons more prone to oxidative damage. Therefore, it is important to intervene with antioxidants to prevent such damage. The more essential antioxidants in overcoming oxidative stress are superoxide dismutase (SOD), glutathione, and catalase. They defend biomembranes from oxidative attacks and prevent lipid peroxidation by scavenging hydrogen peroxide (H_2O_2) and delaying H_2O_2-dependent free radical attacks on lipids. SOD is regarded as a stress protein synthesized to react to oxidative stress. It has a definitive role in overcoming oxidative stress and is part of a protective mechanism. Glutathione is a well-known protective agent against oxidative stress. Glutathione peroxidases are selenoenzymes, which at the expense of glutathione catalyze the reduction in H_2O_2. Catalase, a heme-containing enzyme, plays a vital role in removing H_2O_2 (Behl et al., 2010). MDA, which is one of the natural products of lipid peroxidation in oxidative stress (Ersan et al., 2006), has been reported to be significantly higher in OCD patients and also was indicated to be associated with some amount of tissue damage due to high oxidative stress (Behl et al., 2010). Thus, food rich in antioxidants like SOD, glutathione, and catalase may target MDA production and might have a beneficial effect in overcoming oxidative stress, anxiety, and disorders like OCD.

14.3.2.2 Amino Acids

14.3.2.2.1 Tryptophan

Serotonin plays a major role in the pathophysiology of OCD. It has been observed that selective serotonin reuptake inhibitors (SSRIs) help patients with OCD. It is, therefore, clear that nutrients which increase serotonin levels will decrease OCD symptoms. The amino acid tryptophan, a precursor to serotonin, and its hydroxylated form 5-hydroxytryptophan (5-HTP) as supplements will increase serotonin levels and treat OCD (Lakhan & Vieira, 2008; Camfield et al., 2011). Nonetheless, studies indicate that tryptophan to improve brain serotonin levels is not a highly effective technique as it has been reported that only around 2% of the tryptophan ingested is used for serotonin synthesis in the brain, while most intracellular tryptophan is used for protein synthesis (Brown, 1994; Sirek & Sirek, 1970). Therefore, the hydroxylated form of tryptophan, 5-HTP, has been suggested as a precursor strategy.

14.3.2.2.2 Glycine

Abnormal glutamatergic neurotransmission is a common etiology in OCD (Holton & Cotter, 2018). Higher glutamate levels in the cerebrospinal fluid of OCD patients have been reported (Chakrabarty et al., 2005). Also, the glutamatergic ionotropic NMDA receptor has been associated with the disorder (Arnold et al., 2004). Glycine is an amino acid that modulates glutamatergic function (Clements and Westbrook, 1991) and can reduce OCD symptoms due to its binding effect on NMDA receptors

(Singer et al., 2010). A study by Greenberg et al. (2009) reported a beneficial effect of glycine in OCD patients. However, administering a large dose of glycine (60 g/day) showed side effects such as considerable nausea. Therefore, a lesser dosage in combination with other nutrients might be more helpful.

14.3.2.2.3 N-acetyl cysteine (NAC)

The amino acid–based nutraceutical N-acetyl cysteine (NAC) is a safe and readily available agent, more soluble in water and a better source of cysteine rather than the parenteral cysteine for being less toxic, less reactive, and more oxidizing resistant. Therefore NAC is suggested as a supplement instead of high protein food sources containing cysteine (Karci and Celik, 2020). NAC has been indicated to reduce oxidative stress and inflammation, and therefore it may have a beneficial role in OCD (Ng et al., 2008; Pittenger et al., 2006). Studies indicate that in OCD treatment, NAC has been confirmed to be safe and well-tolerated even at higher dosages (up to 8,000 mg/day) without any major side effects (Atkuri et al., 2007).

14.3.2.3 Vitamins

14.3.2.3.1 Vitamin B Complex

Neurotransmitters (dopamine, serotonin, and melatonin) are necessary for a healthy, balanced mood, emotions, and also sleeping, which is associated with psychiatric health. Vitamin B12 and folic acid, or folate, act as cofactors in the synthesis of these neurotransmitters such as serotonin, norepinephrine, and dopamine. Thus, levels of neurotransmitters may decline due to a deficiency of any of these vitamins (Esnafoğlu and Yaman, 2017). In several studies, increased homocysteine and deficient levels of vitamin B12 and folate have been reported in adult OCD patients (Atmaca et al., 2005; Sharma and Biswas, 2012; Valizadeh and Valizadeh, 2011). Circulating levels of increased homocysteine are well-known as a sensitive marker of folic acid and vitamin B12 deficiency (Folstein et al., 2007). Elevated homocysteine levels may cause damage to deoxyribonucleic Acid (DNA), activation of apoptotic signals, mitochondrial dysfunction, and high oxidative stress (COŞAR et al., 2014). Vitamin B12, folic acid, and homocysteine are associated with carbon transfer metabolism that maintains the methylation of proteins, neural membrane phospholipids, and synthesis of neurotransmitters such as serotonin and other monoamines (Karcı and Celik, 2020). All these studies indicate that vitamin B12, folate and homocysteine may play an essential role in the etiology of OCD.

14.3.2.3.2 Vitamin E

The MDA leads to a decrease in serotonin in the brain in the areas of binding (Britt et al., 1992). Studies indicate MDA levels in OCD patients were significantly higher. In contrast, the concentration of nonenzymatic antioxidant vitamin E, one of the essential antioxidants with a role in the human immune system, was substantially lower in OCD patients. Low vitamin E level in OCD is presumably due to their role in preventing free radical damage that seems more extensive in OCD patients (Ersan et al., 2006). Furthermore, a significant correlation was found between elevated MDA levels and low vitamin E levels.

14.3.2.3.3 Vitamin C

Another important nonenzymatic antioxidant vitamin is vitamin C, or ascorbic acid, which is abundant in foods. Vitamin C level was also reported to be lower in OCD patients. However, the data was not statistically significant, and there are not many studies done to find the impact of vitamin C supplementation on OCD patients (Ersan et al., 2006). Thus, there is a need for research with a broader scope to clarify the use of vitamin C in coping with free radicals in OCD patients.

14.3.2.3.4 Vitamin D

Vitamin D has important roles in immunity modulation, antioxidant processes, inflammatory response, neurotransmission, neuroprotection, and many others (Karcı and Celik, 2020).

Additionally, Vitamin D affects brain serotonin concentrations by regulating tryptophan hydroxylase gene expression, which is a rate-limiting step in serotonin synthesis. Dysregulation of brain serotonergic neurotransmission has been indicated in some psychiatric disorders like OCD (Kaneko et al., 2015). Therefore, one of the ways vitamin D may contribute to OCD development is via the serotonin system. The other possible relationship between vitamin D and OCD may be due to its neuroprotective effect. Vitamin D has antioxidant potential and prohibits an essential enzyme (inducible nitric oxide synthase) from synthesizing nitric oxide (Garcion et al., 1998). Vitamin D levels were reported to be significantly lower in OCD patients (Esnafoğlu and Yaman, 2017). Thus, it explains that vitamin D may play a role in the pathophysiology of OCD.

14.3.2.4 Trace Elements

Trace elements imbalances may be linked with the pathophysiology of OCD. Selenium, an essential trace element present in biological systems as selenoproteins, which characteristically are oxidoreductases, has potential antioxidant properties of glutathione peroxidase (Wołonciej et al., 2016). Selenoproteins have been recognized as modulators of the brain and signaling. Under conditions of selenium deficiency, tissue levels of these enzymes fall, and oxidative stress conditions develop (Ozdemir et al., 2009). A study indicated low serum selenium levels in patients with OCD (Ozdemir et al., 2009). Selenium is a dietary element, so dietary habits may influence the improvement of OCD conditions. Shohag et al., 2012 have also reported low levels of serum zinc, iron, and magnesium in OCD patients. Due to the possible role of the oxidation mechanism in the pathogenesis of OCD, trace elements with antioxidant properties may have the potential to be useful in managing OCD.

14.3.2.5 Carbohydrate Molecule Myoinositol (MI)

Myoinositol is an endogenous glucose isomer and has long been used to treat psychiatric disorders. It is clinically beneficial in depression, OCD, and panic disorder when intervened for a long duration (Kofman et al., 1998). A sustained supply of MI is essential for membrane phospholipid synthesis and the intracellular secondary messenger process (Kim et al., 2005). It is suggested that MI could modulate serotonin reuptake and increase serotonin receptor density, thereby improving OCD symptoms. There are few studies which indicate the positive effect of high-dose MI supplementation (18 g/day) in OCD treatment (Fux et al., 1996; Carey et al., 2004).

14.3.2.6 Dietary Glutamate (Monosodium Glutamate, MSG)

There is abundant evidence indicating disordered glutamatergic neurotransmission may be a key component in the etiology of OCD (Holton and Cotter, 2018). Glutamate, the most ubiquitous neurotransmitter in the body, has been reported to be significantly higher in the cerebrospinal fluid of OCD patients (Chakrabarty et al., 2005; Rosenberg et al., 2000; Bolton et al., 2001). Thus, dietary modulation of glutamate, which has a low risk of side effects, may be a novel, cost-effective intervention for OCD treatment.

14.3.3 Role of Medicinal Plants in Obsessive-Compulsive Disorder (OCD)

There are few medicinal plants with active components which have been found to modulate serotonergic activity and are thus used in the treatment of OCD effectively (Camfield et al., 2011, Karcı and Celik, 2020).

1. St John's wort (SJW) (*Hypericum perforatum*) has several bioactive constituents, including hyperforin, hypericin, and pseduohypericin, which offers a variety of neurobiological effects, including reuptake inhibition of monoamines, enhanced sensitization and serotonin receptor binding, and neuroendocrine modulation with a possible positive impact on OCD.

2. Milk thistle (*Silybummarianum*) is a Mediterranean and Persian plant which has a flavonoid complex silymarin (silibinin, a principal constituent). It has been found to increase cortical serotonin levels, which resulted in obsessional reducing beneficial symptoms in OCD. This may be due to silibinin's inhibition of monoamine oxidase activity.
3. Valerian root (*Valeriana officinalis* L) is a perennial plant containing aleuronic acid, which modulates GABA receptors. It also contains valepotriates, which have been reported to be effective in severe anxiety, insomnia, and migraine. A clinical trial indicated the beneficial effect of valerian extract in OCD treatment (Pakseresht et al., 2011).
4. Turmeric (Curcuma Longa), a perennial herbaceous plant of southern Asia, contains the active component curcumin. Curcumin has anti-inflammatory and antioxidative effects and is beneficial in neuropsychiatric disorders. Although there are no human trials, the administration of curcumin in OCD animal models for over a month has been reported to significantly reduce compulsive behaviors (Chimakurthy and Murthy, 2010).
5. Borage (*Echium amoenum*) is a Persian plant, the flower of which exerts anxiolytic, sedative, and antidepressant effects. A study by Sayyah et al. (2009) indicated a significant improvement in anxiety symptoms after a month of administration of the aqueous extract of the dried flower to OCD patients.

14.3.4 General Dietary and Lifestyle Recommendation for Obsessive-Compulsive Disorder (OCD)

Very few studies have been done to understand the effectiveness of lifestyle interventions such as nutrition, dietary change, and exercise on OCD. However, based on the current information and evidence, in this section of the chapter, we will discuss a few general dietary and lifestyle guidelines that may be implemented to prevent and manage OCD.

a. Antioxidants have been shown to have a vital role in OCD pathophysiology. Therefore, foods rich in particularly enzymatic antioxidants like SOD, catalase, and glutathione might be an effective modulatory factor. SOD, glutathione, and catalase are mostly found in broccoli, cabbage, brussels sprouts, wheatgrass, barley grass, peas, spinach, asparagus, celery, zucchini, kale, cucumber, etc.
b. Amino acid tryptophan and glycine have been shown to have a beneficial effect on OCD. Therefore, include high-protein foods such as meat, fish, beans, milk and milk products, which contain amino acids.
c. Include foods rich in vitamins B12 and folate that help in the synthesis of neurotransmitter like serotonin, which plays a crucial role in neuropsychiatric health. Thus, adding foods like milk, fish, poultry, egg, meat, spirulina, and fortified cereals for vitamin B12 and beans, lentils, broccoli, spinach, asparagus, and dairy products for folate might be beneficial.
d. Vitamin E has been reported to be one of the most effective antioxidants in preventing free radical damage in OCD patients. Therefore, include foods rich in Vitamin E like nuts (almonds, peanut, hazelnuts), seeds (sunflower seeds), dark green leafy vegetables (spinach, kale, lettuce, turnip greens), asparagus, broccoli, beans and peas, and vegetable cooking oil (wheat germ oil, sunflower oil, safflower oil, soybean oil).
e. Include foods rich in vitamin D, which might be beneficial in OCD due to its neurotransmission and neuroprotection abilities. Foods containing a high amount of Vitamin D are fatty fish (salmon, tuna, mackerel), fish liver oil, meat and meat products, egg yolk, and full-fat dairy products like cheese.
f. Glutamate levels are reported to be significantly higher in OCD patients. Therefore, limit dietary glutamate in the form of monosodium glutamate (MSG), commonly found as a

food additive in processed products, soy sauce, and artificial flavors, which otherwise might trigger OCD symptoms. Processed foods containing MSG are labelled in the food packages accordingly.

g. Consume foods rich in myoinositol, like fruits, grains, nuts, beans, and oats, which might be effective in managing OCD. Myoinositol has been shown to improve OCD symptoms by modulating the serotonergic system.
h. Include herbs or spices with anti-inflammatory and antioxidative properties like turmeric, oregano, sage, basil, rosemary, thyme, mint, lemongrass, clove, garlic, ginger, green onion, leek, cinnamon, black and red pepper.
i. Physical activity can help overall psychological well-being. However, no study has been conducted to completely elucidate the effect of regular physical activity on OCD. Furthermore, excessive physical activity has been associated with OCD symptoms (Błachno et al., 2016).

14.4 PERSPECTIVE

With the increasing number of cases worldwide, depression has gained more importance in the past decade. Studies have demonstrated that apart from psychosocial factors, depression is associated with immunological or neurotransmitter abnormalities; therefore, nutrition and diet play an essential role in the prevention and management of depression. The existing evidence suggests that nutrients with potential anti-inflammatory and antioxidant properties may relieve the secretion of inflammatory cytokines, diminish inflammation in the brain, reduce oxidative stress, and upregulate BDNF. Thus, certain nutrients like antioxidants, unsaturated fatty acids, particularly ω-3 PUFA, MUFA; vitamins such as vitamin E, vitamin D, vitamin C, and folate; minerals such as zinc, magnesium, selenium; and polyphenols play a crucial role in preventing or improving depressive symptoms. Superior adherence to a healthy diet which includes higher consumption of more vegetables and fruits, fish, whole grains, and unsaturated fats and a lower intake of saturated fats, red meat, and processed foods, has shown to have beneficial effects. However, at present, sufficient information is not available to recommend daily allowances for all specific nutrients required to maintain mental health. Hence, more dose-dependent epidemiological studies are needed to fill the gaps in this area.

In the case of OCD, the currently available treatment mainly focuses on invasive methods or drugs induced, which have limited effectiveness and many possible adverse effects. Comparatively, very little research focus has been given to lifestyle interventions such as dietary change and physical activity compared to conventional treatments, whereas lifestyle interventions possess less risk and might be an effective alternative strategy or adjunct treatment method. It also means there lies an excellent potential for research in this field to bridge the gap. The existing evidence suggests that nutrients which have the potential in modulating serotoninergic and glutamatergic system and has anti-inflammatory and antioxidant properties may be beneficial in OCD. Studies indicated enzymatic antioxidants like SOD, glutathione, and catalase; amino acid tryptophan, glycine, cysteine (N-acetyl cysteine); vitamins such as vitamin B12, folate, vitamin E, vitamin C, and vitamin D; trace element as selenium, zinc; carbohydrate molecule myoinositol have a significant impact in the pathoetilogy of OCD, which also indicates that these nutrients can be an effective modulatory factor for OCD prevention and management.

REFERENCES

Aan het Rot, M., Moskowitz, D. S., Pinard, G., & Young, S. N. 2006. Social behaviour and mood in everyday life: The effects of tryptophan in quarrelsome individuals. *Journal of Psychiatry and Neuroscience*, *31*(4), 253.

Adam, Y., Meinlschmidt, G., Gloster, A. T., & Lieb, R. 2012. Obsessive–compulsive disorder in the community: 12-month prevalence, comorbidity and impairment. *Social Psychiatry and Psychiatric Epidemiology*, *47*(3), 339–349.

Agnoli, A., Andreoli, V., Casacchia, M., & Cerbo, R. 1976. Effect of S-adenosyl-{l}-methionine (SAMe) upon depressive symptoms. *Journal of Psychiatric Research, 13*(1), 43–54.

American Psychiatric Association. 2000. *Diagnostic and statistical manual of mental disorders.* Fourth Edition. (DSM-IV-TR). American Psychiatric Association, Washington, DC.

American Psychiatric Association. 2013. *Diagnostic and statistical manual of mental disorders (DSM-5).* Fifth Edition. American Psychiatric Pub. ISBN 978-0-89042-554-1.

Arnold, P. D., & Richter, M. A. 2001. Is obsessive–compulsive disorder an autoimmune disease? *CMAJ, 165*(10), 1353–1358.

Arnold, P. D., Rosenberg, D. R., Mundo, E., Tharmalingam, S., Kennedy, J. L., & Richter, M. A. 2004. Association of a glutamate (NMDA) subunit receptor gene (GRIN2B) with obsessive-compulsive disorder: A preliminary study. *Psychopharmacology, 174*(4), 530–538.

Atkuri, K. R., Mantovani, J. J., Herzenberg, L. A., & Herzenberg, L. A. 2007. N-Acetylcysteine—a safe antidote for cysteine/glutathione deficiency. *Current Opinion in Pharmacology, 7*(4), 355–359.

Atmaca, M., Tezcan, E., Kuloglu, M., Kirtas, O., & Ustundag, B. 2005. Serum folate and homocysteine levels in patients with obsessive-compulsive disorder. *Psychiatry and Clinical Neurosciences, 59*(5), 616–620.

Bakian, A. V., Huber, R. S., Scholl, L., Renshaw, P. F., & Kondo, D. 2020. Dietary creatine intake and depression risk among US adults. *Translational Psychiatry, 10*(1), 1–11.

Barr, L. C., Goodman, W. K., & Price, L. H. 1993. The serotonin hypothesis of obsessive compulsive disorder. *International Clinical Psychopharmacology, 8*, 79–82.

Behl, A., Swami, G., Sircar, S. S., Bhatia, M. S., & Banerjee, B. D. 2010. Relationship of possible stress-related biochemical markers to oxidative/antioxidative status in obsessive-compulsive disorder. *Neuropsychobiology, 61*(4), 210–214.

Berk, M., Sanders, K. M., Pasco, J. A., Jacka, F. N., Williams, L. J., Hayles, A. L., & Dodd, S. 2007. Vitamin D deficiency may play a role in depression. *Medical Hypotheses, 69*(6), 1316–1319.

Beydoun, M. A., Shroff, M. R., Beydoun, H. A., & Zonderman, A. B. 2010. Serum folate, vitamin B-12 and homocysteine and their association with depressive symptoms among US adults. *Psychosomatic Medicine, 72*(9), 862.

Bhattacharyya, S., Reddy, Y. J., & Khanna, S. 2005. Depressive and anxiety disorder comorbidity in obsessive compulsive disorder. *Psychopathology, 38*(6), 315–319.

Błachno, M., Bryńska, A., Tomaszewicz-Libudzic, C., Jagielska, G., Srebnicki, T., Wiśniewski, A., & Wolańczyk, T. 2016. Obsessive-compulsive symptoms and physical activity in patients with anorexia nervosa–possible relationships. *Psychiatry, 50*(1), 55–64.

Bo, S., Durazzo, M., Guidi, S., Carello, M., Sacerdote, C., Silli, B., Rosato, R., Cassader, M., Gentile, L., & Pagano, G. 2006. Dietary magnesium and fiber intakes and inflammatory and metabolic indicators in middle-aged subjects from a population-based cohort. *The American Journal of Clinical Nutrition, 84*(5), 1062–1069.

Bolton, J., Moore, G. J., MacMillan, S., Stewart, C. M., & Rosenberg, D. R. 2001. Case study: Caudate glutamatergic changes with paroxetine persist after medication discontinuation in pediatric OCD. *Journal of the American Academy of Child & Adolescent Psychiatry, 40*(8), 903–906.

Bottiglieri, T. 1996. Folate, vitamin B12, and neuropsychiatric disorders. *Nutrition Reviews, 54*(12), 382–390.

Boutcher, S. 2003. *Physical activity and psychological well-being.* Routledge, Taylor & Francis Group, London & New York.

Britt, S. G., Chiu, V. W., Redpath, G. T., & Vandenberg, S. R. 1992. Elimination of ascorbic acid-induced membrane lipid peroxidation and serotonin receptor loss by Trolox-C, a water soluble analogue of vitamin E. *Journal of Receptor Research, 12*(2), 181–200.

Brown, R. R. 1994. Tryptophan metabolism: A review. In: W. Kochen & H. Steinhart (Eds.), *L-Tryptophan: Current Prospects in Medicine and Drug Safety.* Walter de Gruyter, Berlin, pp. 17–30.

Camfield, D. A., Sarris, J., & Berk, M. 2011. Nutraceuticals in the treatment of obsessive compulsive disorder (OCD): A review of mechanistic and clinical evidence. *Progress in Neuro-Psychopharmacology and Biological Psychiatry, 35*(4), 887–895.

Carey, P. D., Warwick, J., Harvey, B. H., Stein, D. J., & Seedat, S. 2004. Single photon emission computed tomography (SPECT) in obsessive–compulsive disorder before and after treatment with inositol. *Metabolic Brain Disease, 19*(1–2), 125–134.

Chakrabarty, K., Bhattacharyya, S., Christopher, R., & Khanna, S. 2005. Glutamatergic dysfunction in OCD. *Neuropsychopharmacology, 30*(9), 1735–1740.

Chimakurthy, J., & Murthy, T. E. 2010. Effect of curcumin on quinpirole induced compulsive checking: An approach to determine the predictive and construct validity of the model. *North American Journal of Medical Sciences, 2*(2), 81.

Clements, J. D., & Westbrook, G. L. 1991. Activation kinetics reveal the number of glutamate and glycine binding sites on the N-methyl-D-aspartate receptor. *Neuron*, *7*(4), 605–613.

Coşar, A., Ipcioğlu, O. M., Özcan, Ö., & Gültepe, M. 2014. Folate and homocysteine metabolisms and their roles in the biochemical basis of neuropsychiatry. *Turkish Journal of Medical Sciences*, *44*(1), 1–9.

De Leo, D. 1987. S-adenosylmethionine as an antidepressant: A double-blind trial versus placebo. *Current Therapeutic Research*, *41*(6), 865–870.

El Mansari, M., Guiard, B. P., Chernoloz, O., Ghanbari, R., Katz, N., & Blier, P. 2010. Relevance of norepinephrine–dopamine interactions in the treatment of major depressive disorder. *CNS Neuroscience & Therapeutics*, *16*(3), e1–e17.

Ellingson, L. D., Meyer, J. D., Shook, R. P., Dixon, P. M., Hand, G. A., Wirth, M. D., Paluch, A. E., Burgess, S., Hebert, J. R., & Blair, S. N. 2018. Changes in sedentary time are associated with changes in mental wellbeing over 1 year in young adults. *Preventive Medicine Reports*, *11*, 274–281.

Ersan, S., Bakir, S., Ersan, E. E., & Dogan, O. 2006. Examination of free radical metabolism and antioxidant defence system elements in patients with obsessive–compulsive disorder. *Progress in Neuro-Psychopharmacology and Biological Psychiatry*, *30*(6), 1039–1042.

Esnafoğlu, E., & Yaman, E. 2017. Vitamin B12, folic acid, homocysteine and vitamin D levels in children and adolescents with obsessive compulsive disorder. *Psychiatry Research*, *254*, 232–237.

Eyles, D., Brown, J., Mackay-Sim, A., McGrath, J., & Feron, F. 2003. Vitamin D3 and brain development. *Neuroscience*, *118*(3), 641–653.

Folstein, M., Liu, T., Peter, I., Buel, J., Arsenault, L., Scott, T., & Qiu, W. W. 2007. The homocysteine hypothesis of depression. *American Journal of Psychiatry*, *164*(6), 861–867.

Fontenelle, L. F., Nascimento, A. L., Mendlowicz, M. V., Shavitt, R. G., & Versiani, M. 2007. An update on the pharmacological treatment of obsessive-compulsive disorder. *Expert Opinion on Pharmacotherapy*, *8*(5), 563–583.

Fux, M., Levine, J., Aviv, A., & Belmaker, R. H. 1996. Inositol treatment of obsessive-compulsive disorder. *American Journal of Psychiatry*, *153*(9), 1219–1221.

Garcion, E., Sindji, L., Montero Menei, C., Andre, C., Brachet, P., & Darcy, F. 1998. Expression of inducible nitric oxide synthase during rat brain inflammation: Regulation by 1, 25-dihydroxyvitamin D3. *Glia*, *22*(3), 282–294.

Gilbody, S., Lightfoot, T., & Sheldon, T. 2007. Is low folate a risk factor for depression? A meta-analysis and exploration of heterogeneity. *Journal of Epidemiology & Community Health*, *61*(7), 631–637.

Goddard, A. W., Shekhar, A., Whiteman, A. F., & McDougle, C. J. 2008. Serotoninergic mechanisms in the treatment of obsessive–compulsive disorder. *Drug Discovery Today*, *13*(7–8), 325–332.

Godos, J., Castellano, S., Ray, S., Grosso, G., & Galvano, F. 2018. Dietary polyphenol intake and depression: Results from the mediterranean healthy eating, lifestyle and aging (meal) study. *Molecules*, *23*(5), 999.

Greenberg, W. M., Benedict, M. M., Doerfer, J., Perrin, M., Panek, L., Cleveland, W. L., & Javitt, D. C. 2009. Adjunctive glycine in the treatment of obsessive-compulsive disorder in adults. *Journal of Psychiatric Research*, *43*(6), 664–670.

Grisham, J. R., Anderson, T. M., & Sachdev, P. S. 2008. Genetic and environmental influences on obsessive-compulsive disorder. *European Archives of Psychiatry and Clinical Neuroscience*, *258*(2), 107–116.

Grosso, G., Micek, A., Marventano, S., Castellano, S., Mistretta, A., Pajak, A., & Galvano, F. 2016. Dietary n-3 PUFA, fish consumption and depression: A systematic review and meta-analysis of observational studies. *Journal of Affective Disorders*, *205*, 269–281.

Gunnell, D., & Ashby, D. 2004. Antidepressants and suicide: What is the balance of benefit and harm. *BMJ*, *329*(7456), 34–38.

Hashimoto, K., Ishima, T., Sato, Y., Bruno, D., Nierenberg, J., Marmar, C. R., Zetterbeg, H., Blennow, K., & Pomara, N. 2017. Increased levels of ascorbic acid in the cerebrospinal fluid of cognitively intact elderly patients with major depression: A preliminary study. *Scientific Reports*, *7*(1), 1–7.

Hoffmann, K., Emons, B., Brunnhuber, S., Karaca, S., & Juckel, G. 2019. The role of dietary supplements in depression and anxiety–A narrative review. *Pharmacopsychiatry*, *52*(06), 261–279.

Holton, K. F., & Cotter, E. W. (2018). Could dietary glutamate be contributing to the symptoms of obsessive-compulsive disorder? *Future Science OA*, *4*(3), FSO277.

Huang, Q., Liu, H., Suzuki, K., Ma, S., & Liu, C. 2019. Linking what we eat to our mood: A review of diet, dietary antioxidants, and depression. *Antioxidants*, *8*(9), 376.

Jacka, F. N., Overland, S., Stewart, R., Tell, G. S., Bjelland, I., & Mykletun, A. 2009. Association between magnesium intake and depression and anxiety in community-dwelling adults: The Hordaland Health Study. *Australian & New Zealand Journal of Psychiatry*, *43*(1), 45–52.

Jacka, F. N., Pasco, J. A., Henry, M. J., Kotowicz, M. A., Nicholson, G. C. & Berk, M. 2004. Dietary omega-3 fatty acids and depression in a community sample. *Nutritional Neuroscience*, *7*(2), 101–106.

Jaisoorya, T. S., Reddy, Y. J., Nair, B. S., Rani, A., Menon, P. G., Revamma, M., Jeevan, C. R., Radhakrishan, K. S., Jose, V., & Thennarasu, K. 2017. Prevalence and correlates of obsessive-compulsive disorder and subthreshold obsessive-compulsive disorder among college students in Kerala, India. *Indian Journal of Psychiatry*, *59*(1), 56.

Jaisoorya, T. S., Reddy, Y. J., Thennarasu, K., Beena, K. V., Beena, M., & Jose, D. C. 2015. An epidemiological study of obsessive compulsive disorder in adolescents from India. *Comprehensive Psychiatry*, *61*, 106–114.

Kaneko, I., Sabir, M. S., Dussik, C. M., Whitfield, G. K., Karrys, A., Hsieh, J. C., Haussier, M. R., Meyer, M. B., Pike, J. W., & Jurutka, P. W. 2015.1, 25-Dihydroxyvitamin D regulates expression of the tryptophan hydroxylase 2 and leptin genes: Implication for behavioral influences of vitamin D. *The FASEB Journal*, *29*(9), 4023–4035.

Karcı, C. K., & Celik, G. G. 2020. Nutritional and herbal supplements in the treatment of obsessive compulsive disorder. *General Psychiatry*, *33*(2).

Khanzode, S. D., Dakhale, G. N., Khanzode, S. S., Saoji, A., & Palasodkar, R. (2003). Oxidative damage and major depression: The potential antioxidant action of selective serotonin re-uptake inhibitors. *Redox Report*, *8*(6), 365–370.

Kheirouri, S., & Alizadeh, M. 2019. Dietary inflammatory potential and the risk of incident depression in adults: A systematic review. *Advances in Nutrition*, *10*(1), 9–18.

Kim, H., McGrath, B. M., & Silverstone, P. H. 2005. A review of the possible relevance of inositol and the phosphatidylinositol second messenger system (PI cycle) to psychiatric disorders—focus on magnetic resonance spectroscopy (MRS) studies. *Human Psychopharmacology: Clinical and Experimental*, *20*(5), 309–326.

Kim, T. H., Choi, J. Y., Lee, H. H., & Park, Y. 2015. Associations between dietary pattern and depression in Korean adolescent girls. *Journal of Pediatric and Adolescent Gynecology*, *28*(6), 533–537.

Kofman, O., Agam, G., Shapiro, J., & Spencer, A. 1998. Chronic dietary inositol enhances locomotor activity and brain inositol levels in rats. *Psychopharmacology*, *139*(3), 239–242.

Kurutas, E. B. 2015. The importance of antioxidants which play the role in cellular response against oxidative/nitrosative stress: Current state. *Nutrition Journal*, *15*(1), 1–22.

Lai, J. S., Hiles, S., Bisquera, A., Hure, A. J., McEvoy, M., & Attia, J. 2014. A systematic review and meta-analysis of dietary patterns and depression in community-dwelling adults. *The American Journal of Clinical Nutrition*, *99*(1), 181–197.

Lakhan, S. E., & Vieira, K. F. 2008. Nutritional therapies for mental disorders. *Nutrition Journal*, *7*(1), 2.

Lewin, A. B., Storch, E. A., Geffken, G. R., Goodman, W. K., & Murphy, T. K. 2006. A neuropsychiatric review of pediatric obsessive-compulsive disorder: Etiology and efficacious treatments. *Neuropsychiatric Disease and Treatment*, *2*(1), 21.

Li, Y., Lv, M. R., Wei, Y. J., Sun, L., Zhang, J. X., Zhang, H. G., & Li, B. 2017. Dietary patterns and depression risk: A meta-analysis. *Psychiatry Research*, *253*, 373–382.

Logan, A. C. 2005. Omega-3 and depression research: Hold the olive oil. *Prostaglandins, Leukotrienes, and Essential Fatty Acids*, *72*(6), 441.

Lotan, A., Fenckova, M., Bralten, J., Alttoa, A., Dixson, L., Williams, R. W., & van der Voet, M. 2014. Neuroinformatic analyses of common and distinct genetic components associated with major neuropsychiatric disorders. *Frontiers in Neuroscience*, *8*, 331.

Lyketsos, C. G. 2006. Lessons from neuropsychiatry. *The Journal of Neuropsychiatry and Clinical Neurosciences*, *18*(4), 445–449.

Madhura, T. K. 2015. Role of Oxidative Stress in the Pathogenesis of OCD. *Biochemistry and Analytical Biochemistry*, 4(4), 219.

Maidment, I. D. 2000. Are fish oils an effective therapy in mental illness—an analysis of the data. *Acta Psychiatrica Scandinavica*, *102*(1), 3–11.

Markus, C. R. 2007. Effects of carbohydrates on brain tryptophan availability and stress performance. *Biological Psychology*, *76*(1–2), 83–90.

McHugh, R. K., & Weiss, R. D. 2019. Alcohol use disorder and depressive disorders. *Alcohol Research: Current Reviews*, *40*(1).

McLean, A., Rubinsztein, J. S., Robbins, T. W., & Sahakian, B. J. 2004. The effects of tyrosine depletion in normal healthy volunteers: Implications for unipolar depression. *Psychopharmacology*, *171*(3), 286–297.

Menon, V., Kar, S. K., Suthar, N., & Nebhinani, N. 2020. Vitamin D and depression: A critical appraisal of the evidence and future directions. *Indian Journal of Psychological Medicine*, *42*(1), 11–21.

Mikolajczyk, R. T., El Ansari, W., & Maxwell, A. E. 2009. Food consumption frequency and perceived stress and depressive symptoms among students in three European countries. *Nutrition Journal*, *8*(1), 31.

Miles, L. 2007. Physical activity and health. *Nutrition Bulletin*, *32*(4), 314–363.

Mizunoya, W., Ohnuki, K., Baba, K., Miyahara, H., Shimizu, N., Tabata, K., Kino, T., Sato, Y., Tatsumi, R., & Ikeuchi, Y. 2013. Effect of dietary fat type on anxiety-like and depression-like behavior in mice. *Springerplus*, *2*(1), 165.

Moret, C., & Briley, M. 2011. The importance of norepinephrine in depression. *Neuropsychiatric Disease and Treatment*, *7*(Suppl 1), 9.

National Institute for Health and Care Excellence (NICE). 2009. *Depression in adults: Recognition and management clinical guideline*. 28 October 2009. www.nice.org.uk/guidance/cg90

Nestadt, G., Samuels, J., Riddle, M., Bienvenu, O. J., Liang, K. Y., LaBuda, M., Walkup, J., Grados, M., Hoehn-Saric, R. 2000. A family study of obsessive-compulsive disorder. *Archives of General Psychiatry*, *57*(4), 358–363.

Ng, F., Berk, M., Dean, O., & Bush, A. I. 2008. Oxidative stress in psychiatric disorders: Evidence base and therapeutic implications. *International Journal of Neuropsychopharmacology*, *11*(6), 851–876.

Ozdemir, E., Cetinkaya, S., Ersan, S., Kucukosman, S., & Ersan, E. E. 2009. Serum selenium and plasma malondialdehyde levels and antioxidant enzyme activities in patients with obsessive–compulsive disorder. *Progress in Neuro-Psychopharmacology and Biological Psychiatry*, *33*(1), 62–65.

Pakseresht, S., Boostani, H., & Sayyah, M. 2011. Extract of valerian root (Valeriana officinalis L.) vs. placebo in treatment of obsessive-compulsive disorder: A randomized double-blind study. *Journal of Complementary and Integrative Medicine*, *8*(1).

Pallanti, S., Grassi, G., Cantisani, A., Sarrecchia, E., & Pellegrini, M. 2011. Obsessive–compulsive disorder comorbidity: Clinical assessment and therapeutic implications. *Frontiers in Psychiatry*, *2*, 70.

Pasco, J. A., Jacka, F. N., Williams, L. J., Evans-Cleverdon, M., Brennan, S. L., Kotowicz, M. A., Nicholson, G. C., Ball, M. J., & Berk, M. 2012. Dietary selenium and major depression: A nested case-control study. *Complementary Therapies in Medicine*, *20*(3), 119–123.

Paul, L. A., Fulton, A. M., & Heppner, G. H. 1989. Reactive oxygen-mediated damage to murine mammary tumor cells. *Mutation Research/Fundamental and Molecular Mechanisms of Mutagenesis*, *215*(2), 223–234.

Pauls, D. L., Alsobrook, J. P., Goodman, W., Rasmussen, S., & Leckman, J. F. 1995. A family study of obsessive-compulsive disorder. *The American Journal of Psychiatry*.

Payne, M. E., Steck, S. E., George, R. R., & Steffens, D. C. 2012. Fruit, vegetable, and antioxidant intakes are lower in older adults with depression. *Journal of the Academy of Nutrition and Dietetics*, *112*(12), 2022–2027.

Penckofer, S., Byrn, M., Adams, W., Emanuele, M. A., Mumby, P., Kouba, J., & Wallis, D. E. 2017. Vitamin D supplementation improves mood in women with type 2 diabetes. *Journal of Diabetes Research*, *2017*. https://doi.org/10.1155/2017/8232863 (Online Published).

Pham, N. M., Nanri, A., Kurotani, K., Kuwahara, K., Kume, A., Sato, M., Hayabuchi, H., & Mizoue, T. 2014. Green tea and coffee consumption is inversely associated with depressive symptoms in a Japanese working population. *Public Health Nutrition*, *17*(3), 625–633.

Pittenger, C., Krystal, J. H., & Coric, V. 2006. Glutamate-modulating drugs as novel pharmacotherapeutic agents in the treatment of obsessive-compulsive disorder. *NeuroRx*, *3*(1), 69–81.

Rayman, M. P. 2000. The importance of selenium to human health. *The Lancet*, *356*(9225), 233–241.

Reddy, M. S. 2010. Depression: The disorder and the burden. *Indian Journal of Psychological Medicine*, *32*(1), 1.

Reichenberg, A., Yirmiya, R., Schuld, A., Kraus, T., Haack, M., Morag, A., & Pollmächer, T. 2001. Cytokine-associated emotional and cognitive disturbances in humans. *Archives of General Psychiatry*, *58*(5), 445–452.

Richardson, T. M., Friedman, B., Podgorski, C., Knox, K., Fisher, S., He, H., & Conwell, Y. 2012. Depression and its correlates among older adults accessing aging services. *The American Journal of Geriatric Psychiatry*, *20*(4), 346–354.

Rosenberg, D. R., MacMASTER, F. P., Keshavan, M. S., Fitzgerald, K. D., Stewart, C. M., & Moore, G. J. 2000. Decrease in caudate glutamatergic concentrations in pediatric obsessive-compulsive disorder patients taking paroxetine. *Journal of the American Academy of Child & Adolescent Psychiatry*, *39*(9), 1096–1103.

Ruhé, H. G., Mason, N. S., & Schene, A. H. 2007. Mood is indirectly related to serotonin, norepinephrine and dopamine levels in humans: A meta-analysis of monoamine depletion studies. *Molecular Psychiatry*, *12*(4), 331–359.

Sagar, R., Dandona, R., Gururaj, G., Dhaliwal, R. S., Singh, A., Ferrari, A., . . . Kumar, G. A. 2020. The burden of mental disorders across the states of India: The global burden of disease study 1990–2017. *The Lancet Psychiatry*, *7*(2), 148–161.

Sánchez-Villegas, A., Delgado-Rodríguez, M., Alonso, A., Schlatter, J., Lahortiga, F., Majem, L. S., & Martínez-González, M. A. 2009. Association of the Mediterranean dietary pattern with the incidence of depression: The Seguimiento Universidad de Navarra/University of Navarra follow-up (SUN) cohort. *Archives of General Psychiatry*, *66*(10), 1090–1098.

Sánchez-Villegas, A., Verberne, L., De Irala, J., Ruiz-Canela, M., Toledo, E., Serra-Majem, L., & Martínez-González, M. A. 2011. Dietary fat intake and the risk of depression: The SUN Project. *PLoS One*, *6*(1), e16268.

Sanchez-Villegas, A., Zazpe, I., Santiago, S., Perez-Cornago, A., Martinez-Gonzalez, M. A., & Lahortiga-Ramos, F. 2018. Added sugars and sugar-sweetened beverage consumption, dietary carbohydrate index and depression risk in the Seguimiento Universidad de Navarra (SUN) Project. *British Journal of Nutrition*, *119*(2), 211–221.

Sartori, S. B., Whittle, N., Hetzenauer, A., & Singewald, N. 2012. Magnesium deficiency induces anxiety and HPA axis dysregulation: Modulation by therapeutic drug treatment. *Neuropharmacology*, *62*(1), 304–312.

Sayyah, M., Boostani, H., Pakseresht, S., & Malaieri, A. 2009. Efficacy of aqueous extract of Echium amoenum in treatment of obsessive–compulsive disorder. *Progress in Neuro-Psychopharmacology and Biological Psychiatry*, *33*(8), 1513–1516.

Sharma, V., & Biswas, D. 2012. Cobalamin deficiency presenting as obsessive compulsive disorder: Case report. *General Hospital Psychiatry*, *34*(5), 578-e7.

Shohag, H., Ullah, A., Qusar, S., Rahman, M., & Hasnat, A. 2012. Alterations of serum zinc, copper, manganese, iron, calcium, and magnesium concentrations and the complexity of interelement relations in patients with obsessive–compulsive disorder. *Biological Trace Element Research*, *148*(3), 275–280.

Singer, H. S., Morris, C., & Grados, M. 2010. Glutamatergic modulatory therapy for Tourette syndrome. *Medical Hypotheses*, *74*(5), 862–867.

Singh, A., Trumpff, C., Genkinger, J., Davis, A., Spann, M., Werner, E., & Monk, C. 2017. Micronutrient dietary intake in Latina pregnant adolescents and its association with level of depression, stress, and social support. *Nutrients*, *9*(11), 1212.

Sirek, A., & Sirek, O. V. 1970. Serotonin: A review. *Canadian Medical Association Journal*, *102*(8), 846.

Slutsky, I., Abumaria, N., Wu, L. J., Huang, C., Zhang, L., Li, B., . . . Tonegawa, S. (2010). Enhancement of learning and memory by elevating brain magnesium. *Neuron*, *65*(2), 165–177.

Stockwell, T., & Chikritzhs, T. 2013. Commentary: Another serious challenge to the hypothesis that moderate drinking is good for health? *International Journal of Epidemiology*, *42*(6), 1792–1794.

Stokes, C. S., Grünhage, F., Baus, C., Volmer, D. A., Wagenpfeil, S., Riemenschneider, M., & Lammert, F. 2016. Vitamin D supplementation reduces depressive symptoms in patients with chronic liver disease. *Clinical Nutrition*, *35*(4), 950–957.

Su, K. P., Huang, S. Y., Chiu, C. C., & Shen, W. W. 2003. Omega-3 fatty acids in major depressive disorder: A preliminary double-blind, placebo-controlled trial. *European Neuropsychopharmacology*, *13*(4), 267–271.

Valizadeh, M., & Valizadeh, N. 2011. Obsessive compulsive disorder as early manifestation of B12 deficiency. *Indian Journal of Psychological Medicine*, *33*(2).

Wołonciej, M., Milewska, E., & Roszkowska-Jakimiec, W. 2016. Trace elements as an activator of antioxidant enzymes. *Postepyhigienyimedycynydoswiadczalnej (Online)*, *70*, 1483–1498.

World Health Organization (WHO). 2000. *Global burden of disease (GBD) estimates*. www.who.int/healthinfo/global_burden_disease/estimates_regional_2000/en/

World Health Organization (WHO). 2008. *The Global Burden of Disease 2004 Update*. World Health Organization, Geneva, Switzerland.

World Health Organization (WHO). 2017. *Depression and other common mental disorder global health estimates*. https://apps.who.int/iris/bitstream/handle/10665/254610/WHO-MSD-MER-2017.2-eng.pdf

Yang, L., Zhao, Y., Wang, Y., Liu, L., Zhang, X., Li, B., & Cui, R. 2015. The effects of psychological stress on depression. *Current Neuropharmacology*, *13*(4), 494–504.

Yary, T., & Aazami, S. 2012. Dietary intake of zinc was inversely associated with depression. *Biological Trace Element Research*, *145*(3), 286–290.

Yary, T., Aazami, S., & Soleimannejad, K. 2013. Dietary intake of magnesium may modulate depression. *Biological Trace Element Research*, *151*(3), 324–329.

15 Phytochemicals and Herbal Medicines

Potential Drug Candidates for Obsessive-Compulsive Disorder Treatment

*Neelakanta Sarvashiva Kiran, Chandrashekar Yashaswini, Gaonkar Lowkesh, Ketaki Range, and Renuka Madhu**
REVA University

15.1 INTRODUCTION

Obsessive-compulsive disorder (OCD) is a serious mental illness accompanied by severe anxiety, high disability, and disturbances in a person's social and work functioning (Angelakis et al., 2015). Obsessive-compulsive disorder is just a widespread, persistent, and long-lasting mental illness in which an individual has uncontrollable, recurrent thoughts (also known as "obsessions") and actions (also known as "compulsions") that they feel compelled to repeat (Wahl et al., 2010). At some point in time in their life, the majority of individuals have irrational thoughts and compulsive behavior, but this does not imply that everyone has "some OCD." This cycle of phobias must be so serious that it occupies a fair bit of time (and over an hour per day), causes severe distress, or interferes with significant activities that the individual values for an OCD diagnosis to be established (Lochner et al., 2011). The personality traits, underlying issues, family histories, and potential to benefit from treatment vary among OCD sufferers. The diversity of OCD symptoms has indeed been researched concerning several features of the disease, such as symptom patterns or temperamental factor clustering and comorbidity trends (Ivarsson et al., 2008). Different approaches to medical and psychological therapy are another significant area of OCD variability. So according to two investigations, the OCD subtype with hoarding differs in its lack of responsiveness. OCD may serve as an example of an anxiety illness when aspects of self-control (emotions and ideas) and world control are emphasized. According to a few theories, OCD is defined by a person's attempts to regulate their thoughts and the way they interact with the outside world through routines, as well as their anxiety over perhaps losing control of their very own thoughts and behavior (Moulding and Kyrios, 2006). Common symptoms or warning signs may include excessively looking for confirmation, opposing change; taking more time than necessary for a child's age doing tasks, getting ready, or having a meal; job redoing; reluctance to use bare hands on anything; excessive body, hand, and/or dishwashing, etc. According to cognitive theories, OCD patients overestimate the significance of typical intrusive thoughts, leading them to use neutralization approaches like compulsions to mitigate the

*Corresponding author: Renuka Madhu, REVA University. renuka.madhu@reva.edu.in

DOI: 10.1201/9781003387893-18

potential negative consequences. This, however, only serves to increase the intrusions' intensity and frequency (Hasler et al., 2005). The most common symptoms of obsessional-compulsive contamination issues and resultant chaos are fears about harming oneself or others when washing following checking.

Obsessions can range on a scale from ego-dystonic emotions to ego-syntonic delusional ideas from a dimensional perspective. According to several research studies, even adult OCD sufferers might have persistent delusional beliefs that their feared consequences would truly materialize (Thomsen, 2013). Obsessive-compulsive disorder (OCD), which is surprisingly widespread in the general population (2.5%–3%) and places a significant health-economic cost on society, may be diagnosed when these tendencies manifest as excessive and life-disrupting obsessions and compulsions. Recent research has revealed that the profile of teenage OCD is distinct, with more indication of learning and memory than of "executive" impairments. This does not mean that adolescent OCD displays less severe cognitive deficiencies than those commonly reported in adult studies using identical assessments (Robbins et al., 2019). Plants contain phytochemicals, and ingesting them often has positive health impacts. Because of their antioxidant and anti-inflammatory properties, phytochemicals may be useful in treating a variety of ailments, according to clinical trials and epidemiologic research. A variety of plants for this condition have been discovered as medicinal plants, which function by altering two key pathways, including the glutamatergic and serotonergic pathways, as possible sources for discovering novel pharmaceutical therapies for psychiatric conditions (Ayati et al., 2020). Polyphenols, which include flavonoids and nonflavonoids, including polyphenolic compounds, stilbene, curcumin, astaxanthin, and sesame lignin, are the primary phytochemicals that provide neuroprotection. Clinical investigations have shown that the high concentrations of flavonoids present in fruits, vegetables, tea, red wine, and chocolate, as well as the resveratrol contained in red wine, increase cognitive performance, albeit this has not yet been proven. By directly scavenging reactive nitrogen and oxygen species (ROS and RNS), upregulating ROS-removing enzymes, and downregulating radical-producing enzymes, flavonoids operate as potent antioxidants, so these are some of the bioactive compounds which help in psychiatric conditions like OCD and so on (Warshaneyan and Srivastava, 2019).

15.2 UNDERSTANDING OBSESSIVE-COMPULSIVE DISORDER (OCD)

OCD cannot presently be diagnosed biologically. Additionally, the pathogenesis and etiology of this illness are in psychopharmacology in OCD: preclinical and clinical data. Therefore, the prognosis of OCD is only established based on evidence that a certain behavioral profile, which defines what comprises OCD, exists. OCD is defined by the presence of obsessions and/or compulsions, according to the current research. Compulsions are repetitive behaviors or mental acts that a person feels compelled to carry out in response to an obsession or following rules that must be followed rigidly, whereas obsessions are repetitive and persistent opinions, urges, or images that are observed as intrusive and unwanted. Three key characteristics of obsessions and compulsions are classified as OCD, and for a broader clinical knowledge of OCD, it may be more beneficial to concentrate on the structure or form of OCD symptoms: (1) The person is overly preoccupied with certain kinds of thoughts (obsessions) and behaviors (compulsions), and this preoccupation disrupts normal functional ability; (2) the preoccupation does not bring the subject any pleasure, but the subject feels compelled to be so engaged; and (3) the person is aware that the weird obsession may appear to others as disconnected from objective reality (Szechtman et al., 2020).

This clinical impression challenged the previously held idea that OCD is resistant to pharmacotherapy, leading to significant and productive research on the use of medication to treat OCD in patients with and without concomitant depression. Like other TCAs, clomipramine inhibits the reuptake of serotonin and norepinephrine, but since it is a tertiary amine TCA, it has a stronger inhibitory effect on serotonin reuptake than secondary amine TCAs like nortriptyline and desipramine, putting out the idea that a serotonin reuptake inhibitor is necessary for a therapeutic outcome in OCD. The development of selective serotonin reuptake inhibitors (SSRIs), which have little to no impact on norepinephrine as well as dopamine reuptake despite obvious structural differences between compounds in this class, provided evidence in favor of this idea. Pharmacotherapy treatment studies identified three aspects of SSRI intervention that must be taken into account in a theory of OCD in addition to the fact that they target serotonin reuptake in OCD. First, unlike, for example, the benefits of morphine on pain reduction, OCD medication does not immediately reduce symptoms. Finally, regardless of the medication, dosage, or length of therapy, a sizable portion of individuals with OCD do not improve with SRI intervention. The hypothesis of OCD signs are mediated by much more than one neurochemical system in a popular interpretation of this study. Clinically, the ineffectiveness of monotherapy has prompted physicians to mix a different medication with an SSRI; this strategy is referred to as "adjunctive" or "augmentation" therapy, and that is what we will discuss next (Szechtman et al., 2020).

OCD has been associated with issues in the neurotransmitter serotonin, according to pathophysiology; however, it may either be a cause or a result of these abnormalities. Serotonin is believed to have a part in controlling anxiety. Serotonin must attach to the specific receptors on the adjoining nerve cell for it to transmit chemical signals from one part to another. The serotonin receptors in OCD patients may be comparatively understimulated, according to a theory. This idea is in line with the fact that many OCD sufferers gain from taking SSRIs, a family of antidepressants that increases the amount of serotonin that is easily accessible to other nerve cells. In addition to the first-line medication for OCD, often in the form of ERP, is also suggested as a first-line therapeutic option for OCD. However, here we focus on its application with SRI pharmacotherapy as an augmentation technique. Considering that just one in three patients undergoing SSRI monotherapy may benefit from antipsychotic augmentation, physicians investigated if pharmacological treatment combined with psychopathology therapy may offer OCD patients more relief. It turns out that doing an empirical analysis of this subject is difficult. Although not a systematic approach like ERP, therapy in clinical practice involves not just giving patients pills but also having personal contact with the healthcare practitioner that may be considered a sort of psychologic intervention. Explaining why ERP augment of clomipramine psychotherapy was clinically equal to ERP standard treatment is crucial from a theoretical viewpoint since it would show whether the effects of clomipramine and ERP are exerted on the same or largely distinct pathways causing OCD. Consequently, a solution was not made possible by the study's design. The patterns in the decline of YBOCS scores did, however, follow a different time course with clomipramine monotherapy compared to ERP monotherapy, which is notable. The empirical evidence from certain patients of a beneficial therapeutic effect with neuroleptic augmentation nevertheless supports this pharmacological action as a component with a potential impact on OCD, even if the required conditions have not yet been established. Dopamine system activity may be involved in the manifestation of the OCD phenotype given that neuroleptics inhibit dopamine receptors (Seeman, 1981). It is essential to employ a basic scientific research methodology that allows potential causes to be experimentally altered to test this hypothesis. Intriguingly, preclinical research has demonstrated that quinpirole, a dopamine agonist, chronically generates compulsion-like behaviors in rats that are likewise seen in OCD patients (Szechtman et al., 2020).

15.3 EPIDEMIOLOGY AND CURRENT SCENARIO OF OCD

The occurrence of obsessive-compulsive disorder (OCD) in the normal community has been estimated in different studies. It is divided into two categories: Previous studies in which only clinical judgement was used to make the diagnosis as well as current research in which the diagnosis was made using structured or semi-structured tools. Roth and Luton conducted the first clinical investigations that examined the incidence of OCD in Tennessee in 1942, and they revealed that approximately 0.3% of the 1,700 participants in their sample had an illness that is consistent with OCD (Roth and Luton, 1943). Using clinical judgement, Brunetti discovered that 1% of the population in rural Vaucluse, France, has OCD (Brunetti, 1975).

According to studies conducted over the past few decades, OCD affects anywhere from 1 to 4% of youth, teenagers, and adults worldwide, frequently having a serious adverse effect on health. Approximately four out of ten people who have OCD go on to develop it as a chronic condition, and many of them try to hide it from others. According to the World Health Organization (WHO), OCD is one of the top ten human disorders. Approximately eight out of every ten people under the age of 18 will develop OCD (Nazeer et al., 2020).

In the US, over 3 million individuals struggle with OCD. Overall life span frequency of a disease among both the adult and pediatric populations in the United States has been estimated to be between 3% and 1%. According to research, between 60% and 50% of all cases begin during infancy or youth. According to epidemiologic data from a nationwide survey in the United States, nearly 21% of OCD patients began before the age of 10, with men and girls experiencing the start of juvenile OCD around ages 9 to 11 and 11 to 13, respectively. In comparison to adults, where the male-to-female OCD prevalence ratio is approximately 1:1, pediatric OCD seems to affect guys more frequently than girls. In the United Kingdom, a 1999 countrywide clinical analysis of 10,438 kids between the ages of 5 and 15 yielded a weighted OCD prevalence estimate of 0.25%. OCD patients were more likely to be low IQ and socioeconomically disadvantaged than the general population (Rosenberg, 2019).

15.4 PHYTOCHEMICALS IN NEUROPSYCHOLOGICAL DISABILITIES

The neurological system of the human body is a complicated one that coordinates its operations as well as sends signals to other parts of the body and the surrounding environment to manage and regulate all critical processes. This system includes the sensory organs, spinal cord, and brain, as well as the nerves that link to the remaining organs. The nervous system is affected by approximately 600 neurological disorders. Neurodegenerative illnesses are among the most common and varied categories of neurological diseases (NDDs). Neurodegeneration is the indefinite impairment of cognitive, motor, sensory, and emotional abilities caused by ageing or environmental factors. The majority of the time, acute neurodegeneration recovers throughout the period, but chronic NDDs worsen over time as the illness develops. The most common chronic NDDs worldwide are amyotrophic lateral sclerosis (ALS), multiple sclerosis, Parkinson's disease, Alzheimer's disease (AD), and Parkinson's disease (PD) (Kavitha et al., 2020). According to Feigin et al., neurological illnesses appear to be the primary and second-leading cause of mortality and disability worldwide, and the extent of the related effect is likely to grow as the population ages (Feigin et al., 2020). As per contemporary estimates, neurological illnesses cause 6.8 million deaths annually (Siuly and Zhang, 2016).

Most neuropathological diseases and brain aging, which is the progressive malfunctioning of specific groups of neurons, include the neurodegenerative process. Pathology in the brain, such as cerebral infarction and neurodegenerative disease, which were widely recognized as major causes of death worldwide, accounted for approximately 2 deaths per 1,000 people and 8% of all deaths (Yadav, 2021). Numerous signaling and neurotransmitter molecules were identified over time and are now being explored as potential targets. On these targets, both traditional and more modern compounds have been tested. By affecting how the key inhibitory neurotransmitter receptors work, phytochemicals derived from medicinal plants play a crucial part in preserving overall chemical equilibrium within the brain. Numerous herbs have been implicated in the treatment of cognitive problems in conventional medicine; these researchers focused on select medicinal plants' important phytochemicals for neuroprotection constituents, such as phenols, saponins, flavonoids, alkaloids, terpenes, fatty acids, and alkaloids (Manjunatha et al., 2022). The activation of specific transcription factors and signaling pathways that increase neurons' tolerance to different stimuli appears to be a promising class of therapies for the treatment of multifactorial disorders. While the inclusion of polyphenol or other phytochemical transporters or receptors in herbal treatments targeting various brain areas is yet unclear (Scassellati et al., 2021).

Additionally, there has been a surge in interest in herbal medicines during the past ten years, whose phytochemical components seem to have long-term health-promoting or therapeutic effects. Several herbal medicines, on the other hand, have specialized medical effects in addition to contributing to human nutrition, which makes them useful resources for short- or long-term periods of time in responding to specific health issues. Fruits and vegetables include phytochemicals that significantly lower the total danger of a number of serious illnesses, especially cardiac disease, cancer, and neurological disorders. Mostly as a result, those who eat more fruits and vegetables might have a lower chance of contracting certain illnesses brought on by neural malfunction (Talebi et al., 2021).

Neural problems have traditionally been treated with herbal medication. Even though the specific modes of action of most medicinal herbs are still unknown, several researchers have demonstrated that they have antioxidant and anti-inflammatory effects in a variety of autonomic nervous systems. Anti-inflammatory herbal medicine and its contents are now being demonstrated to have effective neuroprotection activity against various brain disorders, as accumulating data suggests that persistent inflammatory reactions produced by neuroglia are an important pathogenic function in the central nervous system (Daniela et al., 2022). A useful source of unique compounds targeting effective treatments that have recently been identified through the structural variability of medicinal plants makes them useful for genomics, elevated screening, and proteomics (Srivastava et al., 2019).

15.5 HERBAL MEDICINE AND PHYTOCHEMICALS IN TREATMENT OF OCD

The population is significantly burdened by obsessive-compulsive disorder (OCD), a rather common mental illness. Despite having high efficacy for many individuals, modern conventional drugs can also cause several side effects. Numerous anxiolytic characteristics of plant-based substances have been discovered through their evaluation for various mental diseases (Table 15.1). More thorough research into plant-based treatments, either as potential monotherapies or augmentation strategies, is urgently needed in light of the partial effectiveness, potential side effects, and therapeutic delay in symptom improvement seen with conventional pharmacotherapeutic treatments for OCD. The findings of research on phytochemicals and herbs for treating OCD are promising, despite their early stages. Findings showed that a number of plant remedies, such as *Withania somnifera, Silybum marianum, Hypericum perforatum, Echium amoenum,* and *Crocus sativus*, as well as a number of natural compounds, such as cannabidiol, curcumin, and crocin, had preliminary evidence suggesting serotonin reuptake inhibition, a change in the monoamine pathway, perhaps the key OCD prevention mechanism (Ayati et al., 2020).

Unfortunately, there are no replicated experiments, limited sample numbers, and brief research times. Moreover, animal studies have only been used to evaluate specific herbal remedies, such as *Clitoria ternatea, Colocasia esculenta, Citrus aurantium*, and *Benicasa hispida*, in addition to every phytochemical that has been studied, including crocin, curcumin, and cannabidiol (Murugesan et al., 2022). Among the plants with the greatest research on OCD is *Hypericum perforatum*. In a similar randomized controlled trial (RCT), *Echium amoenum* was assessed, and the results supported its traditional applications by lowering both OCD and anxiety symptoms. Preliminary evidence suggests that some herbal remedies, including *Crocus sativus, Valeriana officinalis,* and *Echium amoenum*, may be effective in treating OCD due to their known psychopharmacological effects, which include neuroendocrine modulation, GABAergic effects, and reuptake inhibition of monoamines like serotonin and dopamine. Certain researched herbs, such as *Silybum marianum*, have anti-OCD mechanisms of action derived from phytomedicine substances. *Cannabidiol*, which

TABLE 15.1
List of Various Plant Species Along with Their Active Phytochemical Components and its Mechanism of Action Against OCD

Active Phytochemical	Plant Species	Mechanism of Action	References
Curcumin	*Curcuma longa*	Neurotransmitter modifications in norepinephrine, serotonin, and dopamine are increased.	Zhang et al., 2020
Carvacrol	*Origanum vulgare*	Modulate neurotransmitter levels to have an impact on neuronal activity.	Lee and Bae, 2017
Acreoline	*Areca catechu*	The active ingredient is known to reverse the effect of monoamines to the normal level, which is responsible for depression and other mental illnesses.	Abbas et al., 2013
Revesterol	*Vitis vinifera*	Boost brain 5-HT, dopamine, as well as noradrenaline levels.	de Oliveira et al., 2018
Ferulic acid	*Ananas comosus*	Reduce stress-related behaviors. Increased brain-derived neurotropic factor mRNA and CREB phosphorylation in the hippocampus.	Lee and Bae, 2017
Quercetin	*Salvia officinalis*	Prevent the HPA axis from becoming overstimulated.	Wang et al., 2021
Eugenol	*Ocimum tenuiflorum*	The monoamine oxidase enzyme, which is responsible for metabolizing the neurotransmitters serotonin and norepinephrine, is inhibited by eugenol.	Helmy et al., 2022
Proanthocyanidin	*Theobroma cacao*	Additionally, proanthocyanidin administration enhanced 5-HT levels in the frontal brain, hippocampus, and hypothalamus. The antidepressant properties of proanthocyanidin have been linked by some authors to the central monoaminergic neuronal function.	Álvarez et al., 2022
L-Theanine	*Camellia sinensis*	L-theanine injection boosts the hippocampus's (HIP) expression of brain-derived neurotrophic factor and modifies the action of brain neurotransmitters like 5-HT, DA, and gamma-aminobutyric acid.	Shen et al., 2019
1,8-cineole	*Lavandula angustifolia*	One of the best-known herbal remedies for mental problems like anxiety and depression is lavender. According to studies, the antidepressant as well as anxiolytic properties of lavender are connected to its impact just on neurotransmitter pathways, particularly because of its antagonistic effects on the N-methyl-D-aspartate receptor (NMDA-receptor) and blockage of the serotonin transporter (SERT).	Patil et al., 2022

is not precisely established and has a wide range of biological effects. It is widely acknowledged that many studies lack replication and have small sample sizes, and that the existing research consulted gives positive, preliminary proof that herbal medicines, including phytochemicals, could be used to cure OCD (Sarris et al., 2011). There may be some gaps in the clinical evidence for the remaining traditional medicines, despite the fact that the majority of them, including such species as *Silybum marianum* and *Crocus sativus*, were assumed to be safe within typical dosages. Regardless of high research rate, the sub chronic as well as the acute toxicity of many medicines and herbs has to be examined in clinical trials, evaluating safety involving humans is as essential as testing in animal models, reliably regarded for routine prescription administration (Ayati et al., 2021).

15.6 EFFECT OF DIET AGAINST OCD

Mental diseases are extremely common, incapacitating, expensive, and poorly managed. Psychiatric disorders are the most common reason for disability among those ages 14 to 45. The public's health depends on raising awareness of psychiatric disorders and expanding treatment choices. An increasing body of research, including the first randomized controlled trial, indicates that eating habits and food preferences may have an impact on the management and preventative measures of brain-based diseases, notably depression. This year saw the publication of the first dietary recommendations to avoid depression. They advise sticking to a conventional eating regimen like the Mediterranean diet, getting enough omega-3 acids, and staying away from processed meals like those packed with refined carbs or sugar (LaChance and Ramsey, 2018). The portion of the turmeric root that gives it its yellowish hue is called curcumin (*Curcuma longa*), and it is utilized as a non-toxic flavoring, preservative, and coloring ingredient in food and other products all over the world (Hewlings and Kalman, 2017). A polyphenol called curcumin has been proven to target numerous signaling molecules and also exhibit function at the cellular and molecular level, supporting its numerous health advantages. Curcumin has also been demonstrated to directly interact with a variety of signaling molecules and to function directly by influencing gene expression. Modern research has concentrated on curcumin's protective properties in many neurological and mental disorders. Volatile oils and the three main curcuminoids, curcumin, demethoxycurcumin, and bisdemethoxycurcumin, are the main components of turmeric roots. These three curcuminoids are commonly found in curcumin extracts, and it is believed that they are all to blame for the pharmacological actions of the extracts. Curcumin pre-administration prevented stress-induced declines in quantities of serotonin and then a serotonin metabolite, 5-hydroxyindoleacetic acid in rats with chronic unpredictable mild stress, as well as forced swimming test models. At large dosages, curcumin also boosted dopamine but, to a lesser extent, noradrenaline. These effects on serotonin seemed to be dose-dependent. Curcumin's impact on the serotonergic system might have been influenced by its interactions with 5-$HT_{1A/1B}$ and 5-HT_{2C} receptors (Seo et al., 2015).

Foeniculum (Fennel), a fruit which in Hindi is called "Saunf," and in Sanskrit, "Madhurika." It has slender leaves and tiny yellow blooms that are grouped. The use of *Foeniculum vulgare* in the food and beverage business, as well as the cosmetic industry, is due to its aromatic properties (Mehra et al., 2021). Although herb leaves and stalks are eaten like vegetables, seeds are thought to be the most valuable part. Prior research indicates that monoamine inhibitors raise the levels of norepinephrine, serotonin, and dopamine in the brain. *Foeniculum vulgare* is a monoamine inhibitor. Patients with depression have been found to have altered levels of several neurotransmitters, including dopamine, norepinephrine, and serotonin. The main phytochemicals of this plant have been identified as phenols, phenolic glycosides, and volatile fragrance compounds such as trans-anethole, estragole, and fenchone. For this widespread condition, many medications are being developed, including tricyclic antidepressants and selective serotonin reuptake inhibitors. Researchers' focus has shifted to natural herbal sources as a result of the negative effects connected with these medications (Abbas et al., 2020).

The herb *B. monnieri* (Linn.) *Penn* (Family- *Scrophulariaceae*), it is popularly referred to as Brahmi and can be found in marshy, moist, and humid areas throughout India and subtropical areas up to 1,000 meters above sea level. It is a plant that thrives in moist, damp, marshy environments and has a variety of medical benefits, including improving memory, cognitive disorders, concentration and alertness, and lowering anxiety. According to numerous accounts, Brahmi contains sterols, alkaloids, glycosides, saponins, and flavonoids. It is used as a nervine tonic and improves memory and learning. Alkaloids such as Brahmi, herpes time, a combination of three additional alkaloids, and saponins like bacoside A and B are the active ingredients. In the forced swim test as well as learned helpless models of depression, the standardized methanolic extract demonstrated considerable antidepressant efficacy, and so this effect was similar to the effects of imipramine (Mannan et al., 2015).

15.7 MECHANISM OF ACTION OF PHYTOCHEMICALS AND HERBAL MEDICINES AGAINST OCD

Different phytoextracts have been used to improve mood in people with behavioral disorders, while the mechanism of anti-OCD remains different for diverse range of phytochemicals. Most of the anti-OCD mechanism have not been clearly understood and deduced yet; subsequently, studies have shown that most of the phytochemical actions against OCD is by blocking the process of serotonin reabsorption (Figure 15.1). The occurrence of OCD has been identified due to hormonal imbalance, especially the decreased levels of serotonin and dopamine. Additionally, oxidative stress induced due to the improper functioning of the nitric oxide synthase [NOS] gene has also been proven to

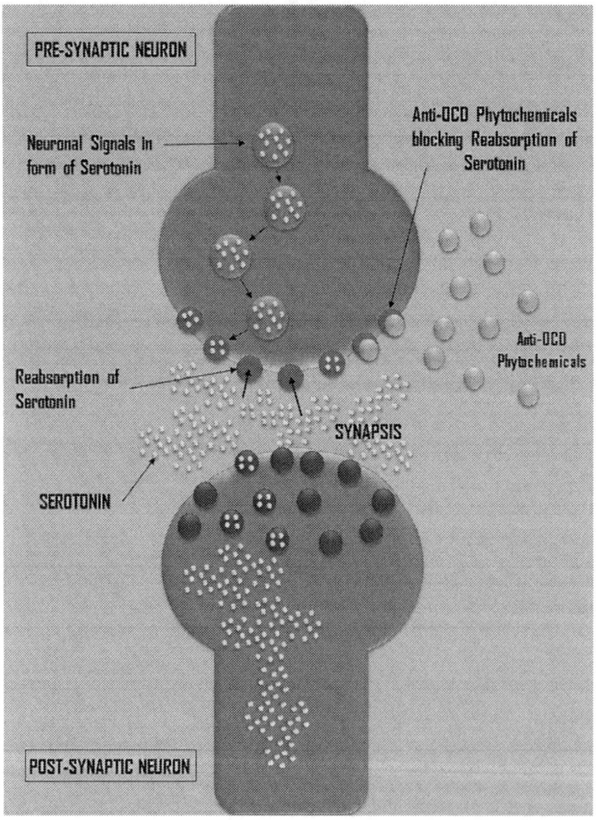

FIGURE 15.1 General mechanism of action of phytochemicals as anti-OCD agents.

trigger the onset of OCD. Recent studies have shown that certain phytoextracts acquired from plants such as *Benincasa hispida*, *Cannabis sativa*, and *Citrus aurantium* have the potential to enhance the serotonergic effects and increasing the rate of serotonin synthesis. Subsequently, these metabolic developments lead to increased consequence of serotonin on the synaptic clefts. This complete phenomenon can gradually lead to the decrease in effects and symptoms of OCD. Consequently, these extracts are also proven to increase the levels of norepinephrine and dopamine, which also play a vital role in its effect as anti-OCD agents.

Another study on *Cannabis sativa* has also been proven for its effects on modulating the expression of the NOS gene by inhibiting the synthesis of ROS-generating NAPDH oxidase, showing its indirect effect as an anti-OCD agent. Certain phytochemicals such as curcumin, silibinin, and hypericin possess anti-OCD property by inhibiting the activity of monoamine oxidase A and B, additionally increasing the levels of dopamine and norepinephrine and subsequently decreasing the levels of γ-aminobutyric acid (GABA) essentially required for the treatment of OCD. Numerous herbal plants and their phytochemical components have been vigorously explored and studied for its neuropsychiatric and anti-OCD properties, some proven to be highly efficient in their therapeutic activity. Though thousands of plants have been identified, their mechanism of anti-OCD activity has not been studied and understood completely.

15.8 FUTURE PROSPECTIVES OF PHYTOCHEMICALS AND HERBAL MEDICINES AGAINST OCD

Numerous fields of study are now included in this field's research, including inflammatory processes, computational psychiatry, neurocognitive endophenotypes, functional imaging, pharmacogenomics, and methods for brain stimulation. The need to constantly reevaluate the significance of research results and also to incorporate these discoveries into novel treatment techniques that are advantageous to patients and physicians as these conditions become more thoroughly understood. Even just the *Diagnostic Statistical Handbook of Mental Disorders, Fifth Edition* (*DSM-5*) is designed to be adaptable and integrate reliable and validated biomarkers as well as discoveries in the field of neuroscience (Prajapati et al., 2011). Comparatively, traditional knowledge-based approaches are proven to be more complex and secure. These natural remedies are believed to be more dependable in India and numerous other nations. This demonstrates that more than 60% of individuals globally place their trust in such antiquated methods. Its affordability is another crucial factor in its applicability. History records the usage of many plant components, such as roots, bark, seeds, leaves, fruits, etc., for medicinal purposes. It has been observed that even today's doctors and scientists are more concerned with how such a product may be employed in medicine (Rizwanullah et al., 2018). Ginkgo biloba, resveratrol, avonoids, and catechins, which have active ingredients that account for 80% of their composition, are only a few of the widely known documented facts that are employed in medicine regularly. There is a growing interest among pharmacologists, microbiologists, biochemists, botanists, and natural product chemists to discover medicinal plants for newer phytochemicals, resulting in the identification of drugs for the treatment of a variety of illnesses, as proven by the immense publications of biomedical research papers (Farooqui, 2013). The majority of these research exercises deal with phytochemicals separation, purification, bioanalytical methods, and characterization of their bioactive principles. Additionally, research into herbal medicine aims to clarify molecular structures, determine mechanisms of action, and identify potential toxicological qualities. This implies that the pharmaceutical industry, medical professionals, and researchers will go to developing nations like India, China, and others for their supply. Because these nations are the top exporters of raw herbal materials and have the widest variety of medicinal plant species. The use of plants in their whole or as pharmaceuticals will continue indefinitely in the future due to their rising appeal as a secure and less expensive alternative to traditional medicinal agents (Khan and Ahmad, 2019).

15.9 CONCLUSION

Several lifestyle choices support neurological system wellness. By placing a slight pressure on brain cells, the system becomes troubled. Phytochemicals found in foods, supplements, and plants and seasonings are a limitless supply of molecules accessible to enhance human health, especially the mental health. Since the beginning of time, herbal plants have been recognized as possible sources for treating illnesses. We must create technology to fundamentally bring and incorporate traditional herbal solutions in our medical services because the cost of healthcare services will only increase. Only the establishment of standardized herbal supplements will make this achievable. So here is a summary of the key anti-anxiety or anti-OCD plants, herbs, and their respective phytochemical constituents that are more effective and have less negative effects.

ACKNOWLEDGMENT

We the authors convey our sincere gratitude to REVA University for providing a platform for research.

CONFLICT OF INTEREST

We the authors declare no potential conflict of interest.

ORCID ID:

- Neelakanta Sarvashiva Kiran: https://orcid.org/0000-0002-2404-3254
- Chandrashekar Yashaswini: https://orcid.org/0000-0001-7415-748X
- Lowkesh Gaonkar: https://orcid.org/0000-0002-1241-8346
- Renuka Madhu: https://orcid.org/0000-0002-6864-8108

REFERENCES

Abbas, A., Ikram, R., Hasan, F., Adil, A., Nisar, U., and ul Ain, Q. 2020. Antidepressant and antiamnesic potential of Foeniculum vulgare. *Journal of Advances in Medicine and Medical Research*, *32*, 131–138.

Abbas, G., Naqvi, S., Erum, S., Ahmed, S., and Dar, A. 2013. Potential antidepressant activity of Areca catechu nut via elevation of serotonin and noradrenaline in the hippocampus of rats. *Phytotherapy Research*, *27*(1), 39–45.

Álvarez, S. A., Rocha-Guzmán, N. E., González-Laredo, R. F., Gallegos-Infante, J. A., Moreno-Jiménez, M. R., and Bravo-Muñoz, M. 2022. Ancestral food sources rich in polyphenols, their metabolism, and the potential influence of gut microbiota in the management of depression and anxiety. *Journal of Agricultural and Food Chemistry*, *70*(4), pp. 944–956.

Angelakis, I., Gooding, P., Tarrier, N., and Panagioti, M. 2015. Suicidality in obsessive compulsive disorder (OCD): A systematic review and meta-analysis. *Clinical Psychology Review*, *39*, 1–15.

Ayati, Z., Emami, S. A., Guillemin, G. J., Karamacoska, D., and Chang, D. 2021. 10 Persian herbal medicines used for brain health. In *Nutraceuticals in Brain Health and Beyond* (pp. 113–123). Academic Press.

Ayati, Z., Sarris, J., Chang, D., Emami, S. A., and Rahimi, R. 2020. Herbal medicines and phytochemicals for obsessive–compulsive disorder. *Phytotherapy Research*, *34*(8), 1889–1901.

Brunetti, P. M. 1975. Rural vaucluse: Two surveys on the prevalence of mental disorders. Summary of data. *Acta psychiatrica scandinavica*, *52*(S263), 12–15.

Daniela, M., Catalina, L., Ilie, O., et al. 2022. Effects of exercise training on the autonomic nervous system with a focus on anti-inflammatory and antioxidants effects. *Antioxidants*, *11*(2), 350.

de Oliveira, M. R., Chenet, A. L., Duarte, A. R., Scaini, G., and Quevedo, J. 2018. Molecular mechanisms underlying the anti-depressant effects of resveratrol: A review. *Molecular Neurobiology*, *55*(6), 4543–4559.

Farooqui, A. A. 2013. Perspective and direction for future research on phytochemicals in neurological disorders. In *Phytochemicals, Signal Transduction, and Neurological Disorders* (pp. 323–340). Springer.

Feigin, V. L., Vos, T., Nichols, E., et al. 2020. The global burden of neurological disorders: Translating evidence into policy. *The Lancet Neurology*, *19*(3), 255–265.

Hasler, G., LaSalle-Ricci, V. H., Ronquillo, J. G., et al. 2005. Obsessive–compulsive disorder symptom dimensions show specific relationships to psychiatric comorbidity. *Psychiatry Research*, *135*(2), 121–132.

Helmy, H., Sadik, N. A. H., Badawy, L., and Sayed, N. H. 2022. Mechanistic insights into the protective role of eugenol against stress-induced reproductive dysfunction in female rat model. *Chemico-Biological Interactions*, *367*, 110181.

Hewlings, S. J., and Kalman, D. S. 2017. Curcumin: A review of its effects on human health. *Foods*, *6*(10), 92.

Ivarsson, T., Melin, K., and Wallin, L. 2008. Categorical and dimensional aspects of co-morbidity in obsessive-compulsive disorder (OCD). *European Child & Adolescent Psychiatry*, *17*(1), 20–31.

Kavitha, R. V., Kumar, J. R., Egbuna, C., and Ifemeje, J. C. 2020. Phytochemicals as therapeutic interventions in neurodegenerative diseases. In *Phytochemicals as Lead Compounds for New Drug Discovery* (pp. 161–178). Elsevier.

Khan, M. S. A., and Ahmad, I. 2019. Herbal medicine: Current trends and future prospects. In *New Look to Phytomedicine* (pp. 3–13). Academic Press.

LaChance, L. R., and Ramsey, D. 2018. Antidepressant foods: An evidence-based nutrient profiling system for depression. *World Journal of Psychiatry*, *8*(3), 97.

Lee, G., and Bae, H. 2017. Therapeutic effects of phytochemicals and medicinal herbs on depression. *BioMed Research International*, *2017*, 6596241.

Lochner, C., Serebro, P., van der Merwe, L., et al. 2011. Comorbid obsessive–compulsive personality disorder in obsessive–compulsive disorder (OCD): A marker of severity. *Progress in Neuro-Psychopharmacology and Biological Psychiatry*, *35*(4), 1087–1092.

Manjunatha, E., Nandeesh, R., and Ahmed, S. M. 2022. Antiulcer and ulcer healing potential of some medicinal plants: A review. *Research Journal of Pharmacognosy and Phytochemistry*, *14*(1), 37–42.

Mannan, A., Abir, A. B., and Rahman, R. 2015. Antidepressant-like effects of methanolic extract of Bacopa monniera in mice. *BMC Complementary and Alternative Medicine*, *15*(1), 1–8.

Mehra, N., Tamta, G., and Nand, V. 2021. A review on nutritional value, phytochemical and pharmacological attributes of Foeniculum vulgare Mill. *Journal of Pharmacognosy and Phytochemistry*, *10*(2), 1255–1263.

Moulding, R., and Kyrios, M. 2006. Anxiety disorders and control related beliefs: The exemplar of obsessive–compulsive disorder (OCD). *Clinical Psychology Review*, *26*(5), 573–583.

Murugesan, V., Govindraj, R., Amarnath Satheesh, M., and Rajagopal, S. 2022. Phytonutrients in neurological disorders. In *Role of Nutrients in Neurological Disorders* (pp. 3–15). Springer.

Nazeer, A., Latif, F., Mondal, A., Azeem, M. W., and Greydanus, D. E. 2020. Obsessive-compulsive disorder in children and adolescents: Epidemiology, diagnosis and management. *Translational Pediatrics*, *9*(Suppl 1), S76.

Patil, S. M., Al-Mutairi, K. A., Firdose, N., Ramu, R., Martiz, R. M., and Ashwini, P. 2022. Pharmacoinformatics based screening discovers swertianolin from Lavandula angustifolia as a novel neuromodulator targeting epilepsy, depression, and anxiety. *South African Journal of Botany*, *149*, pp. 712–730.

Prajapati, R. P., Kalaria, M. V., Karkare, V. P., Parmar, S. K., and Sheth, N. R. 2011. Effect of methanolic extract of Lagenaria siceraria (Molina) Standley fruits on marble-burying behavior in mice: Implications for obsessive–compulsive disorder. *Pharmacognosy Research*, *3*(1), 62.

Rizwanullah, M., Amin, S., Mir, S. R., Fakhri, K. U., and Rizvi, M. M. A. 2018. Phytochemical based nanomedicines against cancer: Current status and future prospects. *Journal of Drug Targeting*, *26*(9), 731–752.

Robbins, T. W., Vaghi, M. M., and Banca, P. 2019. Obsessive-compulsive disorder: Puzzles and prospects. *Neuron*, *102*(1), 27–47.

Rosenberg, D. 2019. Obsessive-compulsive disorder in children and adolescents: Epidemiology, pathogenesis, clinical manifestations, course, assessment, and diagnosis. *Uptodate.com*

Roth Jr, W. F., and Luton, F. H. 1943. The mental health program in Tennessee: I. Description of the original study program, II. Statistical report of a psychiatric survey in a rural county. *American Journal of Psychiatry*, *99*(5), 662–675.

Sarris, J., Panossian, A., Schweitzer, I., Stough, C., and Scholey, A. 2011. Herbal medicine for depression, anxiety and insomnia: A review of psychopharmacology and clinical evidence. *European Neuropsychop harmacology*, *21*(12), 841–860.

Scassellati, C., Galoforo, A. C., Esposito, C., et al. 2021. Promising intervention approaches to potentially resolve neuroinflammation and steroid hormones alterations in Alzheimer's disease and its neuropsychiatric symptoms. *Aging and Disease*, *12*(5), 1337.

Seeman, P. 1981. Brain dopamine receptors. *Pharmacological Reviews*, *32*, 229–313.

Seo, H. J., Wang, S. M., Han, C., et al. 2015. Curcumin as a putative antidepressant. *Expert Review of Neurotherapeutics*, *15*(3), 269–280.

Shen, M., Yang, Y. I., Wu, Y., et al. 2019. L-theanine ameliorate depressive-like behavior in a chronic unpredictable mild stress rat model via modulating the monoamine levels in limbic–cortical–striatal–pallidal–thalamic-circuit related brain regions. *Phytotherapy Research*, *33*(2), 412–421.

Siuly, S., and Zhang, Y. 2016. Medical big data: Neurological diseases diagnosis through medical data analysis. *Data Science and Engineering*, *1*(2), 54–64.

Srivastava, A., Srivastava, P., Pandey, A., Khanna, V. K., and Pant, A. B. 2019. Phytomedicine: A potential alternative medicine in controlling neurological disorders. In *New Look to Phytomedicine* (pp. 625–655). Academic Press.

Szechtman, H., Harvey, B. H., Woody, E. Z., and Hoffman, K. L. 2020. The psychopharmacology of obsessive-compulsive disorder: A preclinical roadmap. *Pharmacological Reviews*, *72*(1), 80–151.

Talebi, M., Talebi, M., Farkhondeh, T., et al. 2021. An updated review on the versatile role of chrysin in neurological diseases: Chemistry, pharmacology, and drug delivery approaches. *Biomedicine & Pharmacotherapy*, *141*, 111906.

Thomsen, P. H. 2013. Obsessive–compulsive disorders. *European child & Adolescent Psychiatry*, *22*(1), 23–28.

Wahl, K., Kordon, A., Kuelz, K. A., et al. 2010. Obsessive-Compulsive Disorder (OCD) is still an unrecognised disorder: A study on the recognition of OCD in psychiatric outpatients. *European Psychiatry*, *25*(7), 374–377.

Wang, G., Li, Y., Lei, C., et al. 2021. Quercetin exerts antidepressant and cardioprotective effects in estrogen receptor α-deficient female mice via BDNF-AKT/ERK1/2 signaling. *The Journal of Steroid Biochemistry and Molecular Biology*, *206*, 105795.

Warshaneyan, S. S., and Srivastava, P. 2019. Analysis and subsequent molecular docking of selected phytochemicals with SLC6A3 and SLC6A4 as potential therapeutic agents for Obsessive-Compulsive Disorder (OCD). *BioRxiv*, 776922.

Yadav, D. K. 2021. Potential therapeutic strategies of phytochemicals in neurodegenerative disorders. *Current Topics in Medicinal Chemistry*, *21*(31), 2814–2838.

Zhang, Y., Li, L., and Zhang, J. 2020. Curcumin in antidepressant treatments: An overview of potential mechanisms, pre-clinical/clinical trials and ongoing challenges. *Basic & Clinical Pharmacology & Toxicology*, *127*(4), 243–253.

16 Nutrition and Obsessive-Compulsive Disorder

Subitha R. and Aparna K.S.*
Department of Biotechnology, Sree Narayana Guru College, Coimbatore, Tamilnadu, India

16.1 INTRODUCTION

Obsessive-compulsive disorder (OCD) is a chronic neuropsychiatric anxiety disorder until about two decades ago characterized by presence of two distinct phenomena: obsessions and compulsions. Obsessions are repetitive thinking, impulses, or mental resemblances causing anxiety. Compulsions are tedious behaviors that a person with OCD feels a craving to do in response to an obsessive thought. The four main grades of OCD behaviors are acting compulsively, seeking reassurance, avoiding triggers, and mental compulsions.[1]

16.1.1 Obsessions

Obsessions are uncontrollable mental recurrences of thoughts, pictures, or impulses. Although you don't want to think these thoughts, you cannot stop them. Sadly, these intrusive thoughts are frequently upsetting and distracting.[2] For example:

- Fear of losing an important thing, being harmed by loved ones, and contamination by the people or environment.
- Extreme worry and concern with order or precision.
- Unwanted intrusive sexual thoughts or images.
- Showing lack of reverence for God or religion.

16.1.2 Compulsions

Compulsions are repetitive activities or rituals that you feel compelled to perform.[3] Compulsions are frequently carried out in an effort to rid oneself of obsessions.[4] For instance:

- Repeated handwashing, bathing, brushing teeth, or toileting.
- Organizing things in a particular way.
- Often cleaning home.
- Recurrent checking of objects and appliances.

*Corresponding author: Subitha R, Assistant Professor, Department of Biotechnology, Sree Narayana Guru College, Coimbatore, TN, India. Mob: +91 8129117289, Email: r.subithabiotech87@gmail.com

DOI: 10.1201/9781003387893-19

- Frequently seeking assurance or approval.
- Ritual concern to numerals.
- Avoiding certain peoples, places, or situations.

16.2 HOW WIDESPREAD IS OCD?

About 2.3% of the world's population, or around 1 in 40 adults and 1 in 100 children, suffer with OCD. In India, prevalence rates of OCD is lower in the range of 2 to 3% (0.6%).[5]

The commonness of this condition is elevated in females (1.8%) than males (0.5%) in a period of 12 smonth.[6] Scientific research revealed that the early onset of OCD is before the age of 19.5 years, wherein males make up majority of the cases before the age of 10.[7, 8] But females are diagnosed during adolescence.

There is a higher risk rate of attention deficit hyperactivity disorder (ADHD) and bipolar disorder in people with early onset of OCD. Molecular study reveals that there is a chance of 25% inheritance in OCD. OCD and its coexistence to mental health includes disorders such as anxiety, panic, phobias, mood disorder, and post-traumatic stress disorder (PTSD) (75.8%). Major depressive disorder, bipolar disorder (63.3%), inability to maintain self-control, including ADHD (55.9%) and SUD (substance use disorders) (38.6%) are the patterns of mood disorders.[9]

16.3 ETIOLOGY OF OCD

The reason of OCD's etiology cannot be pinpointed. Instead, they are the outcome of the interaction of numerous risk factors.

- **Genetics**: OCD is a polygenic disorder with many identified risk loci of small effect.[10, 11] Numerous potentially important genes were discovered through molecular investigations, which contribute to the etiology. It's likely that the disorder has a complicated pattern of inheritance.
- **Psychological factors:** It is marked by chronic intrusive thoughts and images accompanied by ritualistic repetitive behaviors.
- **Neurobiological factors:** OCD results due to incapability and dysfunction of neurotransmitters such as serotonin, dopamine, glutamate, glutamine, and gamma-aminobutyric acid (GABA). Serotonin deficiency, elevated intensity of glutamate, glutamine, GABA, and also the glutamatergic ionotropic N-methyl D-aspartate (NMDA) receptor has been associated with OCD.[12]
- **Illnesses during childhood**: In children, a streptococcus infection can result in sudden onset of OCD.[13]
- **Other illnesses or injuries**: Injury to brain by means of any physical or biological factors such as traumatic brain injuries, certain viruses, and brain tumors correlate to OCD.

Apart from these, structural and immunological factors also lead to etiology and pathophysiology of OCD.

16.4 PREVENTIVE MEASURES OF OCD

Scientists and mental health experts believe that while OCD cannot be prevented from developing in the first place, patients may be able to control their symptoms once they have received a diagnosis. In order to manage their symptoms, they must identify any circumstances or events that may cause them, as well as keep a diary of them. If their symptoms worsen, they must seek medical attention right away. The patient should also continue taking all of their prescriptions as prescribed. It is extremely dangerous for patients to stop taking their medications when their symptoms seem to improve or disappear. Additionally, the patient should keep taking all of their prescriptions as directed. Patients should not stop taking their prescriptions when their symptoms appear to get better or go away since doing so is exceedingly harmful. About 20% of patients no longer respond to medications that once worked for them because they stopped taking them as directed.[14]

16.5 TREATMENTS OF OCD

Although OCD cannot be cured, it can be effectively controlled with therapies that combine medication and psychotherapy, depending on the patient's individual needs and symptoms. Together, these therapies help patients perform better and have less severe symptoms over time.

The mentioned treatments for OCD are listed here:

- **Cognitive-behavioral therapy (CBT)[15] and mindfulness:** While mindfulness helps one examine feelings and ideas objectively, CBT assists in identifying, labelling, and reframing the patient's thoughts.[16]
- **Selective serotonin reuptake inhibitors (SSRIs):** Prozac and Lexapro medications are commonly prescribed as daily medications to OCD patients, and it will take 10 to 12 weeks before there is a visible change in its symptoms. Recovery rate of this drug therapy is about 40 to 60%.
- **Exposure and response prevention (ERP):** Tolerating the anxiety and discomfort that come with obsessions is one of the most effective ways to teach the brain how to react to them differently.

The following treatments had very limited efficacy, and potential adverse effects have prompted researchers to look for substitute tactics. It was also mentioned that patients with mental illnesses had numerous dietary inadequacies found in them. As a result, nutritional and herbal supplements are regarded as one of the most successful OCD treatments.[17, 18]

16.6 NUTRITIONAL THERAPY

According to the Association for Comprehensive Neurotherapy, nutritional diet may benefit in reducing OCD symptoms.[19] A healthy diet, together with medicine, nutritional supplements, and mental health therapy, can help manage the symptoms but may not be able to stop the onset of OCD.[20] Here are a few food groups that have been proven to be helpful in easing OCD symptoms:

- Vitamin B12
- Folic acid
- Homocysteine
- N-acetyl cysteine
- Vitamin D
- Glycine
- Trace elements
- Tryptophan
- Inositol
- L-theanine
- St. John's wort
- Valerian root
- Flax seed oil
- Vitamin B Complex
- Spirulina
- Magnesium

- Lactium
- Whole grains
- Low-fat dairy product
- Fruits and vegetables
- Caffeine
- Probiotics
- Curcumin
- Ashwagandha
- Milk thistle
- Borage Oil
- Saffron

16.6.1 Vitamin B12, Folic Acid, and Homocysteine

Vitamin B12 is a cobalt containing complex organic compound. Folic acid is seen in the human body in a class of water-soluble vitamins and helps in normal daily activities such as sleep, memory, and exercise.[21] Homocysteine is a type of nonessential aminoacid, and it contains sulfur. Low levels of folic acid and vitamin B12 that correlated with elevated levels of homocysteine result in OCD.[22, 23] Previous studies have reported that Vitamin B12, folic acid, and homocysteine take part in the metabolism of carbon transfer, which maintains the protein methylation, phospholipids of neural membrane, and neurotransmitters that include serotonin and monoamines.[24, 25] Due to vitamin B12 and folic acid deficiency, there occurs a drop in neurotransmitter level.[26, 27, 28] Patients with high homocysteine exhibit a wide range of neurological abnormalities that include mental retardation, cerebral atrophy, and seizures. Also, studies reported that high levels cause apoptosis triggered by DNA damage, defective mitochondria due to mitochondrial dysfunction, increased oxidative stress.[29, 30] So OCD can be controlled by having an adequate amount of nutrition. So patients with Vitamin B12 and folic acid deficiency should eat foods rich with folic acid and vitamin B12. Good sources of folic acid are broccoli, brussels sprouts, liver, spinach, asparagus, peas, chickpeas, brownrice, fortified breakfast cereals, meat, salmon, cod, milk, cheese, eggs,. Good sources of Vitamin B12 are meat cod, milk, cheese, eggs, some fortified breakfast cereals, and salmon.

16.6.2 N-Acetyl Cysteine

N-acetyl cysteine is a sulfur-containing aminoacid derived from L-Cysteine. It has antioxidant property, which helps to metabolize glutathione in the body and also interacts with neurotransmitters and hormones such as glutathione and dopamine in the brain.[31, 32] These neurotransmitters in turn carry chemical messages from neurons to various organs of the body. In patients with OCD will have high level of glutamate. So NAC binds to glutamate and reduces its elevated level.[33] Numerous important actions of NAC have been reported, such as antioxidant effects, cytokine activity reduction, modulation of dopamine release, reversal of mitochondrial dysfunction, reductions in apoptosis and ferroptosis, anti-inflammatory activity, increased neurogenesis, and increased glutamate release.[34] Its antioxidant activity, which it achieves through a variety of methods, is perhaps the most crucial of these for the treatment of psychiatric diseases.[35] NAC reduces glutamate transmission by promoting glutathione production, boosting glutathione-S-transferase activity, scavenging free radicals, and activating group II metabotropic glutamate receptors.[36, 37] Glutathione is the important naturally occurring antioxidant in the brain. The availability of L-cysteine regulates its production rate. Supplementing with NAC raises the level of L-cysteine in the body.

Antioxidant reduces oxidative stress, which is related with OCD.[38, 39] Antioxidants reduce oxidative stress, which is linked to the emergence of many psychiatric illnesses. An imbalance between tissue redox defenses and reactive oxygen species like hydrogen peroxide, superoxide, and peroxynitrite is known as oxidative stress. Increased reactive oxygen species, weakened antioxidant defenses, or unrepaired oxidative damage are possible causes.[40, 41] Cellular lipid peroxidation, essential enzyme inactivation, respiratory chain dysfunction, and DNA damage are all brought on by reactive oxygen species.[42, 43] Superoxide dismutase, glutathione reductase, and peroxidase are examples of antioxidant enzymes that convert reactive oxygen species into less harmful molecules, shielding the brain from the negative effects of oxidative stress.[44, 45] Through a rise in glutathione disulfide production, severe sustained oxidative stress can cause glutathione depletion. The antioxidant property of NAC, which lowers stress level by reacting with reactive oxygen species.[2] NAC reduces the inflammatory chemicals such as cytokines, which is the reason for disease like depression which worsen OCD. NAC increases the dopamine levels, which has an important role in movements and emotional responses. OCD can be treated by consuming NAC rich foods. NAC rich foods are lean pork chops, skirt steak beef, lean chicken breast, tuna, lentils, oatmeal, eggs, low-fat yogurt, sunflower seeds, swiss cheese.[34]

16.6.3 Vitamin D

Vitamin D (calciferol), a fat-soluble vitamin obtained from few foods and supplements and also produced endogenously from sunlight.[46] Vitamin D has more roles, including immune distortion, inflammatory responses, antioxidant processes, reduction of inflammation, brain development and functioning, neurotransmission, neuroprotection, proliferation, and differentiation.[47, 48] Origin of synthesis of vitamin D at epidermis is in the form of vitamin D3 (cholecalciferol), which is in inactive form, which should get converted into 25-OH D3 in the liver by hydroxylation.[49] This 25-OH D3 gets into active form called 1,25-dihydroxy-vitamin D3 by rehydroxylation in the kidney.[50] Deficiency of vitamin D causes various neuropsychiatric diseases such as autism, schizophrenia, and OCD.[51, 52] There exists a major relationship between pathophysiology of OCD and vitamin D.[53] 1,25-dihydroxy-vitaminD3 regulates an enzyme called tyrosine/tryptophan hydroxylase, which are the rate-limiting enzymes indopamine, epinephrine, norepinephrine synthesis, and in serotonin synthesis.[54, 55] From the mentioned mechanism, it is clearly understood that deficiency of vitamin D affects the serotonin and catecholamine pathway, which leads to etiology of OCD.[53] Various studies demonstrated that OCD patients have low level of vitamins, which can be controlled through vitamin D supplemented foods such as eggs, mushrooms, milk, fatty fish, fortified yogurt.

16.6.4 Glycine

Glycine is a type of aminoacid obtained through the dietary intake. It is used to synthesize glutathione, an important antioxidant in the body. It can reduce anxiety. Anxiety often rises when NMDA receptors are overworked. N-methyl-D-aspartate (NMDA) is a subcategory of glutamate receptor found in central nervous system.[56, 57] The function of glycine is to bind to these overactive NMDA receptors, which results in reduced anxiety symptoms.[58, 59] Similar to glycine, sarcosine is a different class of amino acid. An endogenous antagonist of glycine transporter-1 is sarcosine. Sarcosine may improve NMDA by increasing synaptic glycine availability and blocking glycine absorption. As a result, the brain will have access to glycine. Therefore, consuming foods high in glycine and sarcosine, such as meat, fish, collagen, and gelatin can lessen OCD.

16.6.5 Trace Elements

16.6.5.1 Selenium

Selenium is a crucial trace element which is naturally present in the body.[60] Numerous enzymes and proteins known as selenoproteins, which help in the synthesis of DNA, protect against cellular damage, and control the metabolism of thyroid hormones, depend on the mineral selenium for their formation. The majority of selenium in the body is stored in muscle tissue, even though the thyroid gland has the highest concentration of selenium due to multiple selenoproteins that support thyroid function. The etiology of OCD may be correlated with imbalances in trace elements. A crucial trace element called selenium contributes to glutathione peroxidase's antioxidant abilities. Researcher found that serum selenium levels were observed to be decreased in OCD patients. They noted an inverse relationship between plasma selenium levels and plasma glutathione peroxidase activity. So they recommended selenium-rich foods such as Brazil nuts, fin fish, shellfish, beef, turkey, chicken, fortified cereals, wholewheat bread, beans, lentils may reduce the OCD symptoms.[61, 62]

16.6.5.2 Zinc

Zinc is a trace element needed for many enzymes to carry out important chemical reactions. It is a main precursor for gene expression, growth of cells, protein building, wound healing, and supportive for immune system. It acts as a signaling molecule in the central nervous system and brain.[63] The modulation of inhibitory and excitatory synaptic transmissions by zinc is significant. Zinc may also be thought of as one of the brain's defenses against oxidative stress. Changes in brain zinc homeostasis have been linked to both some neurologic illnesses (such as epilepsy and Alzheimer's disease) and behavioral abnormalities like anorexia, dysphoria, poor learning, and cognitive performance.

Zinc is an essential mineral for mental health, particularly the one with chronic anxiety. Researchers discovered that 2 billion people worldwide—mostly those with OCD symptoms—are zinc deficient. One study found that taking a zinc supplement reduced obsessions and compulsions without having any negative side effects.[64,65] Oysters, grass-fed beef, pumpkin seeds, cashews, mushrooms, and spinach are some of the finest foods to increase your zinc levels.

16.6.5.3 Iron

In our bodies, every living cell contains traces of iron.[66] All of your body's cells need it to transport oxygen, so low amounts can make you feel worn out, pale, irritated, and foggy. Studies do, however, indicate that iron is essential for the production of serotonin and that low iron levels increase the risk of anxiety disorders. Additionally, researchers discovered that people with mild and moderate OCD have significantly decreased blood iron levels.[67]

16.6.6 Tryptophan

Tryptophan, an essential amino acid is a neurotransmitter needed for growth and maintenance of the body's proteins, molecules, enzymes. Tryptophan encourages the production of serotonin, a signaling chemical that transmits signals between brain and body nerve cells. It is crucial to bodily processes, including regulating mood, sleep, digestion, nausea, wound healing, sexual desire, and anxiety disorders, among others. Tryptophan is the main precursor for the formation of monoaminergic neurotransmitter serotonin (5-hydroxytryptamine, 5-HT). It has an effective role in treating OCD, as serotonin is effective in treating anxiety disorders. Tryptophan is formed from progenitor coenzymes found in VitaminB. Some example of foods with high tryptophan level include elk meat, goat, seaweed, soy, spinach, crabs, halibut, and shrimps.[68]

16.6.7 Inositol

All plants and animals possess the naturally occurring element inositol. It is a simple polyoil second messenger precursor, sugar alcohol, and an isomer of glucose has a vital role in many biological actions.[69] It is found in cell membrane as inositol phospholipids, which is mainly important in the brain for cell signaling and communication and has a role in modulating neurotransmitter signaling and pathways, including serotonin.[70, 71] It is a naturally occurring substance that is related to the water-soluble vitamin B complex and can alter neuronal communication in addition to carrying lipids throughout the body.[72] It is not a necessary nutrient because our bodies can make inositol from glucose on their own.[73] Foods including oranges, nuts, fruits, cantaloupe, beans, and whole wheat grain contain it. However, taking inositol supplements can assist to lessen OCD symptoms. Patients with OCD were given 18 grams of inositol daily for six weeks as part of a trial. At the conclusion of the course time, they discovered a noticeably decreased OCD score.[74, 75]

16.6.8 L-Theanine

L-theanine, an aminoacid, mainly present in tea leaves and mushroom and is used to treat many medical and mental health problems including stress and anxiety. Having L-theanine on a regular basis improves attention, enhances mood, and reduce symptoms of OCD and panic disorder. Additionally, it aids in boosting the alpha wave brain's dopamine, GABA, and serotonin levels, which encourage profound relaxation. L-theanine also possesses qualities that can defend against neurotoxins found in the environment. This amino acid is present in black tea, green tea, and boy boletes. Due to its part in the oxidative damage brought on by "free radicals," it accounts for all the features.

16.6.9 St. John's Wort

Hypericum perforatum, sometimes known as St. John's wort, is a popular herbal remedy for depression.[39, 76] This herb has long been used to treat mental illnesses including depression.[77] Additionally, it contains hypericum, a substance that has been shown to be efficient at modifying serotonin levels.[78] A new, successful open-label research and pharmacokinetic data supporting hypericum's serotonergic, domaminergic, and GABA minergic action raise the possibility that it may be useful for treating OCD.[79]

16.6.10 Valerian Root

A perennial herb called valerian was utilized as a perfume in the sixteenth century. Aleuronic acid, which is found in valerian roots, is linked to GABAA receptor regulation.[80] The substance valepotriates, which is also present in valerian roots, is useful for treating severe anxiety's psychotic side effects. Due to its calming and pain-relieving properties, valerian is most frequently utilized in the treatment of insomnia and migraines. Valerian extract has only been tested in one double-blind, placebo-controlled study for the treatment of OCD.[81] They discovered that valerian extract was superior to a placebo in terms of treating OCD.[82] They also saw that the effects of the extract began to take effect right away (28, 42, 56 days). One significant benefit of this medication was the low prevalence of libido impairment.

16.6.11 Supplements

Natural supplements clearly outperform pharmaceuticals and treatments. They give the brain more starting materials for the production of serotonin. However, if the person is deficient in one or more important nutrients required for the production of serotonin, they may become unable. Several OCD patients have reported success using the following well-known supplements.

16.6.11.1 Flax Seed Oil

It contains tryptophan and omega-3 fatty acids (alpha-linolenic acid ALA), which stimulates serotonin production. According to numerous research, consuming flaxseed oil can help control OCD behavior. It contains a lot of important fatty acids, which improve the connection and growth of brain cells.

16.6.11.2 Vitamin B Complex

OCD is associated mainly with low serotonin levels in the brain.[83] Vitamins intake helps to increase serotonin that reduces the symptoms associated with OCD and found helpful for those suffering from panic disorders. Vitamin B-3 increases tryptophan, precursor in serotonin synthesis. Vitamin B-6 helps in production of serotonin and norepinephrine, which is significant in calming mood and obsessive moods.[84] Vitamin B12 and folic acid work against heart stress for those suffering from anxiety and depression.

16.6.11.3 Spirulina

It is well-known that this dietary supplement contains RNA, which has been found to benefit the nervous system. It aids in nourishing the nerves and myelin sheaths, which help to lessen OCD symptoms.

16.6.11.4 Magnesium

Magnesium—Magnesium is a relaxing mineral that aids in reducing the body's production of stress hormones. Although research on the benefits of magnesium is still in its infancy, numerous studies suggest that it may be useful in the treatment of stress and depression.[85] OCD sufferers frequently fear or worry excessively about leaving the stove on, the taps running, and/or the doors unlocked. Before doing any actions, such as going to work or school, carrying out tasks, running errands, etc., these people may check and recheck that everything is switched off and/or locked to allay their fears.[86] Although the most popular form of treatment for OCD is medication, particularly SSRIs (antidepressants), therapy (such as cognitive-behavioral therapy (CBT), acceptance and commitment therapy (ACT), and exposure-response prevention (ERP) therapy), lifestyle changes, and self-help tools, such as impulse therapy, an online OCD treatment tool, are also effective.[87] Magnesium is one natural remedy that has showed potential in the treatment of OCD.[88] Magnesium has health benefits that extend beyond the body. The different variety of foods rich in magnesium are pumpkin seeds, chia seeds, almonds, flaxseeds, bananas, whole grains, soy milk, spinach, cashew, dark chocolate, avocado, brown rice.

16.6.11.5 Lactium

A supplement called Lactium is made from the casein protein found in milk. It helps to lessen symptoms of stress, such as anxiety and panic attacks.

16.6.11.6 Whole Grains

Whole grains, a dietary staple that is high in carbohydrates, offer more nutrients while having a less dramatic effect on blood sugar levels. Serotonin, a brain chemical that encourages serenity, is produced with its assistance. For reduced OCD symptoms, the Association for Comprehensive Neurotherapy, or ACTN, suggests consuming whole grains. Oats, barley, brown rice, wild rice, quinoa, and popcorn are examples of healthy whole grains.

16.6.12 Low-Fat Dairy Products

Milk and yogurt, two low-fat dairy products, include significant amounts of protein, which helps regulate blood sugar levels, as well as the amino acid tryptophan, which helps people feel calmer. The National Sleep Foundation recommends eating protein-rich foods and carbohydrates before night since carbohydrates increase your brain's ability to absorb tryptophan. Consume balanced meals and snacks, such as yogurt with granola on top or whole grain cereal with low-fat milk, for reduced anxiety and sleep troubles.

16.6.13 Fruits and Vegetables

The best sources of antioxidants, like beta-carotene and vitamin C, which help your body fight against illnesses and infections, are fruits and vegetables. Psychiatrist Daniel K. Hall-Flavin of

MayoClinic.com suggests avoiding overeating and eating a diet high in fruits and vegetables to relieve anxiety symptoms. Fruits and vegetables make people feel fuller between meals and may prevent overeating since they are high in fiber and water. Berries, cherries, citrus fruits, tomatoes, leafy greens, broccoli, bell peppers, brussels sprouts, and squash are a few examples of foods that are very nutrient-rich.

16.6.14 Caffeine

Unexpectedly, caffeine works as a natural OCD treatment. In most cases, it makes people more anxious and stressed. However, studies suggest that it can lessen OCD symptoms. Caffeine dramatically lessens the intensity of OCD symptoms in persons with treatment-resistant OCD, according to two clinical investigations. Caffeine is essentially a stimulant of the central nervous system (CNS). It has a variety of effects, including reducing weariness, boosting alertness, and improving performance on tasks requiring prolonged attention. Lower dosages of caffeine are expected to have less of an impact because the majority of research demonstrating these effects employed relatively high doses (e.g., 100 to 150 mg).[89] A lot of people claim that their anxiety levels rise after consuming caffeine. Additionally, caffeine can improve muscle performance and lessen weariness. There are numerous biological consequences of caffeine. Adenosine, a neurotransmitter that generally inhibits brain activity, is inhibited by it. Tryptophan, serotonin, and noradrenergic pathways are released by the action of adenosine A1 receptors. These neurotransmitters can be released more often when adenosine is inhibited. Additionally, caffeine activates the sympathetic nervous system, which boosts energy metabolism and blood flow.

16.6.14.1 How May Caffeine Alleviate OCD Symptoms?

It is unclear. One explanation is that it might boost the brain's production of dopamine, which might enhance one's capacity to turn one's focus away from obsessions. People may be more motivated to rebuff their invasions after experiencing improvements in their mood and vitality. Additionally, serotonin symptoms, which are key targets for the use of serotonergic antidepressants in the treatment of OCD, may be affected by adenosine inhibition in a variety of downstream ways.

16.6.15 Probiotics

Brain and mental health are greatly influenced by the condition of our gut and the bacteria that live there. Probiotics have thus been demonstrated to be another potential all-natural OCD treatment. Serotonin is a neurotransmitter that is made in the gut, according to studies. We may increase the diversity of bacteria in our gut, improve the conditions for serotonin synthesis, and thus raise serotonin levels and activity in your brain by taking a probiotic pill. Probiotics have also been demonstrated to activate the vagus nerve and lower inflammation, both of which are common in OCD sufferers. Probiotics can enhance psychiatric disorder-related behaviors, including anxiety and OCD, according to a comprehensive evaluation of 38 studies. Probiotics are the microorganisms to improve the various microflora in the gut and to enhance mental health. Several studies show that probiotics reduce the level of cortisol, which is a stress hormone. Probiotics are rich in foods like sauerkraut, pickles, kimchi, cheese, and yogurt. The number of strains will determine OCD apart from number of bacteria. So increase in number of strains will reduce OCD. Besides these probiotics help to reduce stress, overthinking, depression, anxiety, and bipolar disorder.[90]

16.6.16 Curcumin

Curcumin is an active constituent in the spice turmeric. Curcumin relieves OCD by bringing dopamine and serotonin levels back to normal and reducing inflammation in the brain. Several studies in animal models like rats reported that curcumin can reduce OCD symptoms by elevating serotonin and reducing dopamine.[91, 92] Curcumin has a significant impact on the brain's monoamine level.[93] Curcumin can control the levels of norepinephrine, which can treat disorders like epilepsy and depression.[94]

16.6.17 Ashwagandha

It is believed that a problem with the serotonergic system underlies obsessive-compulsive disorder (OCD), which is characterized by anxiety-inducing, recurrent, and intrusive thoughts or obsessions followed by actions or compulsions. SSRIs, which are frequently used to treat OCD, are thought to be only slightly helpful, with reports indicating that 40–60% of patients do not respond to them. Those who do react to SSRI therapy typically have relatively slight reductions in OCD symptoms. Ashwagandha (*Withania somnifera*, Solanaceae) is a herb that is traditionally used in traditional Ayurvedic herbal therapy as a revitalizing agent. Its roots have anxiolytic and antidepressant qualities and have been utilized in the treatment of nerve diseases. They are used to improve both mental and physical health. The phytochemicals withanolides and withanolide glycosides, as well as alkaloids, are thought to be involved in these therapeutic effects (withanine and somniferine). Additionally, ashwagandha may influence serotonergic transmission, according to a prior study conducted on mice. This randomized, double-blind, placebo-controlled trial's objective was to evaluate the effectiveness of ashwagandha root extract as a supplemental treatment for OCD.[95]

16.6.18 Milk Thistle

Over 2,000 years ago, a plant called milk thistle's fruit and seeds were utilized to strengthen the liver. The plant's seeds and fruits are used to make silymarin, one of milk thistle's active components. Silymarin is an intricate blend of flavonolignans with antioxidant qualities. According to studies, silymarin modulates cellular permeability and stabilizes cellular membranes to promote detoxification processes. Patients with OCD can facilitate detoxification using milk thistle, which improves mental wellness. Milk thistle is an effective liver cleanser. Serotonin levels may be abnormal in OCD patients, and milk thistle raises serotonin levels. According to research, using milk thistle for as little as four weeks can reduce obsessions and compulsions.[96]

16.6.19 Borage Oil

A natural treatment called borage oil is created from the seeds of the Borago officinalis plant. Gamma linoleic acid (GLA), an important fatty acid that must be consumed through diet, is present in large quantities in the oil. It is frequently used to lessen inflammation, which is associated to a number of chronic conditions. Perhaps because of its anti-inflammatory properties, it lessens OCD. According to studies, taking 500 mg of borage oil daily can help with anxiety and obsessive-compulsive symptom reduction.[97] Research on animals demonstrates that it can also have anti-anxiety properties comparable to benzodiazepines.

16.6.20 Saffron

Crocus sativus is a plant that produces the spice known as saffron. As a result of the medicinal components it contains, it has a range of health advantages. One of the best vitamins for lowering stress, anxiety, and depression is saffron. Two saffron-derived substances, safranal and crocetin, have been demonstrated to activate GABA receptors and raise serotonin levels in the brain. As a result, researchers have determined that saffron can reduce compulsive behavior.

16.6.21 Exercise

Another all-natural method of managing and treating OCD is exercise. Numerous medical professionals and academics cite exercise as their top piece of advice for achieving optimum mental health because it is convenient and cost-free. It has been demonstrated in five distinct trials that aerobic exercise can lessen OCD symptoms. Participants in the study reported less obsessions and compulsions after exercising for 12 weeks and the advantages last for six months. However, the workout need not be strenuous. Walking alone helps lessen OCD symptoms. For many people, exercising can be a major chore, therefore I advise selecting a sport or aerobic exercise that you like. You won't get bored with it and will work out frequently as a result.

16.6.22 Music

Music has a relaxing impact on the brain and is incredibly restorative. Listening to music raises dopamine and oxytocin levels while naturally lowering cortisol. It has likely helped persons with OCD because of this. In one study, music therapy helped OCD patients with their obsessional, depressive, and anxiety symptoms.

REFERENCES

1. Valizadeh, M., & Valizadeh, N. (2011). Obsessive compulsive disorder as early manifestation of B12 deficiency. *Indian Journal of Psychological Medicine*, *33*(2), 203–204.
2. Behl, A., Swami, G., Sircar, S. S., Bhatia, M. S., & Banerjee, B. D. (2010). Relationship of possible stress-related biochemical markers to oxidative/antioxidative status in obsessive-compulsive disorder. *Neuropsychobiology*, *61*(4), 210–214.
3. Chakrabarty, K., Bhattacharyya, S., Christopher, R., & Khanna, S. (2005). Glutamatergic dysfunction in OCD. *Neuropsychopharmacology*, *30*(9), 1735–1740.
4. Singer, H. S., Morris, C., & Grados, M. (2010). Glutamatergic modulatory therapy for Tourette syndrome. *Medical Hypotheses*, *74*(5), 862–867.
5. Adam, Y., Meinlschmidt, G., Gloster, A. T., & Lieb, R. (2012). Obsessive–compulsive disorder in the community: 12-month prevalence, comorbidity and impairment. *Social Psychiatry and Psychiatric Epidemiology*, *47*(3), 339–349.
6. Fawcett, E. J., Power, H., & Fawcett, J. M. (2020). Women are at greater risk of OCD than men: A meta-analytic review of OCD prevalence worldwide. *The Journal of Clinical Psychiatry*, *81*(4), 13075.
7. Boileau, B. (2022). A review of obsessive-compulsive disorder in children and adolescents. *Dialogues in Clinical Neuroscience*, *13*(4), 401–411.
8. Swedo, S. E., Rapoport, J. L., Leonard, H., Lenane, M., & Cheslow, D. (1989). Obsessive-compulsive disorder in children and adolescents: Clinical phenomenology of 70 consecutive cases. *Archives of General Psychiatry*, *46*(4), 335–341.
9. Fineberg, N. A., & Gale, T. M. (2005). Evidence-based pharmacotherapy of obsessive–compulsive disorder. *International Journal of Neuropsychopharmacology*, *8*(1), 107–129.
10. Mahjani, B., Bey, K., Boberg, J., & Burton, C. (2021). Genetics of obsessive-compulsive disorder. *Psychological* Medicine, *51*(13), 2247–2259.
11. Nestadt, G., Grados, M., & Samuels, J. F. (2010). Genetics of obsessive-compulsive disorder. *Psychiatric Clinics*, *33*(1), 141–158.
12. Bolton, J., Moore, G. J., MacMillan, S., Stewart, C. M., & Rosenberg, D. R. (2001). Case study: Caudate glutamatergic changes with paroxetine persist after medication discontinuation in pediatric OCD. *Journal of the American Academy of Child & Adolescent Psychiatry*, *40*(8), 903–906.
13. Last, C. G., Hersen, M., Kazdin, A., Orvaschel, H., & Perrin, S. (1991). Anxiety disorders in children and their families. *Archives of General Psychiatry*, *48*(10), 928–934.
14. Huppert, J. D., & Roth, D. A. (2003). Treating obsessive-compulsive disorder with exposure and response prevention. *The Behavior Analyst Today*, *4*(1), 66.
15. Ougrin, D. (2011). Efficacy of exposure versus cognitive therapy in anxiety disorders: Systematic review and meta-analysis. *BMC Psychiatry*, *11*(1), 1–13.

16. Sadock, B. J., Sadock, V. A., & Ruiz, P. (2015). *Kaplan & Sadock's synopsis of psychiatry: Behavioral sciences/clinical psychiatry*. Philadelphia, PA: Wolters Kluwer.
17. Karcı, C. K., & Celik, G. G. (2020). Nutritional and herbal supplements in the treatment of obsessive compulsive disorder. *General Psychiatry, 33*(2).
18. Zamani, Z., & Akhondzadeh, S. (2019). Herbal medicine in the treatment of obsessive compulsive disorder: A review. *Journal of Medicinal Plants, 18*(71), 1–5. http://jmp.ir/article-1-2662-en.html
19. Rao, T. S., Asha, M. R., Ramesh, B. N., & Rao, K. J. (2008). Understanding nutrition, depression and mental illnesses. *Indian Journal of Psychiatry, 50*(2), 77.
20. Ng, F., Berk, M., Dean, O., & Bush, A. I. (2008). Oxidative stress in psychiatric disorders: Evidence base and therapeutic implications. *The International Journal of Neuropsychopharmacology, 11*(6), 851–876.
21. Tural, Ü., Çorapçıoğlu, A., Boşgelmez, Ş., Köroğlu, G., Ünver, H., Duman, C., & Önder, E. (2019). Double blind controlled study of adding folic acid to fluoxetine in the treatment of OCD. *Psychiatria Danubina, 31*(1), 69–77.
22. Atmaca, M., Tezcan, E., Kuloglu, M., Kirtas, O., & Ustundag, B. (2005). Serum folate and homocysteine levels in patients with obsessive-compulsive disorder. *Psychiatry and Clinical Neurosciences, 59*(5), 616–620.
23. Esnafoğlu, E., & Yaman, E. (2017). Vitamin B12, folic acid, homocysteine and vitamin D levels in children and adolescents with obsessive compulsive disorder. *Psychiatry Research, 254*, 232–237.
24. Hermesh, H., Weizman, A., Shahar, A., & Munitz, H. (1988). Vitamin B12 and folic acid serum levels in obsessive compulsive disorder. *Acta Psychiatrica Scandinavica, 78*(1), 8–10.
25. Türksoy, N., Bilici, R., Yalçıner, A., Özdemir, Y. Ö., Örnek, I., Tufan, A. E., & Kara, A. (2014). Vitamin B12, folate, and homocysteine levels in patients with obsessive–compulsive disorder. *Neuropsychiatric Disease and Treatment, 10*, 1671.
26. Bottiglieri, T. (2005). Homocysteine and folate metabolism in depression. *Progress in Neuro-Psychopharmacology and Biological Psychiatry, 29*(7), 1103–1112.
27. Folstein, M., Liu, T., Peter, I., Buel, J., Arsenault, L., Scott, T., & Qiu, W. W. (2007). The homocysteine hypothesis of depression. *American Journal of Psychiatry, 164*(6), 861–867.
28. Sharma, V., & Biswas, D. (2012). Cobalamin deficiency presenting as obsessive compulsive disorder: Case report. *General Hospital Psychiatry, 34*(5), 578-e7.
29. Coşar, A., İpçioğlu, O. M., Özcan, Ö., & Gültepe, M. (2014). Folate and homocysteine metabolisms and their roles in the biochemical basis of neuropsychiatry. *Turkish Journal of Medical Sciences, 44*(1), 1–9.
30. Yan, S., Liu, H., Yu, Y., Han, N., & Du, W. (2022). Changes of serum homocysteine and vitamin B12, but not folate are correlated with obsessive-compulsive disorder: A systematic review and meta-analysis of case-control studies. *Frontiers in Psychiatry, 13*.
31. Katiyar, S. K. (2005). Silymarin and skin cancer prevention: Anti-inflammatory, antioxidant and immunomodulatory effects. *International Journal of Oncology, 26*(1), 169–176.
32. Viridian Nutrition. (2019, February 4). *Acetyl-cysteine and obsessive-compulsive disorder*. https://viridian-nutrition.com/blogs/nutrition-articles/nacetylcysteine-and-obsessivecompulsive-disorder
33. Couto, J. P., & Moreira, R. (2018). Oral N-acetylcysteine in the treatment of obsessive-compulsive disorder: A systematic review of the clinical evidence. *Progress in Neuro-Psychopharmacology and Biological Psychiatry, 86*, 245–254.
34. Bonanomi, L., & Toxicology, G. A. (1980). Pharmacokinetics and metabolism of acetylcysteine. *European Journal Respiratory Diseases, 111*, 45–51. [PubMed] [Google Scholar]
35. Bradlow, R. C., Berk, M., Kalivas, P. W., Back, S. E., & Kanaan, R. A. (2022). The Potential of N-Acet³yl-L-Cysteine (NAC) in the treatment of psychiatric disorders. *CNS Drugs*, 1–32.
36. Atkuri, K. R., Mantovani, J. J., Herzenberg, L. A., & Herzenberg, L. A. (2007). N-Acetylcysteine—a safe antidote for cysteine/glutathione deficiency. *Current Opinion in Pharmacology, 7*(4), 355–359.
37. Camfield, D. A., Sarris, J., & Berk, M. (2011). Nutraceuticals in the treatment of obsessive compulsive disorder (OCD): A review of mechanistic and clinical evidence. *Progress in Neuro-Psychopharmacology and Biological Psychiatry, 35*(4), 887–895.
38. Nguyen, T. P., Cribb, L., Ng, C. H., Byrne, G. J., Castle, D., Brakoulias, V., . . . Sarris, J. (2021). Dietary quality and nutrient intake in adults with obsessive–compulsive disorder. *BJPsych Open, 7*(6).
39. Rahman, T., Hosen, I., Islam, M. T., & Shekhar, H. U. (2012). Oxidative stress and human health. *Advances in Bioscience and Biotechnology, 2017*, 1–13.
40. Lafleur, D. L., Pittenger, C., Kelmendi, B., Gardner, T., Wasylink, S., Malison, R. T., . . . Coric, V. (2006). N-acetylcysteine augmentation in serotonin reuptake inhibitor refractory obsessive-compulsive disorder. *Psychopharmacology, 184*(2), 254–256.

41. Van Ameringen, M., Patterson, B., Simpson, W., & Turna, J. (2013). N-acetylcysteine augmentation in treatment resistant obsessive compulsive disorder: A case series. *Journal of Obsessive-Compulsive and Related Disorders*, *2*(1), 48–52.
42. Sarris, J., Oliver, G., Camfield, D. A., Dean, O. M., Dowling, N., Smith, D. J., . . . Ng, C. H. (2015). N-acetyl cysteine (NAC) in the treatment of obsessive-compulsive disorder: A 16-week, double-blind, randomised, placebo-controlled study. *CNS Drugs*, *29*(9), 801–809.
43. Sarris, J., Oliver, G., Camfield, D. A., & Dean, O. M. (2016). Participant characteristics as modifiers of response to N-acetyl cysteine (NAc) in obsessive-compulsive disorder. *Clinical Psychological Science*, *4*(6), 1104–1111.
44. Costa, D. L., Diniz, J. B., Requena, G., Joaquim, M. A., Pittenger, C., Bloch, M. H., . . . Shavitt, R. G. (2017). Randomized, double-blind, placebo-controlled trial of N-acetylcysteine augmentation for treatment-resistant obsessive-compulsive disorder. *The Journal of Clinical Psychiatry*, *78*(7), 12399.
45. Paydary, K., Akamaloo, A., Ahmadipour, A., Pishgar, F., Emamzadehfard, S., & Akhondzadeh, S. (2016). N-acetylcysteine augmentation therapy for moderate-to-severe obsessive–compulsive disorder: Randomized, double-blind, placebo-controlled trial. *Journal of Clinical Pharmacy and Therapeutics*, *41*(2), 214–219.
46. Celik, G., Didem, T. A. Ş., Tahiroğlu, A., Ayşe, A. V. C. I., Yüksel, B., & Perihan, Ç. A. M. (2016). Vitamin D deficiency in Obsessive–Compulsive disorder patients with pediatric autoimmune neuropsychiatric disorders associated with streptococcal infections: A case control study. *Nöro Psikiyatri Arşivi*, *53*(1), 33.
47. Eyles, D. W., Burne, T. H., & McGrath, J. J. (2013). Vitamin D, effects on brain development, adult brain function and the links between low levels of vitamin D and neuropsychiatric disease. *Frontiers in Neuroendocrinology*, *34*(1), 47–64.
48. Holick, M. F. (2015). Vitamin D and brain health: The need for vitamin D supplementation and sensible sun exposure. *Journal of Internal Medicine*, *277*(1), 90–93.
49. Yazici, K. U., Yazici, P., & Ustundag, B. (2017). Vitamin D levels in children and adolescents with obsessive compulsive disorder. *Nordic Journal of Psychiatry*, 1–6. [PubMed] [Google Scholar]
50. Wang, T., Shan, L., Du, L., Feng, J., Xu, Z., Staal, W. G., & Jia, F. (2016). Serum concentration of 25-hydroxyvitamin D in autism spectrum disorder: A systematic review and meta-analysis. *European Child & Adolescent Psychiatry*, *25*(4), 341–350.
51. Cannell, J. J. (2013). Autism, will vitamin D treat core symptoms? *Medical Hypotheses*, *81*(2), 195–198.
52. Chiang, M., Natarajan, R., & Fan, X. (2016). Vitamin D in schizophrenia: A clinical review. *Evidence-Based Mental Health*, *19*(1), 6–9.
53. Parker, G. B., Brotchie, H., & Graham, R. K. (2017). Vitamin D and depression. *Journal of Affective Disorders*, *208*, 56–61.
54. Cui, X., Pertile, R., Liu, P., & Eyles, D. W. (2015). Vitamin D regulates tyrosine hydroxylase expression: N-cadherin a possible mediator. *Neuroscience*, *304*, 90–100.
55. Garcion, E., Sindji, L., Montero-Menei, C., Andre, C., Brachet, P., & Darcy, F. (1998). Expression of inducible nitric oxide synthase during rat brain inflammation: Regulation by 1, 25-dihydroxyvitamin D3. *Glia*, *22*(3), 282–294.
56. Arnold, P. D., Rosenberg, D. R., Mundo, E., Tharmalingam, S., Kennedy, J. L., & Richter, M. A. (2004). Association of a glutamate (NMDA) subunit receptor gene (GRIN2B) with obsessive-compulsive disorder: A preliminary study. *Psychopharmacology*, *174*(4), 530–538.
57. Greenberg, W. M., Benedict, M. M., Doerfer, J., Perrin, M., Panek, L., Cleveland, W. L., & Javitt, D. C. (2009). Adjunctive glycine in the treatment of obsessive-compulsive disorder in adults. *Journal of Psychiatric Research*, *43*(6), 664–670.
58. Cleveland, W. L., DeLaPaz, R. L., Fawwaz, R. A., & Challop, R. S. (2009). High-dose glycine treatment of refractory obsessive-compulsive disorder and body dysmorphic disorder in a 5-year period. *Neural Plasticity*. *2009*, 768398, 1–25.
59. Lack, C. W. (2012). Obsessive-compulsive disorder: Evidence-based treatments and future directions for research. *World Journal of Psychiatry*, *2*(6), 86.
60. Shohag, H., Ullah, A., Qusar, S., Rahman, M., & Hasnat, A. (2012). Alterations of serum zinc, copper, manganese, iron, calcium, and magnesium concentrations and the complexity of interelement relations in patients with obsessive–compulsive disorder. *Biological Trace Element Research*, *148*(3), 275–280.
61. Ozdemir, E., Cetinkaya, S., Ersan, S., Kucukosman, S., & Ersan, E. E. (2009). Serum selenium and plasma malondialdehyde levels and antioxidant enzyme activities in patients with obsessive–compulsive disorder. *Progress in Neuro-Psychopharmacology and Biological Psychiatry*, *33*(1), 62–65.

62. Sayyah, M., Andishmand, M., & Ganji, R. (2018). Effect of selenium as an adjunctive therapy in patients with treatment-resistant obsessive-compulsive disorder: A pilot randomized double blind placebo-controlled clinical trial. *Archives of Psychiatry & Psychotherapy*, *20*(4).
63. Takeda, A. (2000). Movement of zinc and its functional significance in the brain. *Brain Research Reviews*, *34*(3), 137–148.
64. Sayyah, M., Olapour, A., Saeedabad, Y., Parast, R. Y., Malayeri, A. (2012). Evaluation of oral zinc sulfate effect on obsessive-compulsive disorder: a randomized placebo-controlled clinical trial. *Nutrition*, *28*(9), 892–895.
65. Sayyah, M., Olapour, A., shahhosseini Saeedabad, Y., Parast, R. Y., & Malayeri, A. (2012). Evaluation of oral zinc sulfate effect on obsessive-compulsive disorder: A randomized placebo-controlled clinical trial. *Nutrition*, *28*(9), 892–895.
66. Villagomez, A., & Ramtekkar, U. (2014). Iron, magnesium, vitamin D, and zinc deficiencies in children presenting with symptoms of attention-deficit/hyperactivity disorder. *Children*, *1*(3), 261–279.
67. Wołonciej, M., Milewska, E., & Roszkowska-Jakimiec, W. (2016). Trace elements as an activator of antioxidant enzymes. *Postepy higieny i medycyny doswiadczalnej (Online)*, *70*, 1483–1498.
68. Kaneko, I., Sabir, M. S., Dussik, C. M., Whitfield, G. K., Karrys, A., Hsieh, J. C., . . . Jurutka, P. W. (2015). 1, 25-Dihydroxyvitamin D regulates expression of the tryptophan hydroxylase 2 and leptin genes: Implication for behavioral influences of vitamin D. *The FASEB Journal*, *29*(9), 4023–4035.
69. Harvey, B. H., Brink, C. B., Seedat, S., & Stein, D. J. (2002). Defining the neuromolecular action of myo-inositol: Application to obsessive–compulsive disorder. *Progress in Neuro-Psychopharmacology and Biological Psychiatry*, *26*(1), 21–32.
70. Kim, H., McGrath, B. M., & Silverstone, P. H. (2005). A review of the possible relevance of inositol and the phosphatidylinositol second messenger system (PI-cycle) to psychiatric disorders—focus on magnetic resonance spectroscopy (MRS) studies. *Human Psychopharmacology: Clinical and Experimental*, *20*(5), 309–326.
71. Seedat, S., & Stein, D. J. (1999). Inositol augmentation of serotonin reuptake inhibitors in treatment-refractory obsessive–compulsive disorder: An open trial. *International clinical psychopharmacology*, *14*(6), 353–356.
72. Levine, J. (1997). Controlled trials of inositol in psychiatry. *European Neuropsychopharmacology*, *7*(2), 147–155.
73. Fux, M., Levine, J., Aviv, A., & Belmaker, R. H. (1996). Inositol treatment of obsessive-compulsive disorder. *American Journal of Psychiatry*, *153*(9), 1219–1221.
74. Carey, P. D., Warwick, J., Harvey, B. H., Stein, D. J., & Seedat, S. (2004). Single photon emission computed tomography (SPECT) in obsessive–compulsive disorder before and after treatment with inositol. *Metabolic Brain Disease*, *19*(1), 125–134.
75. Fux, M., Benjamin, J., & Belmaker, R. H. (1999). Inositol versus placebo augmentation of serotonin reuptake inhibitors in the treatment of obsessive–compulsive disorder: A double-blind cross-over study. *International Journal of Neuropsychopharmacology*, *2*(3), 193–195.
76. Rahimi, R., Nikfar, S., & Abdollahi, M. (2009). Efficacy and tolerability of Hypericum perforatum in major depressive disorder in comparison with selective serotonin reuptake inhibitors: A meta-analysis. *Progress in Neuro-Psychopharmacology and Biological Psychiatry*, *33*(1), 118–127.
77. Sarris, J., & Kavanagh, D. J. (2009). Kava and St. John's Wort: Current evidence for use in mood and anxiety disorders. *The Journal of Alternative and Complementary Medicine*, *15*(8), 827–836.
78. Kobak, K. A., & Taylor, L. V. (2005). Sky A, Kohlenberg CJ, Greist JH, Tucker P, Warner G, Futterer R, Vapnik T. St John's wort versus placebo in obsessive-compulsive disorder: Results from a double-blind study. *International Clinical Psychopharmacology*, *20*(6), 299–304.
79. Taylor, L. V., & Kobak, K. A. (2000). An open-label trial of St. John's Wort (Hypericum perforatum) in obsessive-compulsive disorder. *The Journal of Clinical Psychiatry*, *61*(8), 575–578.
80. Khom, S., Baburin, I., Timin, E., Hohaus, A., Trauner, G., Kopp, B., & Hering, S. (2007). Valerenic acid potentiates and inhibits GABAA receptors: Molecular mechanism and subunit specificity. *Neuropharmacology*, *53*(1), 178–187.
81. Pakseresht, S., Boostani, H., & Sayyah, M. (2011). Extract of valerian root (Valeriana officinalis L.) vs. placebo in treatment of obsessive-compulsive disorder: A randomized double-blind study. *Journal of Complementary & Integrative Medicine*, *8*, j-jcim.
82. Andreatini, R., Sartori, V. A., Seabra, M. L., & Leite, J. R. (2002). Effect of valepotriates (valerian extract) in generalized anxiety disorder: A randomized placebo-controlled pilot study. *Phytotherapy Research: An International Journal Devoted to Pharmacological and Toxicological Evaluation of Natural Product Derivatives*, *16*(7), 650–654.

83. Goddard, A. W., Shekhar, A., Whiteman, A. F., & McDougle, C. J. (2008). Serotoninergic mechanisms in the treatment of obsessive–compulsive disorder. *Drug Discovery Today, 13*(7–8), 325–332.
84. Dell'Osso, B., Nestadt, G., Allen, A., et al. (2006). Serotonin-Norepinephrine reuptake inhibitors in the treatment of obsessive-compulsive disorder. *Journal of Clinical Psychiatry, 67*, 600–610. doi:10.4088/JCP.v67n0411 [PubMed] [CrossRef] [Google Scholar]
85. Rajizadeh, A., Mozaffari-Khosravi, H., Yassini-Ardakani, M., & Dehghani, A. (2017). Effect of magnesium supplementation on depression status in depressed patients with magnesium deficiency: A randomized, double-blind, placebo-controlled trial. *Nutrition, 35*, 56–60.
86. Hemamy, M., Heidari-Beni, M., Askari, G., Karahmadi, M., & Maracy, M. (2020). Effect of Vitamin D and magnesium supplementation on behavior problems in children with attention-deficit hyperactivity disorder. *International Journal of Preventive Medicine, 11.*
87. Kirkland, A. E., Sarlo, G. L., & Holton, K. F. (2018). The role of magnesium in neurological disorders. *Nutrients, 10*(6), 730.
88. Black, L. J., Allen, K. L., Jacoby, P., Trapp, G. S., Gallagher, C. M., Byrne, S. M., & Oddy, W. H. (2015). Low dietary intake of magnesium is associated with increased externalising behaviours in adolescents. *Public Health Nutrition, 18*(10), 1824–1830.
89. Shams, J., Soufi, E. S., Zahiroddin, A., & Shekarriz-Foumani, R. (2019). Using caffeine on the patients as therapeutic option against treatment-resistant obsessive-compulsive disorder. *Journal of Family Medicine and Primary Care, 8*(5), 1741.
90. Dwyer, A. (2021, April 28). *Could probiotics help with OCD?* https://www.optibacprobiotics.com/professionals/latest-research/general-health/probiotics-and-ocd
91. Chimakurthy, J., & Murthy, T. E. (2010). Effect of curcumin on quinpirole induced compulsive checking: An approach to determine the predictive and construct validity of the model. *North American Journal of Medical Sciences, 2*(2), 81.
92. Daily, J. W. Y., & Park, S. (Agosto de 2016). Efficacy of turmeric extracts and curcumin for alleviating the symptoms of joint arthritis: A systematic review and meta-analysis of randomized clinical trials. *Journal of Medicine Food, 19*(8), 717–729.
93. Mazzio, E. A., Harris, N., & Soliman, K. F. (1998). Food constituents attenuate monoamine oxidase activity and peroxide levels in C6 astrocyte cells. *Planta Medica, 64*(07), 603–606.
94. Lopresti, A. L., Hood, S. D., & Drummond, P. D. (2012). Multiple antidepressant potential modes of action of curcumin: A review of its anti-inflammatory, monoaminergic, antioxidant, immune-modulating and neuroprotective effects. *Journal of Psychopharmacology, 26*(12), 1512–1524.
95. Jahanbakhsh, S. P., Manteghi, A. A., Emami, S. A., Mahyari, S., Gholampour, B., Mohammadpour, A. H., & Sahebkar, A. (2016). Evaluation of the efficacy of Withania somnifera (Ashwagandha) root extract in patients with obsessive-compulsive disorder: A randomized double-blind placebo-controlled trial. *Complementary Therapies in Medicine, 27*, 25–29.
96. Sayyah, M., Boostani, H., Pakseresht, S., & Malayeri, A. (2010). Comparison of Silybum marianum (L.) Gaertn. with fluoxetine in the treatment of obsessive– compulsive disorder. *Progress in Neuro-Psychopharmacology and Biological Psychiatry, 34*(2), 362–365.
97. Sayyah, M., Boostani, H., Pakseresht, S., & Malaieri, A. (2009). Efficacy of aqueous extract of Echium amoenum in treatment of obsessive–compulsive disorder. *Progress in Neuro-Psychopharmacology and Biological Psychiatry, 33*(8), 1513–1516.

17 Effect of Nutrients and Herbal Supplements in the Treatment of OCD

Ramya Kumari B S,[a] *Geetika Pant,*[b] *and Sabyasachi Choudhury*[c]

[a] Department of Chemistry and Biochemistry, M S Ramaiah College of Arts, Science and Commerce, Bengaluru-560054, KA

[b] Department of Biotechnology and Genetics, M S Ramaiah College of Arts, Science and Commerce, Bengaluru-560054, KA

[c] Content Writer, Navigators software Pvt Ltd, Kolkata-700046, West Bengal

17.1 INTRODUCTION

Obsessive-compulsive disorder (OCD) is known to be a very commonly occurring neuropsychiatric disorder. As the name indicates, it is marked by obsessions and compulsions. Generally, OCD begins in childhood or adolescence and gradually progresses as the individual ages. Patients with OCD show comorbid disorders—specifically anxiety, mood swings, mental derangement, and somatoform disorders (Adam et al. 2012). Out of these, anxiety disorder predominates and produces symptoms of fear, worry, and a constant nagging feeling of being overwhelmed. These are triggered by stress conditions, among which are panic attacks, social anxiety, and agoraphobia. Current estimates reveal that 1–3% of the global population are affected with obsessive-compulsive disorder as a chronic neuropsychiatric disease (American Psychiatric Association 2013). This seriously deteriorates the quality of life.

The risk factors contributing to the emergence of OCD may include environmental factors, genetic factors, impaired neurotransmissions, and autoimmune processes and infections. However, stressors or trauma-driven incidents count as being the primary reason behind OCD (Miller and Brock 2017; Brander et al. 2016; Taylor 2011; Swedo 2010; Bhattacharyya et al. 2009). The prime reason why middle-aged women indulge in nutrient/herbal based supplements can be attributed to mental health and psychological concerns. It is due to an increased use of such supplements within the community, it becomes very important to know about the products' efficacy.

The best known pathophysiology of OCD is the oddity and peculiarity of the central nervous system (CNS), mostly in the serotonin, dopamine, and the glutamate pathways (Chakrabarty et al. 2005). The most recommended treatment in case of OCD is cognitive–behavioral therapy (CBT), using exposure and response prevention in conjunction with pharmacotherapy. On the contrary, some specific nutritional and herbal supplements may be effectual in the treatment of OCD. Clinical regulations (Del Casale et al. 2019) have also laid out few other lines of treatment such as exposure and response prevention (ERP), selective serotonin reuptake inhibitor (SSRI), or a combination of CBT, SSRI, and ERP. However, the pre-eminence of either of the aforementioned lines of treatment has not yet been proven. Symptoms of OCD can be managed by using a host of approaches, such as switching over to an alternate SSRI or clomipramine, increasing the dose of SSRI, or furthermore, augmentation with an atypical neuroleptic drug such as risperidone and aripiprazole. Higher dose SSRI treatment comes with its own set of side effects and often, long-term complications such as dizziness and sedation, insomnia, anxiety, diarrhea, constipation, and sexual dysfunctions

 DOI: 10.1201/9781003387893-20

(Papakostas 2008). Despite the fact that these therapeutic methods are very effective, treatment in many cases cannot be started or completed due to the patients' resistance to any of the treatment modalities. The exact etiology and pathophysiology of OCD remain unknown till date. Besides neurotransmitter imbalance, the other causative factor that could lead to OCD points towards oxidative stress. An oxidative stress is caused by an absence of balance between the generation of oxidative free radicals and the neutralizing antioxidants (Hovatta et al. 2010).

Present modes of treatments are limited in their effectiveness and have countless side effects, many of which are long-term. This has led to the search for alternative methods and strategies in combating OCD. Research reveals that various nutritional deficiencies are detected in patients challenged with mental disorders. Therefore, nutritional supplements are thought to be effective in providing relief to patients, by alleviating OCD. It is where the use of complementary medicine comes into play. Some popular herbal products and dietary supplements have the potential to benefit patients suffering from mild to moderate levels of anxiety.

This chapter aims to present the current scientific knowhow about the relationship between the ingestion of specific micro- and macronutrients and mental health. It also particularly deals with the beneficial effect on stress and anxiety, mild cognitive impairment, as well as on neuropsychiatric disorders. Alleviation of OCD directly implies to a significant shift in the quality of life by retaining normal brain functioning and optimal mental well-being. Encouraging subjects to integrate a well-balanced and diversified diet, along with a healthy lifestyle may prove useful in reducing the amount of brain neuroinflammation. This strategy is central and of paramount importance when considering the global human aging populace since the brain significantly endures the lifelong impact of stress-inducing elements.

17.2 PLANT-BASED SUPPLEMENTS

A wholesome dietary regime affects our mental and intellectual fitness by tapping into the antioxidant, anti-inflammatory, neurogenesis, gut-microbiome, and immunity-boosting mechanisms, in addition to epigenetic modifications. A detailed and variegated dietary profile does not only benefit the braincomposition, structure, and functioning but also supports the endogenous hormones, neurotransmitters and peptides, and the gut-brain axis microflora, in turn playing a pivotal role in the management of stress and in safeguarding cognitive function (Adan et al. 2019). Moreover, strengthening and augmenting healthy and balanced diet with plant-based supplements (St. John's wort, Kava, Passionflower, etc.), macronutrients (fatty acids), and micronutrients (vitamins and minerals) can provide a myriad of beneficial effects, thanks to their important biological capacities (Polavarapu and Hasbani 2017).

A number of plant-based/herbal supplements have been studied for improving OCD and treating major depression issues. Of these, St. John's wort, kava, passionflower, milk thistle, valerian root, curcumin, etc. have been studied and are discussed in the following sections.

17.2.1 St John's Wort (*Hypericum Perforatum*)

St John's wort (SJW) is a plant-based derivative with a long history in traditional medicine to treat many disorders. Effectiveness of SJW has primarily been studied in fighting depression. Research by Rahimi et al. (2009) claims SJW to be equivalent to antidepressants in action. SJW has flavonoid that exhibits some anxiolytic effects. *H. perforatum* belongs to the Hypericaceae family with the common name St. John's wort (SJW). The neurobiological effects seen in SJW are neuroprotection (Oliveira et al. 2018), increased sensitization, monoamine reuptake inhibition, neuroendocrine modulation, and binding to receptors such as 5-HT (Sarris and Kavanagh 2009). These monoaminergic effects may have potential benefits in treating OCD symptoms. The evidence for SJW effectiveness against OCD is currently based on only two studies. The first study was conducted in a 12-week open label trial that showed significant improvements after treatment with SJW (900 mg/d) in patients with OCD (n = 12) (Taylor and Kobak 2000). The second study comprised of a placebo-controlled trial wherein a much larger sample size (n = 60) failed to show any marked difference between the SJW (600–1,800 mg/d) treatment and placebo groups (Kobak et al. 2005).

SJW is a known natural antioxidant, an invaluable reservoir of free radical scavenging compounds, and advantageous in the prevention and treatment of oxidative stress-related pathologic disorders (Napoli et al. 2018). The free radical scavenging activity present in SJW is because of Xanthone derivatives and proanthocyanidins naturally found in it (Caccia and Gobbi 2009; Heinrich et al. 2017). In an animal model study performed by Skalisz et al. (2004), it was evaluated that the influence of SJW extract towards the marble-burying behavior of mice at dosages that displayed an anti-depressive-like impact but showed no exponential effect on their locomotor action.

In an investigation conducted by Linde et al. (2008), St. John's wort had been explored for treating depression by analyzing over 40 clinical trials with a wide range of methodological quality. A Cochrane review evaluated 29 trials, which constituted of 5,489 subjects, analyzed 18 comparisons of SJW with placebo and assessed 17 comparisons with antidepressants. It was remarkably revealed that the participating individuals were significantly more likely to respond to the action of St. John's wort than to placebo (relative risk of 1.48, confidence interval 1.23–1.77); however, results from the studies indicated a contrasting response. St. John's wort had a near equivalent effect to the SSRIs (relative risk of response 1.00, CI 0.90–1.15) in the same analysis. During an extensive, prolonged follow-up study on SJW (standardized European extract WS 5570) by Kasper et al. (2008), in treating mild to moderate depression remission rates were measured in 426 responders. These were the ones who either continued with SJW doses or opted for placebo for 26 weeks. It was found that 17 people continuing SJW had a significantly higher time in remission. Similar to what was mentioned previously, Sarris, Nierenberg et al. (2013) recorded that the response to SJW was equivalent to that of SSRIs, with data hinting that an initial partial response is indicative of full response (that may occur within two to four weeks).

Due to the risk of drug interactions (refer Drug interactions with complementary medicines, Aust Prescr 2010;33:177–80), people undergoing other medications should only use SJW with low hyperforin concentration (<4 mg per tablet). The products that are standardized with higher amounts of hypericin and flavonoids must not induce cytochrome enzymes (Whitten et al. 2006). Traces of hypericin are generally identified in SJW products; however, hyperforin quantities are mostly not. It is to be remembered that SJW should not be taken along with antidepressants. This may give rise to serotonin syndrome.

17.2.2 Kava

Kava is a perennial plant that is indigenous to various islands of the South Pacific region (Sarris, LaPorte et al. 2011). In traditional medicine, the roots are prescribed as a water-based beverage which supposedly has medicinal and psychotropic benefits. In Australia, kava is available as water-soluble extracts with maximum daily dose of 250 mg of kavalactones (around four to five tablets/day). While kava has been found to be linked to cases of abuse in the Northern Territory of Australia—most commonly as an admixture with alcohol—current research does not deem it as an addictive when used therapeutically (Sarris, Stough et al. 2013).

It has been generally observed that kava inhibits the enzyme cytochrome P450 in the liver that metabolizes many drugs, thereby reducing their potency to a great extent (Singh 2005). Hence, it is imperative to know the risk of drugs associating with kava. Other side effects that have been reported due to long-term usage include skin rashes or lesion and yellowing of the skin, which is generally reversible, but these have been reported in very few numbers. Despite a lack of reliable, long-term data on kava's efficacy and safety, Singh (2005) and Ernst (2002) provided evidence that highlights its short-term usage (till 24 weeks) can show small-scale betterment in subjects with generalized anxiety (Pittler and Ernst 2003). This is to be mentioned that the short-term risks do not, however, outnumber kava's benefits. Thus, it can be concluded that patients with mild to moderate anxiety levels willing to explore herbal remedies/supplements, without the usage of alcohol and other drugs that are metabolized in the liver, kava seems to be an acceptable short-term choice.

In 2002, the European and UK markets banned the sale of kava since it had reported hepatotoxicity in some of its patients. This certainly could have been due to the earlier concoctions being formulated using acetone or ethanol extracts as base ingredient, from potentially inferior quality material or other less potent parts of the kava plant (Teschke et al. 2013) (Teschke et al. 2012). The use of only peeled roots from noble cultivars by employing a water-solute extraction method is highly recommended (refer kava fact sheet by the Therapeutic Goods Administration www.tga.gov.au/safety/alerts-medicine-kava-050421.htm). While using kava supplements regularly, care has to be taken to perform occasional liver function tests. Use of alcohol and benzodiazepines must be avoided at all costs (Teschke et al. 2011). In spite of all the tests and their corresponding results, practitioners are urged to exercise caution while incorporating kava in their lives in batting anxiety.

17.2.3 Valerian Root

Valerian is a perennial plant with the scientific name *Valeriana officinalis L*, which was used as an ingredient in perfumery during the sixteenth century. These days valerian root extracts are touted to treat nervousness and anxiety disorders by herbalists and few physicians. As per reports by Khom et al. (2007), valerian roots contain aleuronic acid, which is directly related with the modulation of GABAA receptors. The other molecules found in valerian roots are valepotriates, which are effective in combating psychotic symptoms that result due to severe anxiety (Andreatini et al. 2002). Valerian is most commonly used in cases of insomnia and migraine treatment since it acts as a pain reliever and sedative. There is only one available double-blind placebo-controlled trial of valerian extract used in OCD treatment (n = 31). Valerian extract was found to be more effective in OCD treatment as compared with placebo samples. Also, the rapid onset of the effects of valerian extract was found in a period of 28, 42, and 56 days. The main advantage that amplifies the use of this herbal supplement is the low incidence of libido suppression in patients found by researchers (Pakseresht et al. 2011).

17.2.4 Passion Flower

The *Passiflora spp.*, commonly known as passion flower, is a large, showy, delicate blue flower. There exists only a single randomized, double-blind trial wherein 45 drops of passion flower tincture per day were compared to 30 mg per day of oxazepam (belonging to the brand Serax that is currently discontinued in the United States) for 30 days (Akhondzadeh et al. 2001). Investigators also noted a marked anxiety score reduction in both the groups, but without the presence of a placebo group, it was ambiguous whether other environmental factors might have contributed to the effects.

17.2.5 Hawthorn and California Poppy

Only one French study exists in scientific literature which states a combination product called Sympathyl, containing a mixture of 20 mg California poppy (*Eschscholtzia californica*), 75 mg hawthorn (*Crataegus oxyacantha*), and 75 mg elemental magnesium. As per the authors of this study, Sympathyl had a tiny but seemingly positive effect on anxiety symptoms. No further clinical trials were later conducted to pinpoint the efficacy of any individual components in reducing anxiety in patients, with anxiety disorders. Hence, hawthorn and California poppy need more credible research.

17.2.6 Milk Thistle

Milk thistle (*Silybum marianum*), a medicinal plant, that grows well in the Mediterranean and Persian regions (Sayyah et al. 2010) contains a flavonoid-complex silymarin, which according to

Katiyar (2005) has antioxidant, anti-inflammatory, immune-modulating, sedative, and antidepressant properties. Also, silymarin has been seen to enhance the levels of serotonin in the brain cortex and act as monoamine oxidase inhibitor (Mazzio et al. 1998).

The results in only randomized controlled trial (n = 35) using milk thistle for OCD treatment revealed no noticeable difference in effects between milk thistle (600 mg/d) and fluoxetine. It pointed that milk thistle has effects akin to fluoxetine in OCD symptoms. The positive effects, however, begin to show result from the fifth week, without severe side effects (Sayyah et al. 2010).

17.2.7 Turmeric

The Zingiberaceae (ginger) family member turmeric (*Curcuma longa*) has long been utilized as an all-important agent in culinary activities, most specifically in Asian cuisine. Its chief component—curcumin, the yellow pigment, is the major active constituent and a natural polyphenol. It is known to be a dietary safe phytochemical with numerous advantageous effects (Mollazadeh et al. 2019) (47) (Panahi et al. 2018) (Soleimani et al. 2018) (Ghandadi and Sahebkar 2017). Curcumin is known to scavenge various reactive oxygen species (ROSs) such as hydroxyl and nitrogen dioxide radicals. It also shows an excessive capacity as a lipid-solvable antioxidant, thereby inhibiting lipid peroxidation activities in multiple animal models (Reddy and Lokesh 1994).

In an animal model study conducted by Chimakurthy and Murthy (2010), the effect of curcumin was tested in quinpirole-induced model of OCD, and it presented a shielding effect against OCD. Adult Wistar rats were classified into five groups: Group 1 rats were administered peanut oil (0.1 ml/100 mg) as a control; group 2 served as the negative control; in groups 3, 4, and 5, rats were orally administered treatment with 5 and 10 mg/kg of curcumin and 1.8 mg/kg of paroxetine. The entire set of rats, barring the control group, were treated with quinpirole (0.5 mg/kg, PO). It was noted that curcumin lowered obsessive-compulsive symptoms in rats, including repetitive cleansing of the snout and grooming, along with overall length and frequency of stops at objects. Curcumin-treated rats (5 and 10 mg/kg) displayed decreased dopamine concentrations, but occurrence of an advancement in serotonin concentrations only at 10 mg/kg of curcumin were observed.

17.3 NUTRITION: A KEY PLAYER IN OCD

Obsessive-compulsive disorder (OCD) is a well-known neuropsychiatric disorder, and as the name goes, it is distinguished by obsessions and compulsions. A healthy and balanced diet is very important for well-being and positive state of mind. Over dietary profile and nutrient intake entirely affects the metabolic rate, hormone secretion, release of neurotransmitters, and overall growth and development of body. Some nutritional supplements such as vitamin D, vitamin B12, folic acid, homocysteine, N-acetyl cysteine, glycine, myoinositol, and trace elements, along with plant-based products like St. John's wort, valerian root, milk thistle, curcumin and borage have been researched in OCD treatment studies (Adam Y et al. 2012; Boileau B 2011). The intake of proper diet not only modulates the stress hormone but also plays a significant role on mood swings and mental health. Research has proved that selective serotonin reuptake inhibitors (SSRIs) reduce the level of stress in OCD. Hence, nutrient diets rich in serotonin, or an amino acid tryptophan, which is a precursor for serotonin, has improved the OCD patients to a great extent (Fontenelle LF et al. 2007; Yaryura-Tobias and Bhagavan 1977). The effectiveness of herbal and nutritional supplements in the treatment of OCD must be supported with more concrete and conclusive evidences. Here are some of the important nutrient supplements which directly or indirectly influences the disorder at various levels.

17.3.1 Vitamins, Folic Acid, and Homocysteine

Vitamin D has some important roles in immunity modulation, inflammatory response, and antioxidant processes. It is also critical for normal brain development and functioning and plays pivotal roles in nervous system processes, as in neurotransmission, neuroprotection, neural cell proliferation, and cell differentiation.

Earlier research has illustrated that vitamin D deficiency is deeply associated with multiple neuropsychiatric diseases; some of which are autism, major depressive disorders, schizophrenia, and OCD (Eyles DW et al. 2013). Studies showed that people with OCD have low levels of vitamins (B12), minerals such as Zn, Fe, and Mg when tested against normal healthy patients.

Vitamin B12, folic acid, and homocysteine all play an important role in carbon transfer metabolism, which is useful in the upkeep and maintenance of methylation of proteins, neural membrane phospholipids, and neurotransmitters such as serotonin and other monoamines. Thus, incidences of dropping down of neurotransmitter levels with the deficiency of vitamin B12 and folic acid may be observed (Cannell JJ 2013).

17.3.2 Zinc

Zinc is the antioxidant trace element, indispensable in many molecular processes such as gene expression, protein synthesis, and enzymatic catalysis. Zinc is widely present in the central nervous system, the cerebral cortex, pineal gland, and hippocampus and actively modulates inhibitory and excitatory transmissions.

17.3.3 N-Acetyl Cysteine

N-acetyl cysteine (NAC), when orally taken, gets converted to cysteine in the liver and is further used for glutathione synthesis. It has been seen that cysteine may cross the blood-brain barrier. Hence, cysteine prodrugs such as NAC inhibit release of synaptic glutamate by cysteine-glutamate exchange and may contribute in treating of psychiatric disorders such as OCD (Arnold P et al. 2004).

17.3.4 Glycine

Meat, fish, dairy products, and beans are some of the high-protein foods that contain glycine. A case report demonstrated benefits of glycine in patients suffering with OCD.

17.3.5 Myoinositol

Myoinositol (MI) being an endogenous isomer of glucose has been used in the treatment of psychiatric disorders since a considerable time. High amounts of MI have been isolated from vegetarian sources like fruits, grains, nuts, and beans.

17.3.6 Valerian Root (*Valeriana officinalis L*)

The root extract of the valerian plant is most commonly used in patients suffering with insomnia and migraine due to its sedative and pain-relieving effects. There is so far only one double-blind placebo-controlled trial in study of valerian extract in OCD treatment (Fineberg and Gale 2005).

17.3.7 Curcumin

Curcumin is an active constituent of the popular rhizome, turmeric. Both turmeric and curcumin are believed to strengthen the overall energy of the body, relieve gas, cast worms out, improve digestion, regulate menstruation cycles in women, dissolve gallstones, and provide relief in arthritis (Goddard AW et al. 2008). Curcumin is most commonly used in arthritis-associated pain alleviation along with other musculoskeletal disorders due to its anti-inflammatory effect. However, there are no human studies that extort the benefits of curcumin in the treatment of OCD.

17.3.8 Ashwagandha Root Extract

Ashwagandha (*Withania somnifera*, Solanaceae family) is widely considered as a rejuvenating and revitalizing herb in traditional Ayurvedic medicine. Its potent roots, which are used to enhance mental and physical health, also have anxiolytic and antidepressant properties that have been used in the treatment of nervous disorders since centuries.

17.3.9 Flax Seed Oil

There are several evidences that indicate that ingesting flax seed oil can help manage OCD behavior and symptoms. It is invariably rich in essential fatty acids that nourishes and strengthens the brain cells, its communication and development.

17.4 CONCLUSION

By the end of this chapter, it is a well-established fact that mental health is an essential and integral constituent of human health. Poor mental health is very much proportional to an unhealthy lifestyle. It has been well-proved through scientific findings that an amalgamation of micro and macronutrients in a well-balanced and holistic diet, accompanied by a healthy and active lifestyle, is indispensable in the preservation of optimum brain functioning and overall well-being. This approach is of prime importance when considering the pressing problem of global human aging and the significant ensuing impact on our brain, which suffers at the hands of a long-standing impact of stress factors.

REFERENCES

Adam, Yuki, Gunther Meinlschmidt, Andrew T. Gloster, and Roselind Lieb. "Obsessive–compulsive disorder in the community: 12-month prevalence, comorbidity and impairment." *Social Psychiatry and Psychiatric Epidemiology* 47, no. 3 (2012): 339–349.

Adan, Roger A. H., Eline M. van der Beek, Jan K. Buitelaar, John F. Cryan, Johannes Hebebrand, Suzanne Higgs, Harriet Schellekens, and Suzanne L. Dickson. "Nutritional psychiatry: Towards improving mental health by what you eat." *European Neuropsychopharmacology* 29, no. 12 (2019): 1321–1332.

Akhondzadeh, Shahin, H. R. Naghavi, M. Vazirian, A. Shayeganpour, H. Rashidi, and M. Khani. "Passionflower in the treatment of generalized anxiety: A pilot double-blind randomized controlled trial with oxazepam." *Journal of Clinical Pharmacy and Therapeutics* 26, no. 5 (2001): 363–367.

Andreatini, Roberto, Vânia A. Sartori, Maria L. V. Seabra, and José Roberto Leite. "Effect of valepotriates (valerian extract) in generalized anxiety disorder: A randomized placebo-controlled pilot study." *Phytotherapy Research: An International Journal Devoted to Pharmacological and Toxicological Evaluation of Natural Product Derivatives* 16, no. 7 (2002): 650–654.

Arnold, Paul D., David R. Rosenberg, Emanuela Mundo, Subi Tharmalingam, James L. Kennedy, and Margaret A. Richter. "Association of a glutamate (NMDA) subunit receptor gene (GRIN2B) with obsessive-compulsive disorder: A preliminary study." *Psychopharmacology* 174, no. 4 (2004): 530–538.

Bhattacharyya, Sagnik, Sumant Khanna, Koushik Chakrabarty, Anita Mahadevan, Rita Christopher, and S. K. Shankar. "Anti-brain autoantibodies and altered excitatory neurotransmitters in obsessive–compulsive disorder." *Neuropsychopharmacology* 34, no. 12 (2009): 2489–2496.

Boileau, Bernard. "A review of obsessive-compulsive disorder in children and adolescents." *Dialogues in Clinical Neuroscience* 13, no. 4 (2011).

Brander, Gustaf, Ana Perez-Vigil, Henrik Larsson, and David Mataix-Cols. "Systematic review of environmental risk factors for obsessive-compulsive disorder: A proposed roadmap from association to causation." *Neuroscience & Biobehavioral Reviews* 65 (2016): 36–62.

Caccia, Silvio, and Marco Gobbi. "St. John's wort components and the brain: Uptake, concentrations reached and the mechanisms underlying pharmacological effects." *Current Drug Metabolism* 10, no. 9 (2009): 1055–1065.

Cannell, John Jacob. "Autism, will vitamin D treat core symptoms?" *Medical Hypotheses* 81, no. 2 (2013): 195–198.

Chakrabarty, Kaushik, Sagnik Bhattacharyya, Rita Christopher, and Sumant Khanna. "Glutamatergic dysfunction in OCD." *Neuropsychopharmacology* 30, no. 9 (2005): 1735–1740.

Chimakurthy, Jithendra, and Talasila E. G. K. Murthy. "Effect of curcumin on quinpirole induced compulsive checking: An approach to determine the predictive and construct validity of the model." *North American Journal of Medical Sciences* 2, no. 2 (2010): 81.

Del Casale, Antonio, Serena Sorice, Alessio Padovano, Maurizio Simmaco, Stefano Ferracuti, Dorian A. Lamis, Chiara Rapinesi et al. "Psychopharmacological treatment of obsessive-compulsive disorder (OCD)." *Current Neuropharmacology* 17, no. 8 (2019): 710–736.

DSM-5 American Psychiatric Association. "Diagnostic and statistical manual of mental disorders." *Arlington: American Psychiatric Publishing* 10 (2013).

Ernst, Edzard. "The risk–benefit profile of commonly used herbal therapies: Ginkgo, St. John's Wort, Ginseng, Echinacea, Saw Palmetto, and Kava." *Annals of Internal Medicine* 136, no. 1 (2002): 42–53.

Eyles, Darryl W., Thomas H. J. Burne, and John J. McGrath. "Vitamin D, effects on brain development, adult brain function and the links between low levels of vitamin D and neuropsychiatric disease." *Frontiers in Neuroendocrinology* 34, no. 1 (2013): 47–64.

Fineberg, Naomi A., and Tim M. Gale. "Evidence-based pharmacotherapy of obsessive–compulsive disorder." *International Journal of Neuropsychopharmacology* 8, no. 1 (2005): 107–129.

Fontenelle, Leonardo F., Antonio L. Nascimento, Mauro V. Mendlowicz, Roseli G. Shavitt, and Marcio Versiani. "An update on the pharmacological treatment of obsessive-compulsive disorder." *Expert Opinion on Pharmacotherapy* 8, no. 5 (2007): 563–583.

Ghandadi, Morteza, and Amirhossein Sahebkar. "Curcumin: An effective inhibitor of interleukin-6." *Current Pharmaceutical Design* 23, no. 6 (2017): 921–931.

Goddard, Andrew W., Anantha Shekhar, Aaron F. Whiteman, and Christopher J. McDougle. "Serotoninergic mechanisms in the treatment of obsessive–compulsive disorder." *Drug Discovery Today* 13, no. 7–8 (2008): 325–332.

Heinrich, Miriam, Peter Lorenz, Rolf Daniels, Florian C. Stintzing, and Dietmar R. Kammerer. "Lipid and phenolic constituents from seeds of Hypericum perforatum L. and Hypericum tetrapterum Fr. and their antioxidant activity." *Chemistry & Biodiversity* 14, no. 8 (2017): e1700100.

Hovatta, Iiris, Juuso Juhila, and Jonas Donner. "Oxidative stress in anxiety and comorbid disorders." *Neuroscience Research* 68, no. 4 (2010): 261–275.

Kasper, S., H. P. Volz, H. J. Möller, A. Dienel, and M. Kieser. "Continuation and long-term maintenance treatment with Hypericum extract WS® 5570 after recovery from an acute episode of moderate depression—A double-blind, randomized, placebo controlled long-term trial." *European Neuropsychopharmacology* 18, no. 11 (2008): 803–813.

Katiyar, Santosh K. "Silymarin and skin cancer prevention: Anti-inflammatory, antioxidant and immunomodulatory effects." *International Journal of Oncology* 26, no. 1 (2005): 169–176.

Khom, S., I. Baburin, E. Timin, A. Hohaus, G. Trauner, B. Kopp, and S. Hering. "Valerenic acid potentiates and inhibits GABAA receptors: Molecular mechanism and subunit specificity." *Neuropharmacology* 53, no. 1 (2007): 178–187.

Kobak, Kenneth A., Leslie VH Taylor, Alexander Bystritsky, Cary J. Kohlenberg, John H. Greist, Phebe Tucker, Gemma Warner, Rise Futterer, and Tanya Vapnik. "St John's wort versus placebo in obsessive–compulsive disorder: Results from a double-blind study." *International Clinical Psychopharmacology* 20, no. 6 (2005): 299–304.

Linde, Klaus, Michael M. Berner, and Levente Kriston. "St John's wort for major depression." *Cochrane Database of Systematic Reviews* 4 (2008).

Mazzio, Elizabeth A., Natholyn Harris, and Karam FA Soliman. "Food constituents attenuate monoamine oxidase activity and peroxide levels in C6 astrocyte cells." *Planta Medica* 64, no. 07 (1998): 603–606.

Miller, Michelle L., and Rebecca L. Brock. "The effect of trauma on the severity of obsessive-compulsive spectrum symptoms: A meta-analysis." *Journal of Anxiety Disorders* 47 (2017): 29–44.

Mollazadeh, Hamid, Arrigo FG Cicero, Christopher N. Blesso, Matteo Pirro, Muhammed Majeed, and Amirhossein Sahebkar. "Immune modulation by curcumin: The role of interleukin-10." *Critical Reviews in Food Science and Nutrition* 59, no. 1 (2019): 89–101.

Napoli, Edoardo, Laura Siracusa, Giuseppe Ruberto, Alessandra Carrubba, Silvia Lazzara, Antonio Speciale, Francesco Cimino, Antonella Saija, and Mariateresa Cristani. "Phytochemical profiles, phototoxic and antioxidant properties of eleven Hypericum species–A comparative study." *Phytochemistry* 152 (2018): 162–173.

Oliveira, Ana Isabel, Cláudia Pinho, Pedro Fonte, Bruno Sarmento, and Alberto CP Dias. "Development, characterization, antioxidant and hepatoprotective properties of poly (Ɛ-caprolactone) nanoparticles loaded with a neuroprotective fraction of Hypericum perforatum." *International Journal of Biological Macromolecules* 110 (2018): 185–196.

Pakseresht, Siroos, Hatam Boostani, and Mehdi Sayyah. "Extract of valerian root (Valeriana officinalis L.) vs. placebo in treatment of obsessive-compulsive disorder: A randomized double-blind study." *Journal of Complementary & Integrative Medicine* 8 (2011): J-jcim.

Panahi, Yunes, Yasin Ahmadi, Manouchehr Teymouri, Thomas P. Johnston, and Amirhossein Sahebkar. "Curcumin as a potential candidate for treating hyperlipidemia: A review of cellular and metabolic mechanisms." *Journal of Cellular Physiology* 233, no. 1 (2018): 141–152.

Papakostas, George I. "Tolerability of modern antidepressants." *Journal of Clinical Psychiatry* 69, no. Suppl E1 (2008): 8–13.

Pittler, Max H., and Edzard Ernst. "Kava extract versus placebo for treating anxiety." *Cochrane Database of Systematic Reviews* 1 (2003).

Polavarapu, Aparna, and Daphne Hasbani. "Neurological complications of nutritional disease." In *Seminars in pediatric neurology*, vol. 24, no. 1, pp. 70–80. WB Saunders, 2017.

Rahimi, Roja, Shekoufeh Nikfar, and Mohammad Abdollahi. "Efficacy and tolerability of Hypericum perforatum in major depressive disorder in comparison with selective serotonin reuptake inhibitors: A meta-analysis." *Progress in Neuro-Psychopharmacology and Biological Psychiatry* 33, no. 1 (2009): 118–127.

Reddy, A. Ch, and Belur R. Lokesh. "Studies on the inhibitory effects of curcumin and eugenol on the formation of reactive oxygen species and the oxidation of ferrous iron." *Molecular and Cellular Biochemistry* 137, no. 1 (1994): 1–8.

Sarris, Jerome, and David J. Kavanagh. "Kava and St. John's Wort: Current evidence for use in mood and anxiety disorders." *The Journal of Alternative and Complementary Medicine* 15, no. 8 (2009): 827–836.

Sarris, Jerome, Andrew A. Nierenberg, Isaac Schweitzer, Jonathan E. Alpert, Jerrold F. Rosenbaum, Nadia Iovieno, Jennifer Covino, Maurizio Fava, and David Mischoulon. "Conditional probability of response or nonresponse of placebo compared with antidepressants or St John's wort in major depressive disorder." *Journal of clinical Psychopharmacology* 33, no. 6 (2013): 827–830.

Sarris, Jerome, C. Stough, Rolf Teschke, Zahra T. Wahid, Chad A. Bousman, Greg Murray, Karen M. Savage, Peter Mouatt, Chee Ng, and Isaac Schweitzer. "Kava for the treatment of generalized anxiety disorder RCT: Analysis of adverse reactions, liver function, addiction, and sexual effects." *Phytotherapy Research* 27, no. 11 (2013): 1723–1728.

Sarris, Jerome, Emma LaPorte, and Isaac Schweitzer. "Kava: A comprehensive review of efficacy, safety, and psychopharmacology." *Australian & New Zealand Journal of Psychiatry* 45, no. 1 (2011): 27–35.

Sayyah, Mehdi, Hatam Boostani, Siroos Pakseresht, and Alireza Malayeri. "Comparison of Silybum marianum (L.) Gaertn. with fluoxetine in the treatment of Obsessive– Compulsive Disorder." *Progress in Neuro-Psychopharmacology and Biological Psychiatry* 34, no. 2 (2010): 362–365.

Singh, Yadhu N. "Potential for interaction of kava and St. John's wort with drugs." *Journal of Ethnopharmacology* 100, no. 1–2 (2005): 108–113.

Skalisz, L. L., V. Beijamini, and R. Andreatini. "Effect of Hypericum perforatum on marble-burying by mice." *Phytotherapy Research: An International Journal Devoted to Pharmacological and Toxicological Evaluation of Natural Product Derivatives* 18, no. 5 (2004): 399–402.

Soleimani, Vahid, Amirhossein Sahebkar, and Hossein Hosseinzadeh. "Turmeric (Curcuma longa) and its major constituent (curcumin) as nontoxic and safe substances." *Phytotherapy Research* 32, no. 6 (2018): 985–995.

Swedo, Susan E. "Streptococcal infection, Tourette syndrome, and OCD: Is there a connection? PANDAS: Horse or zebra?" *Neurology* 74, no. 17 (2010): 1397–1399.

Taylor, Leslie vH, and Kenneth A. Kobak. "An open-label trial of St. John's Wort (Hypericum perforatum) in obsessive-compulsive disorder." *The Journal of Clinical Psychiatry* 61, no. 8 (2000): 575–578.

Taylor, Steven. "Etiology of obsessions and compulsions: A meta-analysis and narrative review of twin studies." *Clinical Psychology Review* 31, no. 8 (2011): 1361–1372.

Teschke, Rolf, Jerome Sarris, Xaver Glass, and Johannes Schulze. "Kava, the anxiolytic herb: Back to basics to prevent liver injury?" *British Journal of Clinical Pharmacology* 71, no. 3 (2011): 445–448.

Teschke, Rolf, Jerome Sarris, and Vincent Lebot. "Contaminant hepatotoxins as culprits for kava hepatotoxicity–fact or fiction?" *Phytotherapy Research* 27, no. 3 (2013): 472–474.

Teschke, Rolf, Jerome Sarris, and Isaac Schweitzer. "Kava hepatotoxicity in traditional and modern use: The presumed Pacific kava paradox hypothesis revisited." *British Journal of Clinical Pharmacology* 73, no. 2 (2012): 170–174.

Whitten, D. L., S. P. Myers, J. A. Hawrelak, and H. Wohlmuth. "The effect of St John's wort extracts on CYP3A: A systematic review of prospective clinical trials." *British Journal of Clinical Pharmacology* 62, no. 5 (2006): 512–526.

Yaryura-Tobias, Jose A., and Hemmige N. Bhagavan. "{l}-tryptophan in obsessive-compulsive disorders." *The American Journal of Psychiatry* 134, no. 11 (1977): 1298–1299

18 Impact of Micronutrients on Obsession-Compulsion Disorders During Adolescence

Dhanalakshmi Thiyagarajan,[a,*,†] *Anitha Arumugam,*[b] *and Rohini Venugobal*[c]

[a] Associate Professor, Nehru Institute of Health Sciences, Coimbatore, Tamil Nadu, India

[b] Women Scientist, Department of Botany, Bharathiar University, Coimbatore, Tamil Nadu, India

[c] Associate Professor of Physics, Department of Science and Humanities, Nehru Institute of Engineering and Technology, Coimbatore, Tamil Nadu, India

18.1 INTRODUCTION

Adolescence is a point of conversion which consists of physical, psychological, cognitive, and social changes. Physically, they are entering puberty, and there is a change in brain organization. They are embarking on the stage of sexual interest. Furthermore, psychologically, adolescents' cognitive abilities are budding. Finally, adolescents are receiving social conversion through the school environment and other transitions like roles to play in the family and community (National Research Council [NRC], 2002). These evolutions happen concurrently and at various paces for every adolescent within each sexual category. Adolescents' development is being impacted frequently by structural and environmental factors (Victor, 2005).

Adolescents' disappointment to not having expectations met, and as they are not accustomed to these transitions, direct to pessimistic health behaviors. Adolescence is the most challenging stage of life, and it is quite common for teens to cultivate their own traits and eccentricities as they make an effort to steer the turmoil. These habits and uncommon attitudes are signs of a possible psychological health issue among young adults, and one of these is obsessive-compulsive disorder (OCD). This psychiatric mess can lead to noteworthy anguish, and OCD may become devastating in more serious cases of the disorder.

Perhaps these changes will create stress and anxiety among them, and they will get obsessive thoughts and compulsion behaviors. For example, nail biting, skin picking, hair pulling, writing and rewriting assignment, checking and rechecking, and etc. The keystone of physical health is good nutrition; however many researches report diet's upshot on psychological health as well. One of the special sections in clinical psychological science indicates the various approaches that psychological studies are taking to know some ways in which nutrition and mental health interconnect (Asle & Sverre, 2020).

*Corresponding author.

†This is the first author.

 DOI: 10.1201/9781003387893-21

There are lots of evidence that have assessed wide-ranging micronutrients formulas to get rid of OCD (Rucklidge, 2009; Kaplan et al., 2002), mood anxiety disorders (Kaplan et al., 2001, 2002, 2004; Rucklidge et al., 2010; Gately & Kaplan, 2009; Erazier et al., 2009; Popper, 2001; Simmons, 2001; Rucklidge & Harrison, 2010), attention deficit hyperactivity disorder (ADHD) (Rucklidge & Harrison 2010; Rucklidge et al., 2011), mood and tetchiness symptoms associated with autism (Mehl-Madrona et al., 2010), and aggressive or antisocial behaviors in cases not medically referred, for example, schoolchildren (Schoenthaler & Bier, 2000) and young delinquents (Gesch et al., 2002). Hence, this chapter focuses on how micronutrients are used to get rid of obsession behaviors and compulsive thoughts during adolescent period of human development.

18.1.1 OCD

Obsessive-compulsive disorder (OCD) comprises of two main parts, the "obsession thoughts" and the "compulsion behaviors."

18.1.1.1 Obsession

Obsessions are relentless, invasive and unnecessary, unacceptable, or senseless thoughts, metaphors, or urges. These thoughts trigger prejudiced misery such as anxiety, fright, and uncertainty of negative effect of some physical and mental issues. Nevertheless, they are not only everyday fears about subjects like job, health, dealings, or economic conditions. Obsessions usually relay to a restricted number of subjects, which are described in Table 18.1 (Jonathan et al., 2015).

The obsessions can be segregated into two classes, autogenous and reactive, on the basis of the way they provoke and are experienced in recent research (Lee & Kwon, 2003; Lee & Telch, 2010; Lee et al., 2005). Autogenous obsessions are feelings and images that enter into the consciousness with or without particular triggers. In practice they are extremely disgusting and stressful, and the person afflicted experiences intolerable sexual, antagonistic, or wicked ideas, misgivings, images, or desires. The other class of reactive obsessions are triggered by identifiable circumstances such as when driving, in the bathroom, and might seem as rational behaviors, such as checking, washing compulsions. As seen in Table 18.1, many people with OCD show multiple types of obsessions (Jonathan et al., 2015).

18.1.1.2 Compulsive Behaviors

Compulsions are unrelenting, extreme, recurring behaviors such as checking, washing, ordering and psychological mannerisms such as counting, praying, repeating words silently. The aim of

TABLE 18.1
Common Obsessions (Jonathan et al., 2015)

Categories	Example
Contamination	Anxiety of germs, virus, bacteria from body fluids
Responsibility for harm	Dread of losing resistor of themself by doing on thoughts or instincts, (hurting robbery, ejaculating out swear utter words or becoming ferocious
Sex and morality	feelings or metaphors about children, relatives, lesbian, and/or violence
Violence	feeling of shooting someone with the breadknife, missing adored ones from their lives in dreadful coincidence
Religion	Fear about upsetting Divinity or about living a ethical life
Symmetry and order	A fascination on firm numbers and/or colors being worthy/right or evil/improper

TABLE 18.2
Common Compulsive Rituals (Jonathan et al., 2015)

Categories	Example
Decontamination	Frequently washing, cleaning and wiping hands and other tilings and even over and again bathing
Checking	Checking which not hurt others and themselves, not make any eixor, often checking some parts of electronic goods, house appliances and etc.
Repeating routine activities	Rewriting and rereading assignments and repeating actions such as hair pulling, tapping, touching, getting up and sit on the chair.
Ordering/arranging	Saying the word yes whenever hearing the word correct and setting figures on the wall until they are dangled.
Mental rituals	Substituting awful feeling by thinking of a great thought and often prays the almighty until it is said right.

compulsions is to avoid grief or foil alarmed incidence or circumstances; nevertheless, the compulsions frequently show unwanted and quixotic behaviors associated to the obsessions. Not all rituals or types of repetitive attitudes are compulsions, such as bedtime habits, religious practices, and learning a new ability. Compulsions are all reactions to obsessions that decrease anxiety, manage or oppose unnecessary thoughts, or ease uncertainties that are consequent characteristics of obsessions. Compulsive rituals, evading, and counterbalancing all function as anxiety-reduction approaches. Table 18.2 shows some common obvious rituals in OCD. (Jonathan et al., 2015).

18.1.2 Development of Adolescents

Adolescence is the most swift stage of transition in the human development. Physiological development leads to psychological maturity; however individual and situation attributes have been influencing the transitions of adolescence. This phase has health consequences not only in teenage years but the entire life span and which is an acknowledged phase of life in many more societies across the globe. The teenage period begins at approximately the onset of puberty, and then further into adult roles of job and marital stage (Schlegel & Barry, 1991). There are some controversies among the people as to how to accurately explain adolescence; it is connected to the second decade of life though. Many reports differ regarding the start of adolescence; it may be 11 or 12 to 14 years, mid-teenage is 15 to 17 years, and late youth is 18 to 21 years. One statistical report says that around 10.5 million people are between the ages of 10 and 19, more than 20% of the inhabitants in the world (Statistics South Africa, 2011; Richter, 2006).

On the view of development, adolescence is both a thrilling and an exigent time. Biologically, teenagers' bodies are maturing, and cognitively, adolescent start to think about the different paths they can take in the world. Socially, converting dealings with family relationships, acquaintances, and colleagues play a main role in carving their experiences (Lauren & Sharlene, 2012). Developmental targets of the teenage period in Western societies are the following: creating a triumphant transition to high school education, improving academic skills which are wanted for higher education level/work, accomplishing emotional freedom, figuring out intimate friendships with same and opposite gender, and developing an intelligence of uniqueness (Cicchetti & Rogosch, 2002). Nevertheless, cultural, historical, and political occasions may shape adolescents and how society thinks about them as well through the demands and stress it keeps on them, and by what civil liberties and occasions it offers them as they embark on their voyage into their teenage years (Lauren & Sharlene, 2012).

18.1.3 Psychosocial Development of Adolescents

Psychosocial development is another vital change that takes place in adolescence (Derman, 2013), and "self-definition and personality development" arise during this time (Michael et al., 2013). In connection with hormonal and neurodevelopmental transitions that lead to psychosocial and emotional changes and gradually increasing cognitive and intellectual capabilities that develop into rigid reasoning talents, common sense, and ethical thinking. Furthermore, capacity of conceptual thinking and creating lucid sentence structures develop. Age-specific duties and deeds that show adult roles are viewed in self-definition. The adolescent increasingly becomes an individual who is accustomed to social tasks, aims to live their life on their own desires, presumes their levels of responsibility, and determines their own personality by resuming new relationships (Kurtman, 2005). Teens' desires and interests changes from family to outside members, such as acquaintances and other relationships. Their interest in their studies declines and gets disrupted, which can lead to failing in schooling, and adolescents provide negative responses to their parents. This relationship moves from dependence to freedom (AkçanParlaz et al., 2012; Güler et al., 2009). Adolescents moving away from family can head over to sensation of misery, solitude, and uncertainty (Karadamar et al., 2014). There are many ideas (ego, uniqueness, temperament, and character) used to elucidate personality development and psychosocial changes during teen years (Kurtman, 2005).

18.1.4 Psychological Risk of Adolescents

Adolescence is a crucial and cautious transition period of human growth, and many more changes take place physiologically and psychologically, according to their environments and satisfaction of their expectations. Youths have particular expectations from the family, the self, and the society, and some of these are social recognition, expectation of adoration from family members and others, self-respect, interest in success, and having biological stamina and prettiness (Akman et al., 2012; Karadamar et al., 2014). The adolescents' disappointment should they fail to meet these anticipations or become accustomed to these changes can directly lead to harmful health issues and manners. Unhappily, for a small proportion of adolescents who fail to improve their own habits and abnormal behaviors, this could be a sign of a severe psychological health issues, such as obsessive-compulsive disorder, and cause considerable distress. Recent studies report that technology intervention among the young generation, such as social media, can affect psychology deeply. This study emphasized that persons with high OCD symptoms seemed to place a larger importance to social media, and this can be make OCD people hypothetically even more susceptible to the adverse consequences of social media (Guazzini et al., 2022).

OCD may make everyday life traumatic for teens, and compulsion habits consume a lot of time, which can make one late for school or other commitments. This can lead to anxiety or arguments with family. Especially at school, OCD influences attention and focus, affecting things like finishing tasks and school attendance through the checking, writing, rewriting, erasing, redoing assignment. Adolescents can become distressed that they are mad and make an effort to camouflage their OCD from others, which can get exhausting throughout the day. Therefore, all this stress leads them to become gloomy, annoyed, or bursting at risk condition. Risk activities may be defined as any behavior that keeps someone at peril for negative physical, psychological, or social effects, and it can be for short term or long term. This can tend to go high in the late youth and near the beginning twenties (Arnett, 1999). According to the World Health Organization (2011), virtually two-thirds of hasty fatalities are related with circumstances or behaviors that begin in adolescence. Hence, OCD-affected adolescents, and adolescents in general, should be given counseling with psychiatric medication, and the best treatment is giving them some nutrients and micronutrient supplements. This may be helpful in recovering them without any side effects and they become healthy persons as well.

18.1.5 Diagnostic Assessment

An accurate diagnosis has been found for OCD after many years (Stengler et al., 2013). It took many years to detect in teens during development, and many signs of awkwardness and attempts to camouflage symptoms made it difficult to distinguish real OCD from normal rituals (Evans et al., 1997). Besides, even as OCD is frequently connected with an attribute set of symptoms such as overwashing and repeated checking, this disorder is noticeably heterogeneous; two persons with

OCD may have exclusively nonoverlapping symptom summaries, which makes it more difficult to diagnose. However, the six-question screening method used to identify most of the pediatric OCD cases and the short screener (SOCS) is suggested by the National Institute for Health and Clinical Excellence. The short OCD screener is as follows and which has 97% sensitivity (Uher et al., 2007).

1. Do you wash or clean many times?
2. Do you ensure things a lot?
3. Is there any thought that makes you anxious and that you would want to get away from but cannot?
4. Does it take you a long time to finish your daily activities (e.g. getting ready for school)?
5. Are you worried about arranging things in a special order or are you distressed by muddles?
6. Do these issues trouble you?

As it is not used as a diagnostic instrument, additional appraisal is needed in persons who tested positive, including getting a detailed background history of obsessions and compulsions and developmental details and an individual interview with the adolescent (Barton & Heyman, 2009). And last, especially important given that there are general "taboo" obsessions (e.g., sexual obsessions), and the adolescent may be loathed to open up about them in front of family members (Fernández et al., 2013).

Another method of diagnosis is differential diagnosis, and after admission in a hospital, the person had a conditional diagnosis of "not otherwise specified" (NOC)/OCD/borderline intellectual functioning. Pediatric autoimmune neuropsychiatric disorder associated with strep was included to the person's list of diagnosis, as well as generalized anxiety disorder (GAD) and social anxiety disorder (Megan et al., 2012).

18.1.6 Treatments for OCD

Neuropsychiatric disorder is a kind of obsessive-compulsive disorder that commonly presents in adolescence, and comorbid disorders (e.g., mood, anxiety, psychotic and somatoform disorder) are usual in patients with OCD (Adam et al., 2012). Structural, neurochemical, hereditary, and immunological factors take part in the etiology and pathophysiology of OCD (Boileau, 2011). Changes in certain neurotransmitters are complicated in OCD symptoms and serotonin dearth, and the abnormal function in the accessibility of serotonin transporter (5-HIT) have been implicated in OCD symptomatology (Goddard et al., 2008). Besides, increased glutamate, glutamine, and gamma-aminobutyric acid (GABA) levels in the caudate nucleus (Bolton et al., 2001) and augmented glutamate levels in the cerebrospinal fluid have been detected in patients with OCD (Chakrabarty et al., 2005), and also, the glutamatergicionotropic N-methyl D-aspartate (NDA) receptor has been linked with OCD (Arnold et al., 2004).

The pandemic situation created a huge space in the availability of psychotherapy, and especially good quality treatment like CBT for OCD (Storch et al., 2021). CBT is the top treatment for OCD. Besides this dearth of obtainability and approachability being intrinsically challenging, hence treatments that are unapproachable to the common people are not able to have a significant public psychological effect. So Lundström et al. (2022) determined the main research questions to determine whether internet-based CBT (ICBT) must be suggested for incorporation in healthcare settings. But there are some drawbacks in ICBT, such as cost.

The suggested treatments for OCD are CBT (cognitive-behavioral therapy) using experience reaction prevention (Ougrin, 2011) with pharmacotherapy with selective serotonin reuptake suppressors (venlafaxine or duloxetine) (Dell'Osso et al., 2006; Fineberg & Gale, 1999). The results were inadequate, with probable side effects of current treatments, so more studies need to be conducted to find out substitute remedies. The patients with a mental disorder have been diagnosed with several nutritional deficiencies, therefore, nutritional supplements may be an effective treatment (Kuygun & Gül Celik, 2020), and let's discuss further this concept in following section.

18.1.7 Effects of Nutrients and Micronutrients on OCD Adolescents

"Let food be thy medicine, and medicine be thy food" is a saying by Hippocrates (Lucock, 2004). Perhaps he was referring to using food to rear both physical and psychological fitness, and many evidences provide information about everything, from guidelines to be concerned of with livestock to treatment of TB (tuberculosis) and heart diseases (Ritter, 1910). It is remarkable that this handy source of information affirmed that mental sickness was known to be the effect of "imperfect nutrition." In the beginning of the mid-nineteenth century, there was a change in the development of psychiatric medications for curing of the "mental ill." Many medical journal articles from the nineteenth century onwards reported that symptom developed as a consequence of interferences with nutrients. There were lots of case studies (Goggans, 1984) and controlled trials (Wyatt et al., 1999) showing betterment of psychiatric symptom with B vitamins in the late twentieth century. Correspondingly, in the twenty-first century, there were many case series (Heiden et al., 1999) and controlled trials signifying amelioration in mood stability with dietary minerals (magnesium). It is noteworthy, nevertheless, that until Bell's report in 1992, most interventions for mental illness consisted of a single nutrient, and in his study, a combination of vitamins B1, B2, and B6 was detected to enlarge the therapeutic impact of tricycUc antidepressants in 14 seniors suffering from depression (Bell et al., 1992). Consequently, some more research made evident that nutrients which augment serotonin levels would decrease the symptoms of OCD. The amino acid tryptophan is a precursor to serotonin, and treatment of OCD by giving them with tryptophan supplements would boost serotonin levels (Yaryura-Tobias & Bhagavan, 1977).

18.1.8 Multiple Micronutrients for Treatment of OCD and Other Mental Disorders

The general treatment of delusions and illusions is usually psychotic medications, and these substances have been linked with major adverse events in the short-term, such as sleepiness, stiffness, constipation, obesity, etc., and furthermore, long-term increased peril to health includes diabetics, heart diseases, etc. (McIntyre & Jerrell, 2008). There are many research and case study reports that have estimated broad-spectrum micronutrient formulas for mental illnesses like OCD (Rucklidge, 2009; Kaplan et al., 2002), mood and anxiety disorders (Kaplan et al., 2002; Rucklidge et al., 2010; Gately & Kaplan, 2009; Erazier et al., 2009; Kaplan et al., 2004; Kaplan et al., 2001; Popper, 2001; Simmons, 2003; Rucklidge & Harrison, 2010), as well as antagonistic or antisocial behaviors in samples (schoolchildren and adolescent criminals) (Schoenthaler & Bier, 2000; Gesch et al., 2002).

Conventional logical methodology needs the treatment of only one independent variable at a time, however medical research with nutrients has started to create much evolution by incorporating complex independent variables, normally formulas having impartial amount of micronutrients such as vitamins, minerals, amino acids, and essential fatty acids (omega-3). This formula solves many human problems and can be taken in balanced combinations supplemented naturally by foods (Mitchell et al., 2003). Even though some other formulas are reported and found literature has a single multimicronutrient supplement which makes use of a 36-ingredient formula called "EMPowerplus," Gesch and his colleagues (2002) have formulated more resembled formula, which consists of 26 micronutrients to treat 172 teens prisoners (Gesch et al., 2002).

As seen in Table 18.3, the EMPowerplus (EMP+) consists of amino acids, minerals, vitamins, and antioxidants, and many independent research groups and research studies are targeting psychological health of humans with this multimicronutrient formula. Recently, there have been 17 psychological health articles published on EMP+, reproduced by scientists at various academic and research institutions as well as clinicians in personal practice. The researchers have reported that EMP+ benefited three countries in the cure of mood and anxiety symptoms in children and adults by using many studies, such as case-control studies, within-subject case studies, open-label case series, case reports with general chronological treatment information,

TABLE 18.3
EMPowerplus Capsules (Kaplan & Leung, 2011)

Components	Quantity
Vitamin A	1536IU
Vitamin C	160 mg
Vitamin D	384 IU
Vitamin E	96 IU
Vitamin B1	4.8 mg
Vitamin B2	3.6 mg
Vitamin B3	24 mg
Vitamin B5	5.8 mg
Vitamin B6	9.6 mg
Vitamin B9	384 meg
Vitamin B12	240 meg
Vitamin H	288 meg
Calcium	352 mg
Phosphorous	224 mg
Magnesium	160 mg
Potassium	64 mg
Iodine	54.4 meg
Zinc	12.8 mg
Selenium	54.4 meg
Copper	1.9 mg
Manganese	2.6 mg
Chromium	166.4 meg
Molybdenum	38.4 meg
Iron	3.7 mg
Proprietary Blend	444.1 mg
dl-phenylalanine, glutamine, citrus bioflavonoids, grape seed extract, chohnebitartrate,, inositol, ginkgo biloba, methionine, germanium ses qui oxide, boron, vanadium, nickel	Trace
Other Ingredients for encapsulations: gelatin, silicon, magnesium stearate	

and huge database analyses (Frazier et al., 2009, 2012; Mehl-Madrona et al., 2010; Rucklidge et al., 2012; Rucklidge & Blampied, 2011; Rucklidge, Taylor et al., 2011; Rucklidge, Johnstone et al., 2011; Rucklidge, Harrison et al., 2011; Rucklidge et al., 2010; Rucklidge & Harrison, 2010; Rucklidge, 2009; Gately & Kaplan, 2009; Kaplan et al., 2004; Kaplan et al., 2002; Kaplan et al., 2001; Simmons, 2003; Popper, 2001).

EMPowerplus formula has been shown to ameliorate the targeted variety of symptoms, as shown in the the results of the published 13 treatment studies. However, the aim was on unsteady mood, such as bipolar symptoms and explosive rage, in the early period, the publications from 2001 to 2002, and currently there have been reports of reduced anxiety and ameliorated OCD symptoms, besides the abolition of symptoms of mental illness such as illusions (Kaplan & Leung, 2011). The reports on EMPowerplus have delivered many considerable support for using this formula to cure a plethora of psychological health disorders. The strong point of these studies have comprised the level of particular on the patient history, including results of psychiatric diagnosis and symptoms, medication use for how many numbers and types of medications, and how long should have follow up observations with help of standardized psychological methods (Kaplan & Leung, 2011). Thus, it is obvious that micronutrients which ameliorate serotonin levels would decrease the symptoms of OCD in adolescents and other age groups.

18.1.9 Perspectives

Obsessive-compulsive disorder is an apprehension disorder which makes repetitive stressful thoughts and obsessions, which is followed by uncontrollable compulsive behaviors. These kinds of disorders will occur particularly during the adolescent stage of human development. This transition period is crucial and swiftly develops from childhood to adulthood and to become mature examples for the next generation in the society. They play a key role to create a healthy, wealthy, and peaceful society without any criminal offence. Adolescents' conversion phase is entirely dependent on the biological, psychoscocial, and emotional environments to mold them in perfection. Any disappointment and desperate happening for the teens/youth may lead them into perilous conditions, to themselves and to society as well. The young teens may develop anxiety, distress, and other psychological strain, which can lead to mental health disorders. These disorders should be properly treated by psychiatric counseling with medications; however, the pharmacological therapy can lead to some health illnesses, such as drowsiness, obsession, stomach obstruction, diabetes, and heart diseases, as side effects. Many research and case studies reported and replaced with micronutrients instead of pharmacotherapy. These micronutrients have the capability to induce serotonin levels and reduce the symptoms of OCD among OCD adolescents and improve their mental health issues. In my perspective, the adolescence phase is perilous but can create good circumstances. Society, parents, teachers, tutors, governments, and NGOs should consider to lead them in the proper path and give them counseling by a psychiatrist with perfect nutrient supplements as a traditional and natural treatment without any side effects. Therefore, this chapter clearly mentioned that nutrients and micronutrients would augment serotonin levels, which diminishes the symptoms of OCD (Kuygun & Gül Celik, 2020).

REFERENCES

Adam Y., Meinlschmidt G., Gloster A. T. 2012. Obsessive–compulsive disorder in the community: 12-month prevalence, comorbidity and impairment. *Soc Psychiatry PsychiatrEpidemiol,* 4, 339–349.

AkçanParlaz E., Tekgül N., Karademirci E., Öngel K. 2012. Adolescence period: Physical growth, psychological and social development process. *Turkish Family Physician*, 3, 10–16.

Akman M., Tüzün S., Ünalan P. C. 2012. Healthy eating patterns and physical activity status of adolescents. *Nobel Medicus,* 8(1), 24–29.

Arnett J. J. 1999. Adolescent storm and stress, reconsidered. *American Psychologist*, 54, 317–326.

Arnold P., Rosenberg D., Mundo E. 2004. Association of a glutamate (NMDA) subunit receptor gene (GRIN2B) with obsessive-compulsive disorder: a preliminary study. *Psychopharmacology* 174, 530–538.

Asle H., Sverre U. J. 2020. Latent trait, latent-trait state, and a network approach to mental problems and their mechanisms of change. *J. of Clin. Psychol. Sci*, 8(4), 595–613.

Barton R., Heyman I. 2009. Obsessive–compulsive disorder in children and adolescents. *Paediatr Child Health* 19, 67–72.

Bell I., Edman J., Morrow E., Marby B., Perrone G., Kayne H., Greenwald M. 1992. Cole J. Vitamin Bl, B2, and B6 augmentation of tricyelie antidepressant treatment in geriatric depression with cognitive dysñinction./,4m. *CollNutr*11, 159–163.

Boileau B. 2011. A review of obsessive-compulsive disorder in children and adolescents. *Dialogues Clin Neurosci* 13, 401–411.

Bolton J., Moore G. J., MacMillan S. 2001. Case study: Caudate glutamatergic changes with paroxetine persist after medication discontinuation in pediatric OCD. *J Am Acad Child Adolesc Psychiatr* 40, 903–906.

Chakrabarty K., Bhattacharyya S., Christopher R. 2005. Glutamatergic dysfunction in OCD. *Neuropsychopharmacol* 30, 1735–1740.

Cicchetti D., Rogosch F. A. 2002. A developmental psychopathology perspective on adolescence. *J Consult Clin Psychol* 70, 6–20.

Dell'Osso B., Nestadt G., Allen A. 2006. Serotonin-Norepinephrine reuptake inhibitors in the treatment of obsessive-compulsive disorder. *J Clin Psychiatr* 67, 600–610.

Derman O. 2013. Adolescents physical, sexual development. *Turkey Clinics J Fam Med-Special Topics* 4(1), 7–11.

Erazier E. A., Eristad M. A., Arnold L. E. 2009. Multinutrient supplement as treatment: Literature review and ease report of a 12-year-old boy with bipolar disorder. *J Child Adolesc Psychopharmacol* 19, 453–460.

Evans D. W., Leckman J. F., Carter A. R. 1997. Habit, and perfectionism: The prevalence and development of compulsive-like behavior in normal young children. *Child Dev* 68, 58–68.

Fernández de la Cruz L., Barrow F., Bolhuis K. 2013. Sexual obsessions in pediatric obsessive-compulsive disorder: Clinical characteristics and treatment outcomes. *Depress Anxiety* 30, 732–740.

Fineberg N. A., Gale T. M. 1999. Evidence-Based pharmacotherapy of obsessive–compulsive disorder. *Int J Neuropsychopharmacol* 8, 107–129.

Frazier E. A., Fristad M. A., Arnold L. E. 2009. Multinutrient supplement as treatment: Literature review and case report of a 12-year-old boy with bipolar disorder. *J Child Adolesc Psychopharmacol* 19, 453–460.

Frazier E. A., Fristad M. A., Arnold L. E. 2012. Feasibility of a nutritional supplement as treatment for pediatric bipolar spectrum disorders. *J Altern Complement Med* 18, 678–685.

Gately D., Kaplan B. J. 2009. Database analysis of adults with bipolar disorder consuming a micronutrient formula. *Clin Med Insights: Psychiatr* 4, 3–16.

Gesch C. B., Hammon S. M., Hampson S. E., Eves A., Crowder M. J. 2002. Influence of supplementary vitamins, minerals and essential fatty acids on the antisocial behaviour of young adult prisoners. *Br J Psychiatr* 181, 22–28.

Goddard A. W., Shekhar A., Whiteman A. F. (2008). Serotoninergic mechanisms in the treatment of obsessive–compulsive disorder. *Drug Discov Today* 13, 325–332.

Goggans F. A. (1984). case of mania secondary to vitamin B12 deficiency. *Am J Psychiat* 141, 300–301.

Guazzini A., Gursesli M. C., Serritella E., Tani M., Duradoni M. 2022. Obsessive-compulsive disorder (OCD) types and social media: Are social media important and impactful for OCD people? *Eur J Investig Health Psychol Educ* 12, 1108–1120. https://doi.org/10.3390/ejihpe12080078.

Güler Y., Gönener H. D., Altay B., Gönener A. 2009. Adolescents obesity and nursing care. *Firat Health Ser J* 4(10), 165–182.

Heiden A., Erey R., Presslich O., Blasbichler T., Smetana R., Kasper S. 1999. Treatment of severe mania with intravenous magnesium sulphate as a supplementary therapy. *Psychiatry Res* 89, 239–246.

Jonathan S. A., Jacoby R. J. 2015. Obsessive-compulsive and related disorders: A critical review of the new diagnostic class. *Annu Rev Clin Psychol* 11, 165–186.

Kaplan B. J., Crawford S. G., Gardner B., Earrelly G. 2002. Treatment of mood lability and explosive rage with minerals and vitamins: Two ease studies in children. *J Child Adolcsc Psychopharmacol* 12, 205–219.

Kaplan B. J., Fisher J. E., Crawford S. G. 2004. Improved mood and behavior during treatment with a mineral-vitamin supplement: An open-label case series of children. *J. Child Adolesc Psychopharmacol* 14, 115–122.

Kaplan B. J., Simpson J. S., Eerre R. C., Gorman C. P., McMullen D. M., Crawford S. G. 2001. Effective mood stabilization with a chelated mineral supplement: An open-label trial in bipolar disorder. *J Clin Psychiatry* 62, 936–944.

Kaplan F. B., Leung B. 2011. Micronutrient treatment of mental disorders. *Integrat Med* 10, 3–4.

Karadamar M., Yİğİt R., Sungur M. A. 2014. Evaluation of healthy lifestyle behaviours in adolescents. *Anatolian J Nurs Health Sci* 17(3), 131–137.

Kurtman E. 2005. *Our behavior development and learning* (4th ed.). Samsun: Eser of set Matbaacılık Ltd. Şt., pp. 28–32.

Kuygun Karcı C., Gül Celik G. 2020. Nutritional and herbal supplements in the treatment of obsessive compulsive disorder. *General Psychiatr* 33, e100159. doi:10.1136/gpsych-2019–100159.

Lauren W., Sharlene S. 2012. Adolescence, Chapter_7_CAD.indd, pp. 213–244.

Lee H. J., Kwon S. M. 2003. Two different types of obsession: Autogenous obsessions and reactive obsessions. *Behav Res Ther* 41(1), 11–29.

Lee H. J., Kwon S. M., Kwon J. S., Telch M. J. 2005. Testing the autogenous-reactive model of obsessions. *Depress Anxiety* 21(3), 118–129.

Lee H. J., Telch M. J. 2010. Differences in latent inhibition as a function of the autogenous-reactive OCD subtype. *Behav Res Ther* 48(7), 571–790.

Lucock M. 2004. Is folic acid the ultimate functional food component for disease prevention? *BMJ* 328, 211–214.

Lundström L., Flygare O., Andersson E. 2022. Effect of internet-based vs face-to-face cognitive behavioral therapy for adults with obsessive-compulsive disorder: A randomized clinical trial. *JAMA Netw Open* 5(3), e221967. doi:10.1001/jamanetworkopen.2022.1967.

McIntyre R. S., Jerrell J. M. 2008. Metabolic and cardiovascular adverse events associated with antipsychotic treatment in children and adolescents. *Arch Pediatr Adolesc Med* 162, 929–935.

Megan R., Annette V., Amany W., Helen L., Elske B., Bonnie J. K. 2012. Efficacy and cost of micronutrient treatment of childhood psychosis. *BMJ Case Reports*, 1–7, bcr2012007213.

Mehl-Madrona L., Leung B., Kennedy C., Paul S., Kaplan B. J. 2010. Micronutrients versus standard medication management in autism: A naturalistic case-control study. *Child Adoksc Psychopharmaeol* 20, 95–103.

Michael E. L., Bornstein M. H. 2013. *The Oxford handbook of social cognition*. New York: Oxford University Press, pp. 1–15.

Mitchell B. L., Ulrich C. M., McTiernan A. 2003. Supplementation with vitamins or minerals and immune function: Can the elderly benefit? *Nutr Res* 23, 1117–1139.

National Research Council. 2002. *Community programs to promote youth development*. Washington, DC: National Academies Press.

Ougrin D. 2011. Efficacy of exposure versus cognitive therapy in anxiety disorders: Systematic review and meta-analysis. *BMC Psychiatr* 11, 200.

Popper C. W. 2001. Do vitamins or minerals (apart from lithium) have mood-stabilizing effects? *J Clin Psychiatr* 62, 933–935.

Richter L. M. 2006. Studying adolescence. *Sci* 312, 1902–1905.

Ritter T. J. 1910. *The people's home medical book*. Cleveland, OH: The R. C. Barnum and Co., pp. 142–150.

Rucklidge J. J. 2009. Successful treatment of OCD with a micronutrient formula following partial response to cognitive behavioral therapy (CBT): A case study *J Anxiety nísoríí* 23, 836–840.

Rucklidge J. J., Andridge R., Gorman B. 2012. Shaken but unstirred? Effects of micronutrients on stress and trauma after an earthquake: RCT evidence comparing formulas and doses. *Hum Psychopharmacol* 27, 440–454.

Rucklidge J. J., Blampied N. M. 2011. Post-earthquake psychological functioning in adults with attention-deficit/hyperactivity disorder: Positive effects of micronutrients on resilience. *J Psychol* 40, 51–57.

Rucklidge J. J., Gately D., Kaplan B. J. 2010. Database analysis of children and adolescents with bipolar disorder consuming a micronutrient formula. *BMC Psychiatr* 10, 74.

Rucklidge J. J., Harrison R. 2010. Successful treatment of bipolar disorder II and ADHD with a micronutrient formula: A case study. *CNS Spectr* 15, 289–295.

Rucklidge J. J., Harrison R., Johnstone J. 2011. Can micronutrients improve neurocognitive functioning in adults with ADHD and severe mood dysregulation? A pilot study. *J Altern Complement Med* 17(12), 1125–1131.

Rucklidge J. J., Johnstone J., Harrison R., Boggis A. 2011. Micronutrients reduce stress and anxiety in adults with attention-deficit/hyperactivity disorder following a 7.1 earthquake. *Psychiatry Res* 189(2), 281–287.

Rucklidge J., Taylor M., Whitehead K. 2011. Effect of micronutrients on behavior and mood in adults with ADHD: Evidence fi-om an 8-week open label trial with natural extension. *J Attend Discorder* 15, 79–91.

Schlegel A., Barry H. 1991. *Adolescence: An anthropological enquiry*. New York, NY: Free Press. ISBN 0029278953, pp. 215–228.

Schoenthaler S. J., Bier I. D. 2000. The effect of vitamin-mineral supplementation on juvenile delinquency among American schoolchildren: A randomized, double-blind placebo-controlled trial. *J Altern Complement Med* 6, 7–17.

Simmons M. 2001. Nutritional approach to bipolar disorder. *J Clin Psychiatr* 2003, 64, 338; author reply 338–9., Popper CW. Do vitamins or minerals (apart from lithium) have mood-stabilizing effects? J *Clin Psychiatr* 62, 933–935.

Simmons M. 2003. Nutritional approach to bipolar disorder. *Clin Psychiatry* 64, 338; author reply 338–339.

Statistics South Africa. 2011. *Mid-year population estimates 2011* (Statistical Release P0302). Pretoria: Statistics South Africa.

Stengler K., Olbrich S., Heider D. 2013. Mental health treatment seeking among patients with OCD: Impact of age of onset. *Soc Psychiatry PsychiatrEpidemiol* 48, 813–819.

Storch E. A., Sheu J. C., Guzick A. G. 2021. Impact of the COVID-19 pandemic on exposure and response prevention outcomes in adults and youth with obsessive-compulsive disorder. *Psychiatry Res* 295, 113597. doi:10.1016/j.psychres.2020.113597.

Uher R., Heyman I., Mortimore C. 2007. Screening young people for obsessive compulsive disorder. *Br J Psychiatry* 191, 353–354.

Victor P. K. 2005. Adolescent development: Perspectives and frameworks-A Discussion Paper, United Nations Children's Fund (UNICEF), New York, pp. 1–2.

World Health Organization. 2011. *Young people: Health risks and solutions* (Fact Sheet No. 345). Geneva: World Health Organization.

Wyatt K. M., Dimmock P. W., Jones P. W., O'Brien P. M. S. 1999. Efficacy of vitamin B-6 in the treatment of premenstrual syndrome: Systematic review. *BMJ* 318, 1375–1381.

Yaryura-Tobias J. A., Bhagavan H. N. 1977. L-tryptophan in obsessive compulsive disorders. *Am J Psychiatry* 134(11), 1298–1299.

19 Immune System and Obsessive-Compulsive Disorder

Intervention Through Immunotherapy

*Neelakanta Sarvashiva Kiran, Gorthi Vaishnavi, Gaonkar Lowkesh, Chandrashekar Yashaswini, and Senthilkumar Rajagopal**
Department of Biotechnology, School of Applied Sciences, REVA University, Bengaluru—560064, KA, India

19.1 INTRODUCTION

Obsessive-compulsive disorder has been defined either by obsessions or compulsive behaviors that causes the person suffering from it to feel extremely distressed. OCD is characterized by a pattern of unpleasant thoughts and anxieties (obsessions) that cause you to engage in repetitive actions (compulsions). These compulsive thoughts and behaviors disrupt daily life and cause severe distress. According to statistics, 1.160% to 1.80% of the world's population suffers from OCD (Wahl et al., 2010). Although only 1.2% of OCD cases are officially diagnosed, the prevalence of OCD is approximately 2.20% in the USA. Obsessions and compulsions are repetitive ritualistic activities that people with OCD, or those with increased risk of getting OCD, engage in to try to avoid, lessen, or completely remove distress or the feared effects of the obsessions. Typically, the person views the intrusions as unwarranted or excessive and makes an effort to resist those (Rapinesi et al., 2019). A small percentage of obsessions are thought to be exaggerated beliefs and, very rarely, illusions. Because of its frequency, accompanying expenses, and challenges in diagnosing the condition, OCD is a serious public health concern. In general, covert and mental compulsions are more difficult to control or resist than overt ones since they are more "portable" and convenient to carry out (Overduin et al., 2012). OCD has significant diagnostic comorbidity with some other mental diseases, which makes measuring it difficult. Concerns about contamination along with washing and maybe even cleaning, worries about harming oneself or others along with checking, intrusive aggressive as well as sexual thoughts along with mental rituals, and worries about symmetry along

*Corresponding author: Dr. R. Senthilkumar, PhD, Department of Biotechnology, School of Applied Sciences, REVA University, Bengaluru—560064, KA, India. Email: senthilanal@yahoo.com

 DOI: 10.1201/9781003387893-22

with ordering or counting are common examples of sequences of obsessive-compulsive disorder in OCD patients. Obsessions are unrelated to daily worries, which are present in generalized anxiety disorder, perceived physical flaws, which are present in body dysmorphic disorder, or worry about developing a serious illness, which is present in health anxiety (Stein et al., 2019). There are OCD-specific treatment plans that combine psychotherapy, medication, or neuromodulatory therapies that can lessen symptoms and induce remission. However, despite the benefits of these treatments, symptoms frequently recur or fail to improve with first-line medications. When a condition first manifests in childhood or adolescence, young adults may shy away from peer interaction or lose the ability to live independently (Freedman and Richter, 2021). OCD is one of the top ten disorders that have the most negative impact on quality of life and lost income, according to the World Health Organization. OCD occurs everywhere, though the specifics may vary depending on cultural factors. (For instance, in some cultures, religious concerns are more prevalent). The prevalence of compulsive washing is more common in women, while sexual obsessions, magical figures, or obsessional slowness are more common in men, according to epidemiological surveys conducted around the world (Rehn et al., 2018).

Our knowledge of the specific components that mediate interactions seen between the brain, infectious pathogens, and the immune system has significantly improved as a result of recent developments in neuroimmunology. As these polypeptides are recognized to play a crucial role inside the brain and to broadly mediate inflammatory and immunological responses, the activity of cytokines is also of special interest in this regard (Rocha et al., 2008). The pathophysiology of several neuropsychiatric illnesses, including depression, OCD, dementia, schizophrenia, and post-traumatic stress disorder, has been linked by recent studies to cytokines. Plasma cytokine levels as well as other immunological markers have been investigated in a few prior research, but the findings have been conflicting (Lamothe et al., 2018). The predominance of negative results does not rule out an autoimmune etiology for all cases of OCD or a specific subtype of OCD. Several theories attempt to explain these observations. First off, autoantibodies might not be linked to adult OCD. Second, the OCD-causing antibodies may only be produced briefly at the time of the illness' inception. It is plausible that transient autoantibodies with disease-causing potential could cause neurologic malfunction and cause either immediate or delayed OCD symptoms (Veale and Roberts, 2014). Third, immune cell characteristics may change as a result of serotonin reuptake inhibitors (SRIs), the most commonly given medications for OCD. However, numerous investigations on OCD patients' immunological measures after therapy have not found any appreciable changes. Once more, the high frequency of reports of higher peripheral cytokine levels in major depression (MD) patients, particularly interleukin-6, appears to be linked to mental distress and hyperactivity (Robbins et al., 2019). We expected that interleukin generation would be decreased in OCD patients because of the presence of MD because OCD is a prevalent anxiety disease that is associated with MD. Cytokines are chemicals that allow immune cells to communicate with one another or with non-immune cells. Addressing the mechanism and routes of probable immunologic disruption in OCD can be achieved by studying cytokines (Pauls et al., 2014). Except for a positive and significant link with IL-6 (interleukin-6) or solubilized IL-6 receptor plasma levels as well as the severity of compulsive behaviors, the initial investigations on cytokine changes in OCD patients comprised very few individuals and were negative. Since IL-6 levels are elevated in autoimmune disorders, elevated IL-6 stages in OCD may support the presence of an autoimmune etiological factor (Novellino et al., 2020). Furthermore, tocilizumab, an antibody that neutralizes the IL-6 receptor, is suggested as a treatment for some OCD cases since it has consistently been demonstrated to be useful in autoimmune diseases that involve IL-6. Thus, it is possible to speculate that tocilizumab or other particular immunological therapies may be able to assist some OCD patients who may have an immunological etiological element (Bennett and Molofsky, 2019).

19.2 UNDERSTANDING OF THE IMMUNE SYSTEM

19.2.1 Innate Immunity

The complex connections involving innate immunity as well as brain illnesses have drawn more attention to the whole range of neuropsychiatric and neurodegenerative disorders. Through the activity of numerous distinct cytokines, barriers including the blood-brain barrier and innate immune cells like microglia, astrocytes, macrophages, and mast cells are implicated in triggering illness events in these groups (Dominguez-Andres and Netea, 2019). Immune system cells originate from hematopoietic lineage precursor cells in the bone marrow and thus are found in specific lymphoid organs such as the lymph nodes, spleen, as well as thymus, as well as skin, and gut. When activated, innate immune system cells frequently go to a nearby lymphoid organ because they act as sentinels in areas where foreign organisms are likely to be encountered. Both inducible and non-inducible pathways are used by innate immunity. Natural morphological and physiological barriers in the body function broadly to ward off infections (Biron, 2016). When these barriers are broken, inducible innate mechanisms are triggered by patterns of identification that are either mediated by soluble substances released in the extracellular environment or by receptors present on the surface of innate immunologic leucocytes. Recent studies have discovered that innate immunity creates resistance by maintaining long-lasting nonspecific responses and sensitivity to pathogenic pathogens (Madsen et al., 2014). Although the host's defense is generally benefited by the innate immunity activity, recent research indicates that this isn't always the case. In fact, under risk factors, innate immunity may support a neuro-inflammatory loop and contribute to human disorders marked by an unchecked rise in inflammation. Almost all cells can participate in innate immunity by manufacturing specific innate cytokines, especially type 1 IFNs, and by reacting to these cytokines by inducing new and enhanced intracellular molecular defense mechanisms against infections (McDonald and Levy, 2019). However, the primary immune cellular components in charge of innate responses are macrophages, dendritic cells, and natural killer (NK) cells. Although eukaryotic organisms frequently lack these motifs, macrophages and DCs have pattern recognition receptors (PRRs) that react to pathogen-associated molecular pattern molecules (PAMPs), which are motifs shared by many different classes of infectious pathogens. The quick array of protections provided by innate immunity is established within hours, whilst the antigen-specific innate immune responses are developed within the first few weeks following infection (Schenten and Medzhitov, 2011). Additionally, several specific innate responses—including DC activation, antigen production process, displacement to lymphoid cells, up-regulation of co-stimulatory cells, and the composition of early cytokine profiles to shape subsequent adaptive responses—are crucially important in the early stages of the induction of acquired immunity. The synthesis of prepared effector molecules by the innate immune system's response was long thought to be the mechanism by which nonspecific protection was mediated. It is now known, however, that the Toll-like receptor (TLR) family of pattern recognition receptors (PRR) and the signaling pathways utilized by innate immune cells to carry out a defensive response have relative selectivity for pathogen-associated molecular patterns (PAMPs), which are representative of various microbe categories (Koenderman et al., 2014).

19.2.2 Acquired Immunity

An immunity that arises after a person acquires antibodies from some other source or when their immune system reacts to foreign particles or bacterium. Adaptive and passive immunity are the two varieties of acquired immunity. Pathogen defense is most effectively achieved through adaptive immunity, which is facilitated by T and B cells. However, some immune cell subgroups of lymphatic origin (lymphocytes), whether or not they have an antigenic receptor encoded by arranging DNA segments, resemble innate cells (Iwasaki and Medzhitov, 2015). They are significant contributors to human immune responses and have distinctive characteristics in terms of formation, phenotype, localization, and functions. Following a microorganism infection or vaccination, an individual develops adaptive immunity. The immunological response the body produces can help to stop the microbe from infecting others in the future. Instead of producing antibodies on its own, the immune system of a person with passive immunity acquires antibodies to an illness or toxin. Immature B and T cells go through antigen receptor gene reconfigurations during development to produce an incredibly wide repertoire with a collection of antigen-specific receptors that can recognize any conceivable antigen (Flajnik, 2018). Long-lived memory lymphocytes that are unique to the pathogen are produced following pathogen destruction and cellular contraction. When the invading pathogen is encountered again, these memory lymphocytes react quickly and aggressively, producing additional effector cells to destroy the infection (Bonilla and Oettgen, 2010). B cells perform a variety of critical functions in the immune response, including antigen presentation, cytokine production, and setting up the structure of lymphoid organs, although they are most well-known as the adaptive immune system's antibody-producing cells. B cells can respond to a range of foreign antigen molecules by making highly specific antibodies. We will outline the levels of B-cell formation, the fundamental procedures involved in producing B-cell receptors, and the procedures involved in B-cell stimulation and antibody production in this review (Cyster and Allen, 2019). T cells play a key role in controlling the immune response. They do this by altering the way that other immune cells function and by changing how endothelial as well as parenchymal cells behave. They regulate antibody responses, stimulate innate immune cells, and also lyse target cells by expressing membrane-bound compounds and secreting soluble mediators. Specific T-cell subsets inhibit immunological responses and shorten their duration. Therefore, poor T-cell activity has far-reaching effects on the immunological response (Gaud et al., 2018). T-Cells mature in the thymus, wherein they go through a rigorous selection procedure called central tolerance. Only cells with minimal affinity for self-antigens and the ability to interact also with major histocompatibility complex (MHC) molecules are permitted to leave the thymus. This makes sure that the repertoire has very few self-reactive T cells. Due to the fact that this mechanism is unaffected by lupus, lupus in humans and mice does not result in a larger frequency in self-reactive T cells than in healthy individuals (Brudno and Kochenderfer, 2019).

19.3 COMPREHENSION OBSESSIVE-COMPULSIVE DISORDER

Obsessive-compulsive disorder (OCD), a once-ignored condition, has been acknowledged as a prevalent, severely incapacitating, and perhaps curable early detection of cognitive issue (Fineberg et al., 2020). People who suffer from obsessive-compulsive disorder (OCD) have intrusive thoughts and must do routines to overcome them (Ambramowitz, 2021). Over the past ten years, there have been significant advancements in terms of physiology, therapy, and the phenomenology of the deep brain response to OCD, which have resulted in therapeutic treatments, innovative nosological conceptualizations, and service offerings (Horesh et al., 2021).

Obsessive-compulsive and related disorders (OCRDs) have recently been added to the International Classification of Diseases 11th (ICD-11) clinical categorization system, making OCD currently the most common of the obsessive-compulsive spectrum disorders. OCRDs also include olfactory reference disorders, hair-pulling, hypochondriasis, skin picking, hoarding, and bodily dysmorphic disorders (Fineberg et al., 2020). The mainstay of modern OCD treatment consists of available serotonin reuptake inhibitors, which include drugs like clomipramine (Del Casale et al., 2019). Stimulus response prevention is a component of cognitive behavioral therapy (CBT), and newly published research indicates that early treatment improves outcomes (Foa, 2022). The development of therapeutic treatments and specialized services was a significant study topic for treatment-resistant OCD globally.

19.3.1 Occurrence and Epidemiology

Over the course of their lives, people with OCD typically experience other mental problems. OCD sufferers typically report depressive symptoms, despite the fact that major depressive disorder

(MDD) and OCD illnesses seem to be distinct (Tadayonnejad et al., 2020). According to Howes et al. (2022), OCD patients had a tenfold higher prevalence of MDD compared to people throughout the wider community. Several researchers state that the majority of OCD patients had at least one-third of contemporaneous MDD at the time of examination, even though approximately 60% of patients with OCD suffer a period of depression during each other's life span (Youssef et al., 2021).

According to Matsunaga et al. 2020 study, a number of people experienced the anxiety and stress seen in the population during the pandemic in addition to OCD symptoms, such as irrational hand washing, inspecting things or surfaces to see if they could be infected or not, phobia of strangers, worry about the socioeconomic effects of the pandemic, and worry about getting sick. According to Walitza et al. (2020), early adulthood during preadolescence is affected by OCPD by up to 3%, which can be debilitating and stunt a child's overall development, but any disease or infection may set off an immune-mediated process that results in the development of the condition which provided clinical estimates for OCD in the general population that ranged from a 2% rise up to 19%. In society, people were most worried about separation, social phobia, as well as obsessions with the death or injury of friends or family members (Menzies et al., 2021). In young age who are beginning to acquire individuation and autonomy. Separation anxiety and the need for reassurance are also prevalent compulsions. Children must learn fundamental academic skills during their school years. Reading and comprehension are two activities that may become monotonous as a result of compulsions brought on by a lack of confidence and excellence (Castel, 2020). Teenagers should negotiate their associated moral and sexual growth, which can manifest as obsessions with things like spirituality, culture, ethics, or sexuality (Coimbra-Gomes, 2021).

19.3.2 Mechanism of Onset of Obsessive-Compulsive Disorder

Dopamine, glutamate, and serotonin are a few essential neurotransmitter networks that are present in cortico-striato-thalamo-cortical (CSTC) circuits or might play a key role in the underlying causes of OCD symptoms (Yin et al., 2022). Because of the initial finding of a specific serotonin reuptake inhibitor (SRI) response to OCD, serotonergic system research has received a lot of attention (Derksen et al., 2020). Individuals' responses towards the addition of dopamine D2 receptor antagonists to SRIs, among other data, brought attention to the involvement of the dopaminergic system (Coşkun, 2019). In contrast, current findings have mostly concentrated on the glutamatergic system. Despite the fact that the specific effectiveness of OCD individuals' use of SRIs may have provided substantial fuel to an understanding of serotonin, very little is known about a lack of natural serotonin that is important to OCD development. There is limited data; however, several studies have found that there are modifications in serotonin as well as its metabolites within the cerebrospinal fluid of OCD patients and that they return to normal following effective therapy (van Roessel et al., 2022).

In certain investigations, modified serotonin transporter receptor affinity has already been described in various regions, including the brain stem. Animal models' stereotypical actions, including such grooming, are significantly influenced by dopamine (Chamberlain et al., 2021). OCD may have an impact on this neurotransmitter, which also plays an important role in several behavioral and emotional functions, including the perception of rewards. A significant study indicates that the dopaminergic system plays an essential function in neurodevelopmental disorders, which most people regard as compulsive and related disorders (OCRDs) that are among the most significant (Scherma et al., 2019).

Some studies have found a connection between OCD and variations in catecholaminergic genes, including COMT, or even molecular diagnostic data suggesting alterations leading to some reduction in striatal dopamine D2 receptors among other dopaminergic receptors. These findings indicate that there is dopamine in OCD (Kaur et al., 2020). Furthermore, dopamine D2 receptor blockers may be useful in the treatment of OCD. According to these synapses, transmitting toward the striatum, glutamatergic prefrontal cortical neurons connect at CSTCs (Casado-Sainz et al., 2021).

DLGAP3, also known as SAPAP3, is an important part of a mouse model of OCD and is expressed in the striatum. Mice with SAPAP3 deactivation not only display obsessive grooming behaviors that are diminished by such a particular SRI but also do not have abnormalities in corticosteroid connections (Hadjas et al., 2020). OCD has been linked to several neurotransmitters and neuropeptides. Even though OCD research is still in its infancy, the role that frequent mental diseases play in inflammatory and immunological processes is significantly becoming more and more clear (Halverson and Alagiakrishnan, 2020).

19.3.3 Treatment of Obsessive-Compulsive Disorder

OCD is linked to genetic variations, though these only have a minor impact on epigenetic processes like DNA methylation. Considering serotonin reuptake inhibitors seem to be the preferred therapeutic therapy, the serotonin transporter (SLC6A4) gene has been widely studied (Gruenblatt et al., 2018). The use of selective serotonin reuptake inhibitors (SSRIs) as a crucial component for care with a wide range of mental illnesses spread rapidly. It functions primarily by preventing serotonin from reaching the serotonin transporter (SERT) and from being reabsorbed synoptically, which causes the postsynaptic neuron's serotonin levels to increase inside the serotonergic synapse. According to researchers, six important SSRI biomolecules have had a similar mode of action until now: paroxetine, sertraline, fluvoxamine, citalopram, escitalopram, and fluoxetine (Edinoff et al., 2021). Several of the typical psychophysiological processes that the serotonergic system regulates, like aggression, hunger, sleeping, mental distress, emotional cognition, and impulsive behavior, are disrupted in the depressed state (Srivastava et al., 2022). Raphe nuclei were the primary sites for serotonin synthesis inside the brain (Kim et al., 2019). The serotonin receptor is uniquely distributed across the brain and participates in a variety of physiological functions. In relation to stress-related reactions, it appears that the majority of receptor subtypes are engaged for depressive symptoms or antidepressant efficacy (Scherma, 2019).

The 5-hydroxytryptamine (5-HT) transporter's activity is selectively inhibited by SSRIs, which have little to no sensitivity towards transporting dopamine and norepinephrine. Due to their strong and specific sensitivity towards SERT, most therapies prevent SERT and 5-HT from attaching and being taken up by presynaptic cells. Over activation of the postsynaptic receptors is caused by excess 5-HT in the synaptic cleft. Throughout the period, postsynaptic and presynaptic receptors are subsequently down regulated as a consequence, resulting in an overall decline in the quantity of 5-HT generated inside the central nervous system and, moreover, a decrease primarily in SERT's expression. The SERT is down regulated after lengthy SSRI usage, but not the norepinephrine transporter.

19.4 EFFECTS OF OCD ON IMMUNE SYSTEM

Obsessive-compulsive disorder (OCD) is indeed a prevalent mental illness with an estimated lifetime incidence of 2.5% in the general adult population (Ruscio et al., 2010). Growing evidence suggests that immunological pathways may have a direct role in the pathogenesis of specific OCD subtypes, such as pediatric autoimmune neuropsychiatric disorders associated with streptococcal infections (PANDAS) (Stein, 2002). PANDAS is a subtype of acute-onset OCD that is thought to result from an autoimmune reaction to a group A streptococcal infection (Sigra et al., 2018). Specifically, antibodies against group A-hemolytic *Streptococcus* (GABHS) inter with self—antigens in the basal ganglia and in cortical structures, causing behavioral and motor disorders (Aron et al., 1965; Husby et al., 1976). It is crucial to check for any background of inflammatory disorders in pregnancy since OCD has been seen to occasionally be connected to the maternal transmission of serum neuronal antibodies following an inflammation event in pregnancy. It is crucial to investigate pregnancy histories and history of losses since immune factors have been linked to the pathophysiology of repeated, spontaneous miscarriages (Hansen et al., 2021; Matthiesen et al., 2012).

Group A beta-hemolytic *Streptococcus* (GAS) infections in kids have been linked to autoimmune reactions, the emergence of rheumatic fever (RF), neurological issues [including Sydenham chorea (SC)], and mental disorders like OCD. Even though pediatric GAS infections are relatively common, it is interesting that only a small percentage of affected children exhibit these neuropsychiatric symptoms; this clearly shows that there is a genetic disposition to OCD. Only acute occurrences of RF appear to increase the likelihood of developing OCD (Murphy et al., 2006). Family studies indicate that this connection may run in the family since first-degree members of RF pro-bands have higher aggregate rates of OCD-related and anxiety disorders compared to control individuals, including generalized anxiety disorders and separation anxiety (Dale, 2005; Ravindran et al., 1999). The connection of two polymorphisms in the tumor necrosis factor (TNF)-α gene promoter region with both OCD and RF is an intriguing result in the hunt for candidate genes because TNF-α is a proinflammatory cytokine implicated in RF and numerous other autoimmune illnesses (Fluitman et al., 2010; Hounie et al., 2007). Two decades ago, the National Institutes of Mental Health OCD study group introduced the name "pediatric autoimmune neuropsychiatric diseases linked with *Streptococcus* (PANDAS)" for those illnesses exhibiting both choreiform movements and OCD symptoms, notably obsessive and compulsive behaviors (Murphy and Pichichero, 2002).

Numerous studies have suggested that the immune system may be involved in the pathogenesis of adult OCD. It should be noted right away that the facts that are now available are without a doubt intriguing, albeit being limited. In fact, some clinical data suggests that OCD is much more likely than other mental disorders to be accompanied with immune-related illnesses, like thyroid issues and systemic lupus erythematous (Rocha et al., 2008; Hoekstra and Minderaa, 2005). The existence of anti-basal ganglia antibodies as well as the lower peripheral T-cell counts which have been linked to the degree of symptom severity imply that autoimmunity processes may play a role in adult OCD (Konuk et al., 2007; Sayyah et al., 2011). In adult OCD patients, increased CD8+ and reduced CD4+ lymphocytes as well as reduced natural killer (NK) cell activities have been noted, and these changes have been regarded as stress indicators (Nicholson et al., 2012; Marazziti et al., 1999). Intriguingly, immune cell abnormalities in people with adult OCD who have responded well to therapy appear to return to normal (Marazziti et al., 2009).

First, autoantibodies may not be linked to OCD in adulthood. Second, OCD-related antibodies may only be generated briefly at the time of the disease' development; it is feasible that brief autoantibodies with a pathogenic potential might startle the nervous system and cause instantaneous or prolonged OCD symptoms (Dale et al., 2005; Teixeira, 2003). The age of beginning of OCD was recently found to be negatively connected with tumor necrosis factor (TNF)-alpha levels in a sample of 31 OCD patients, while the length of disease was found to be marginally correlated with IL-6 levels (Konuk et al., 2007). However, several investigations on OCD patients were unable to find any appreciable changes in immunological parameters following therapy. Severe stress is associated with significant deterioration in OCD patients' quality of life, which leads to anomalies in the hypothalamic-pituitary, which may be the cause of the changes identified in the encouraging studies (Rapaport et al., 2005). Many research teams have proposed a link between major depression (MD) and weakened immunity. Recently, it has also been proposed that depression is linked to cellular immune activation, which increases the release of proinflammatory cytokines, such interleukin-1, interleukin-6, and tumor necrosis factor-alpha (Schiepers et al., 2005). Yet again, psychological stress and HPA hyperactivity appear to be linked to the large number of reports of higher peripheral cytokine production in MD patients, with interleukin-6 being the most often reported (Schiepers et al., 2005).

Furthermore, the age of onset for OCD is bimodal, peaking at 12–14 years old (early signs) and 20–22 years old (late onset), with early childhood symptoms appearing in 30–50% of individuals. It has been hypothesized that early onset OCD may differ from late-onset types of the condition in its pathophysiological processes (Dell'Osso et al., 2016). According to available data, immunological changes may be involved in the pathophysiology of OCD. Additionally, Sydenham's chorea, a very well-known neuropsychiatric side effect of streptococcal infections, is known to co-occur with OCD (Swedo et al., 1989).

19.5 IMMUNOTHERAPY AS A POSSIBLE INTERVENTION FOR TREATMENT OF OCD

OCD symptoms brought on by infections can be successfully treated with immunotherapy using immunoglobulin or plasma exchange. These encouraging outcomes are only seen in a small percentage of kids with PANDAS-related OCD (Millichap, 2000). Clinical improvement was associated with antineuronal antibody titters, and treatment with plasmapheresis significantly decreased OCD symptoms. Depending on antineuronal antibody titters, the sufferers' OCD symptoms also manifested in several remission-relapse events (Swedo et al., 1994). More lately, OCD patients who received IV-IG treatment exhibited a reduction in movement abnormalities of around 45% and a 30% drop in OCD symptoms (Swedo, 1995). The therapeutic benefits of IV-IG and plasmapheresis, which are used to treat autoimmune illnesses, would suggest that autoimmunity plays a role in the pathogenesis in OCD (Hanna et al., 1996). Advancement in the primary PANDAS symptomatology has been reported in several investigations when immunotherapy, such as corticosteroids, NSAIDs, intravenous immunoglobulin (IVIG), or plasma exchange, has been used (Frankovich et al., 2015; Brown et al., 2017). Although the exact mechanisms of action are uncertain, intravenous immunoglobulin (IVIG), an immunomodulatory treatment, has been proven to be beneficial for each of these conditions (Heitink-Pollé et al., 2014; Wong and White, 2016). A previous double-blind, placebo-controlled experiment in PANDAS patients found that plasma drug apheresis and intravenous immunoglobulin (IVIG) were both helpful in lowering obsessive-compulsive symptoms (by 45% and 58%, respectively), but a placebo infusion had no discernible impact (Perlmutter et al., 1999). Apheresis and IVIG have demonstrated no benefit in the treatment of non-PANDAS OCD, therefore these benefits are thought to be attributable to the immunomodulatory effects of IVIG (Murphy et al., 2012; Swedo et al., 2012). In addition, in PANDAS children receiving IVIG therapy, biochemical indicators of therapeutic response just haven't been thoroughly assessed (Williams et al., 2016). Regardless of the kind of infusion (placebo or IVIG) that individuals got in the blind phase of the experiment, a substantial mean decrease in OCD symptoms was seen in those who received an open-label IVIG infusion. A subtype of OCD with inflammatory or autoimmune origins has been supported by a number of independent studies. But this randomized controlled study was unable to give unambiguous evidence in favor of the therapeutic usage of IVIG in PANDAS. Consequently, when we wrap up future investigations of immunomodulatory therapies for PANDAS, we may need to include only individuals who do not react to antibiotic therapy, and they should be controlled to achieve a more modest therapeutic intervention from IVIG than the current trial (Pearlman et al., 2014).

19.6 MECHANISM OF ACTION OF IMMUNOTHERAPY ON OCD

Immunotherapy has always been a boon to medical sciences, proving its effectiveness in various diseases. Several immunomodulatory drugs have also been approved and marketized in various countries that show potential as anti-OCD agents (Table 19.1). Diverse range of investigations have been done to analyze the potential of immunotherapy against the progression and cure for OCD. Although the mechanism of action of immune-stimulating drugs against OCD have not been clearly understood yet, the probable pathway of action deduced in recent studies has always been targeting the mitochondria of the microglial cells and attempting to better express the malfunctioned iNOS gene.

The immunotherapy drugs induced for treatment generally trigger the calcium [Ca^{2+}] and potassium [K^{+}] pumps, decreasing its efficiency. This action causes an occurrence of oxidative stress in the mitochondria, leading to the deterioration of the mitochondrial membrane potential, resulting in the release of mitochondrial-reactive oxygen species [mt-ROS]. These mt-ROS act as inflammatory cytokines that bind with cell specific receptors, such as pattern recognition receptors [PRR], NOD-like receptors [NLR], and Toll-like receptors [TLR]. This cytokine and receptor complex, though released by the cell, trigger the assembly of Caspase-1 and NLRP3 resulting in the formation

TABLE 19.1
List of Approved Anti-OCD Drugs and its Mechanism of Action as OCD Suppressors

Anti-OCD Drug	Country of Origin	Mechanism of Action Reported for OCD	Therapeutic Drug Dosage	Half-life of Drug	References
Clomipramine	Switzerland	Noradrenaline reuptake inhibitor, alpha-1 receptor antagonist, sodium channel and NDMA receptor down-regulator.	75–300 mg/day	32 hours	Paxos, 2022
Fluvoxamine	Belgium	Serotonin reuptake pump suppression at the neuronal membrane inhibits serotonin neuronal absorption and boosts the effects of serotonin on $5HT1_A$ auto-receptors.	50–300 mg/day	15.6 hours	Laux, 2021
Paroxetine	Denmark	Serotonin receptor suppression of presynaptic reuptake of serotonin. Clinically negligible affinity exists for the alpha-1, alpha-2, and beta-adrenergic receptors, the dopamine D1 and D2 receptors, the histamine H1 receptors, and the $5\text{-}HT_{1A}$, $5\text{-}HT_{2A}$, and $5\text{-}HT_{2C}$ receptors of serotonin.	40–60 mg/day	21 hours	Harika-Germaneau et al., 2022; Chou et al., 2022
Sertraline	United States	Inhibiting serotonin (5-HT) reuptake at the presynaptic neuronal membrane as well as activity of norepinephrine receptors in the brain. It has affinity for sigma-1 and sigma-2 receptor binding sites; however, it binds to sigma-1 binding sites more strongly. In vitro, sertraline exhibits minimal to no affinity for benzodiazepine, GABA, dopaminergic, serotonergic (5 $HT1_A$, 5 $HT1_B$, and 5 HT2), or dopaminergic receptors.	50–200 mg/day	26 hours	Comer et al., 2004
Citalopram	Denmark	Binding to histamine, acetylcholine, and norepinephrine receptors prevents 5-HT from absorbing serotonin in the synaptic cleft of the central nervous system.	20–60 mg/day	35 hours	Middleton et al., 2019

Escitalopram	Denmark	Increasing the serotonergic activity by attaching to the major binding site on the serotonin transporter, to which endogenous 5-HT binds, and preventing the re-uptake of serotonin into the presynaptic neuron. Desensitization of 5-HT_{1A} auto-receptors, which normally shut down endogenous 5-HT release in the presence of excess 5-HT, is eventually brought on by sustained elevation of synaptic 5-HT; this desensitization may be required for SSRIs to have their full clinical effect and may explain why they typically take longer to start working.	10–20 mg/day	27–32 hours	Rubin-Kahana et al., 2019
Fluoxetine	United States	Preventing the neurotransmitter serotonin from being reabsorbed presynaptically. The outcome is a rise in 5-hydroxytryptamine (5-HT) levels throughout the brain. Additionally, fluoxetine is 5-HT selective, as evidenced by its strong affinity for 5-HT transporters, limited affinity for noradrenaline transporters, and lack of affinity for dopamine transporters.	20–60 mg/day	1 to 3 days	Acharya et al., 2021
Mirtazapine	Netherlands	The central nervous system experiences an increase in noradrenergic and serotonergic activity as well as antagonist activity at presynaptic 2-adrenergic inhibitory auto receptors and heteroreceptors. Antagonist of 5-HT_2 and 5-HT_3 serotonin receptors. Although it indirectly boosts 5-HT_{1A} transmission, it has not been observed to bind appreciably to the serotonin 5-HT_{1A} and 5-HT_{1B} receptors.	30–60 mg/day	20–40 hours	Lugo-Radillo and Cortes-Lopez, 2020

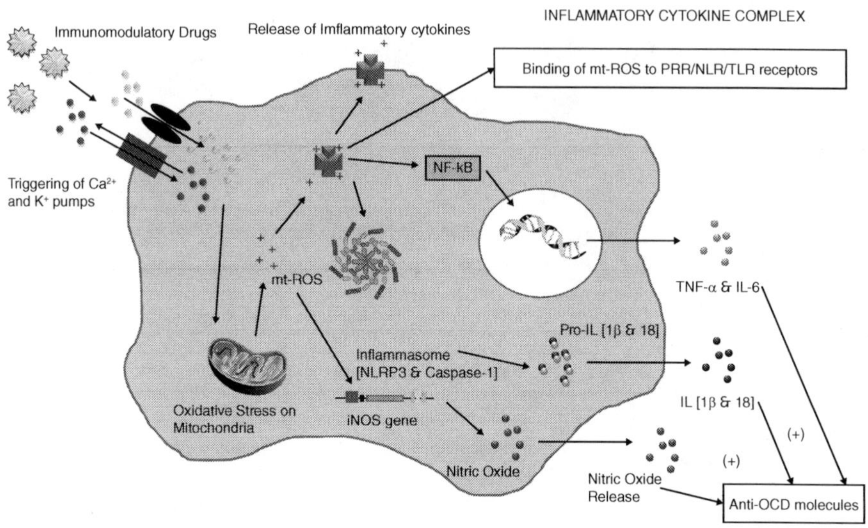

FIGURE 19.1 Mechanism of action of immunotherapy towards suppression of OCD progression.

of inflammasome. This results in formations and release of IL-18 and IL-1β by processing of pro-IL-18 and pro-IL-1β. Additionally, IL-6 and TNF-α are also released by the action of inflammasome activating the nuclear protein NF-κB-dependent transcriptional factor. Subsequently, the mt-ROS also causes iNOS gene to oxidise L-arginine initiating the release of nitric oxide [NO], essential for the suppression of OCD progression (Culmsee et al., 2019) (Figure 19.1). Many immunomodulatory drugs have been fabricated synthetically and naturally discovered, showing it's potential to decrease the symptoms of OCD, while its mechanism of action still remains to be investigated.

19.7 FUTURE PROSPECTIVE OF UTILIZATION OF IMMUNOTHERAPY FOR OCD TREATMENT

Further investigations should focus on the prevalence of antibodies against neurons in OCD patients and the diagnostic approaches that most effectively target autoimmune OCD therapies, among other things, by creating precise treatment algorithms and therapeutic standards. As a result of better knowledge of the immunological underpinnings of the condition, new OCD patients' treatment options may become available by improving their reaction to therapy and reducing recurrence (Endre et al., 2022). Encouragement for earlier detection or identification of pediatric neuroimmunological illness to increase the benefits of early intervention. Faster response times and direct therapy via therapeutic plasma exchange (TPE) will be future developments in understanding the processes of neuroimmunological diseases. Immunoadsorption, in which the body receives separated plasma that has been cleaned of a specific pathogenic molecule and returned to be renewed rather than replaced, has attracted renewed interest. TPE and immunoadsorption are equally effective in adults, and immunoadsorption correlates with being able to offer more safety features (Eyre et al., 2018).

19.8 CONCLUSION

As we know, obsessive-compulsive disorder (OCD) is a severe mental health issue that is defined by a pattern of unpleasant thoughts. Obsessions can lead to compulsive behaviors that disrupt everyday life and cause great distress. Symptoms in OCD patients frequently take the form of repetitive behaviors like worrying about contamination while washing and possibly even cleaning, worrying about harming oneself or others while checking, having intrusive sexual and aggressive thoughts while performing mental rituals, and worrying about symmetry while ordering or counting. In fact, obsessive-compulsive disorder (OCD) affects a large group of adults worldwide on a lifetime basis, making it one of the most common mental illnesses. There is mounting evidence that some OCD subtypes, such as pediatric autoimmune neuropsychiatric disorders associated with streptococcal infections, may be directly impacted by immunological pathways. The immune system may be implicated in the etiology of adult OCD, according to several research. In fact, according to certain clinical data, immune-related diseases like thyroid problems and systemic lupus erythematosus are far more likely to coexist with OCD than they are with other mental disorders. Through this chapter, the research community would further gain insight of the immunological pathway related to OCD. Also, we can possibly see how OCD symptoms are brought on by infections and further development of immunotherapy technology for treatment and management of this mental disorder.

ACKNOWLEDGMENT

We, the authors, convey our sincere gratitude to REVA University for providing a platform for research.

CONFLICT OF INTEREST

We, the authors, declare no potential conflict of interest.

REFERENCES

Abramowitz, J. S. 2021. *The family guide to getting over OCD: Reclaim your life and help your loved one.* Guilford Publications.

Acharya, P., Baral, R., Manandhar, S., Shahi, P., and Gurung, B. 2021. Formulation and in-vitro characterization of fluoxetine hydrochloride loaded fast dissolving oral film using HPMC 15CPS and HPMC K4M. *Journal of Pharmaceutics and Drug Development*, *8*(1), 101.

Aron, A. M., Freeman, J. M., and Carter, S. 1965. The natural history of Sydenham's chorea: Review of the literature and long-term evaluation with emphasis on cardiac sequelae. *The American Journal of Medicine*, *38*(1), 83–95.

Bennett, F. C., and Molofsky, A. V. 2019. The immune system and psychiatric disease: A basic science perspective. *Clinical & Experimental Immunology*, *197*(3), 294–307.

Biron, C. A. 2016. Innate immunity: Recognizing and responding to foreign invaders—no training needed. In *Viral pathogenesis* (pp. 41–55). Academic Press.

Bonilla, F. A., and Oettgen, H. C. 2010. Adaptive immunity. *Journal of Allergy and Clinical Immunology*, *125*(2), S33–S40.

Brown, K. D., Farmer, C., Freeman Jr., et al. 2017. Effect of early and prophylactic nonsteroidal anti-inflammatory drugs on flare duration in pediatric acute-onset neuropsychiatric syndrome: An observational study of patients followed by an academic community-based pediatric acute-onset neuropsychiatric syndrome clinic. *Journal of Child and Adolescent Psychopharmacology*, *27*(7), 619–628.

Brudno, J. N., and Kochenderfer, J. N. 2019. Recent advances in CAR T-cell toxicity: mechanisms, manifestations and management. *Blood Reviews*, *34*, 45–55.

Casado-Sainz, A., Gudmundsen, F., Baerentzen, S. L., et al. 2021. Nigro-striatal dopamine activation lowers behavioral and neuronal phenotypes associated with obsessive-compulsive disorder. *Preprint] bioRxiv*, *11*(10.1101), 2021–02.

Castel, P. H. 2020. Guilty cognitions, faulty brains: Obsessive-compulsive disorders in the age of the condition-of-autonomy (1980–2010). In *Obsessional neurosis* (pp. 1–33). Routledge.

Chamberlain, B. L., and Ahmari, S. E. 2021. Animal models for OCD research. *The Neurobiology and Treatment of OCD: Accelerating Progress*, 55–96.

Chou, P. H., Sack, A. T., and Su, K. P. 2022. Targeting three brain regions (Bilateral SMA, left and right DLPFC) sequentially in one session using combined repetitive transcranial magnetic stimulation and intermittent theta-burst stimulation in treatment-refractory obsessive-compulsive disorder: A case report.

Coimbra-Gomes, E. 2021. *Language & normativity in sexuality & gender-related obsessive-compulsive disorder: A netnographic-informed corpus-assisted discourse study* (Doctoral dissertation, Queen Mary, University of London).

Comer, J. S., Kendall, P. C., Franklin, M. E., Hudson, J. L., and Pimental, S. S. (2004). Obsessing/worrying about the overlap between obsessive-compulsive disorder and generalized anxiety disorder in youth. *Clinical Psychology Review, 24*, 663–683

Coşkun, M. 2019. Successful treatment of obsessive-compulsive disorder with aripiprazole in preschool children: Case series. *Psychiatry and Clinical Psychopharmacology, 29*, 151–152.

Culmsee, C., Michels, S., Scheu, S., Arolt, V., Dannlowski, U., and Alferink, J. 2019. Mitochondria, microglia, and the immune system—how are they linked in affective disorders? *Frontiers in Psychiatry, 9*, 739.

Cyster, J. G., and Allen, C. D. 2019. B cell responses: Cell interaction dynamics and decisions. *Cell, 177*(3), 524–540.

da Rocha, F. F., Correa, H., and Teixeira, A. L. 2008. Obsessive–compulsive disorder and immunology: A review. *Progress in Neuro-Psychopharmacology and Biological Psychiatry, 32*(5), 1139–1146.

Dale, R. C. 2005. Post-streptococcal autoimmune disorders of the central nervous system. *Developmental Medicine & Child Neurology, 47*(11), 785–791.

Dale, R. C., Heyman, I., Giovannoni, G., and Church, A. J. 2005. Incidence of anti-brain antibodies in children with obsessive–compulsive disorder. *The British Journal of Psychiatry, 187*(4), 314–319.

Del Casale, A., Sorice, S., Padovano, A., et al. 2019. Psychopharmacological treatment of obsessive-compulsive disorder (OCD). *Current Neuropharmacology, 17*(8), 710–736.

Dell'Osso, B., Benatti, B., Hollander, E., et al. 2016. Childhood, adolescent and adult age at onset and related clinical correlates in obsessive–compulsive disorder: A report from the international college of obsessive–compulsive spectrum disorders (ICOCS). *International Journal of Psychiatry in Clinical Practice, 20*(4), 210–217.

Derksen, M., Feenstra, M., Willuhn, I., and Denys, D. 2020. The serotonergic system in obsessive-compulsive disorder. In *Handbook of behavioral neuroscience* (Vol. 31, pp. 865–891). Elsevier.

Dominguez-Andres, J., and Netea, M. G. 2019. Long-term reprogramming of the innate immune system. *Journal of Leukocyte Biology, 105*(2), 329–338.

Edinoff, A. N., Akuly, H. A., Hanna, T. A., et al. 2021. Selective serotonin reuptake inhibitors and adverse effects: A narrative review. *Neurology International, 13*(3), 387–401.

Endres, D., Pollak, T. A., Bechter, K., et al. 2022. Immunological causes of obsessive-compulsive disorder: Is it time for the concept of an "autoimmune OCD" subtype? *Translational Psychiatry, 12*(1), 1–14.

Eyre, M., Hacohen, Y., Barton, C., Hemingway, C., and Lim, M. 2018. Therapeutic plasma exchange in pediatric neurology: A critical review and proposed treatment algorithm. *Developmental Medicine & Child Neurology, 60*(8), 765–779.

Fineberg, N. A., Hollander, E., Pallanti, S., et al. 2020. Clinical advances in obsessive-compulsive disorder: A position statement by the international college of obsessive-compulsive spectrum disorders. *International Clinical Psychopharmacology, 35*(4), 173.

Fineberg, N. A., Van Ameringen, M., Drummond, L., Hollander E., Stein, D., Geller, D., et al. 2020. How to manage obsessive-compulsive disorder (OCD) under COVID-19: A clinician's guide from the international college of obsessive-compulsive spectrum disorders (ICOCS) and the obsessive-compulsive and related disorders research network (OCRN) of the European college of neuropsychopharmacology. *Comprehensive Psychiatry, 100*, 152174.

Flajnik, M. F. 2018. A cold-blooded view of adaptive immunity. *Nature Reviews Immunology, 18*(7), 438–453.

Fluitman, S. B., Denys, D. A., Heijnen, C. J., and Westenberg, H. G. 2010. Disgust affects TNF-α, IL-6 and noradrenalin levels in patients with obsessive–compulsive disorder. *Psychoneuroendocrinology, 35*(6), 906–911.

Foa, E. B. 2022. Cognitive behavioral therapy of obsessive-compulsive disorder. *Dialogues in Clinical Neuroscience, 12*(2), 199–207.

Frankovich, J., Thienemann, M., Rana, S., and Chang, K. 2015. Five youth with pediatric acute-onset neuropsychiatric syndrome of differing etiologies. *Journal of Child and Adolescent Psychopharmacology, 25*(1), 31–37.

Freedman, D. E., and Richter, M. A. 2021. A narrative review of exercise and obsessive-compulsive disorder. *General Hospital Psychiatry, 71*, 1–10.

Gaud, G., Lesourne, R., and Love, P. E. 2018. Regulatory mechanisms in T cell receptor signalling. *Nature Reviews Immunology, 18*(8), 485–497.

Gruenblatt, E., Marinova, Z., Roth, A., et al. 2018. Combining genetic and epigenetic parameters of the serotonin transporter gene in obsessive-compulsive disorder. *Journal of Psychiatric Research*, *96*, 209–217.

Hadjas, L. C., Schartner, M. M., Cand, J., et al. 2020. Projection-specific deficits in synaptic transmission in adult Sapap3-knockout mice. *Neuropsychopharmacology*, *45*(12).

Halverson, T., and Alagiakrishnan, K. 2020. Gut microbes in neurocognitive and mental health disorders. *Annals of Medicine*, *52*(8), 423–443.

Hanna, G., Singh, V. K., Curtis, G., Himle, J., Weidmer-Mikhail, E., and Koram, D. 1996. Proceedings of the 43rd annual meeting of the American academy of child and adolescent psychiatry, Philadelphia, PA, p. 109, Abstract #NR-80.

Hansen, N., Luedecke, D., Malchow, B., et al. 2021. Autoantibody-associated psychiatric syndromes in children: Link to adult psychiatry. *Journal of Neural Transmission*, *128*(6), 735–747.

Harika-Germaneau, G., Langbour, N., Patri, S., et al., 2022. Impact of brain-derived neurotrophic factor Val66Met polymorphism and response to escitalopram or paroxetine in obsessive–compulsive disorder. *CNS Spectrums*, *27*(5), 645–651.

Heitink-Pollé, K. M., Nijsten, J., Boonacker, C. W., de Haas, M., and Bruin, M. C. 2014. Clinical and laboratory predictors of chronic immune thrombocytopenia in children: A systematic review and meta-analysis. *Blood, the Journal of the American Society of Hematology*, *124*(22), 3295–3307.

Hoekstra, P. J., and Minderaa, R. B. 2005. Tic disorders and obsessive-compulsive disorder: Is autoimmunity involved? *International Review of Psychiatry*, *17*(6), 497–502.

Horesh, D., Hasson-Ohayon, I., and Harwood-Gross, A. 2021. The contagion of psychopathology across different psychiatric disorders: A comparative theoretical analysis. *Brain Sciences*, *12*(1), 67.

Hounie, A. G., Pauls, D. L., do Rosario-Campos, M. C., et al. 2007. Obsessive-compulsive spectrum disorders and rheumatic fever: A family study. *Biological Psychiatry*, *61*(3), 266–272.

Howes, O. D., Thase, M. E., and Pillinger, T. 2022. Treatment resistance in psychiatry: State of the art and new directions. *Molecular Psychiatry*, *27*(1), 58–72.

Husby, G. V., Van de Rijn, I., Zabriskie, J. B., Abdin, Z. H., and Williams Jr, R. C. 1976. Antibodies reacting with cytoplasm of subthalamic and caudate nuclei neurons in chorea and acute rheumatic fever. *The Journal of Experimental Medicine*, *144*(4), 1094–1110.

Iwasaki, A., and Medzhitov, R. 2015. Control of adaptive immunity by the innate immune system. *Nature Immunology*, *16*(4), 343–353.

Juul-Madsen, H. R., Viertlböeck, B., Härtle, S., Smith, A. L., and Göbel, T. W. 2014. Innate immune responses. In *Avian immunology* (pp. 121–147). Academic Press.

Kaur, S., Singh, S., Jaiswal, G., et al. 2020. Pharmacology of dopamine and its receptors. In *Frontiers in pharmacology of neurotransmitters* (pp. 143–182). Springer.

Kim, M., Kwak, S., Yoon, Y. B., et al. 2019. Functional connectivity of the raphe nucleus as a predictor of the response to selective serotonin reuptake inhibitors in obsessive-compulsive disorder. *Neuropsychopharmacology*, *44*(12), 2073–2081.

Koenderman, L., Buurman, W., and Daha, M. R. 2014. The innate immune response. *Immunology Letters*, *162*(2), 95–102.

Konuk, N., Tekın, I. O., Ozturk, U., et al. 2007. Plasma levels of tumor necrosis factor-alpha and interleukin-6 in obsessive compulsive disorder. *Mediators of Inflammation*, *2007*.

Lamothe, H., Baleyte, J. M., Smith, P., Pelissolo, A., and Mallet, L. 2018. Individualized immunological data for precise classification of OCD patients. *Brain Sciences*, *8*(8), 149.

Laux, G. 2021. Serotonin reuptake inhibitors: Citalopram, escitalopram, fluoxetine, fluvoxamine, paroxetine, and sertraline. In *NeuroPsychopharmacotherapy* (pp. 1–13). Springer International Publishing.

Lugo-Radillo, A., and Cortes-Lopez, J. L. 2020. Long-term amelioration of OCD symptoms in a patient with chronic consumption of psilocybin-containing mushrooms. *Journal of Psychoactive Drugs*, *53*(1), 1–3.

Marazziti, D., Mungai, F., Masala, I., et al. 2009. Normalisation of immune cell imbalance after pharmacological treatments of patients suffering from obsessive-compulsive disorder. *Journal of Psychopharmacology*, *23*(5), 567–573.

Marazziti, D., Presta, S., Pfanner, C., et al. 1999. Immunological alterations in adult obsessive-compulsive disorder. *Biological Psychiatry*, *46*(6), 810–814.

Matsunaga, H., Mukai, K., and Yamanishi, K. 2020. Acute impact of COVID-19 pandemic on phenomenological features in fully or partially remitted patients with obsessive–compulsive disorder. *Psychiatry and Clinical Neurosciences*, *74*(10), 565.

Matthiesen, L., Kalkunte, S., and Sharma, S. 2012. Multiple pregnancy failures: An immunological paradigm. *American Journal of Reproductive Immunology*, *67*(4), 334–340.

McDonald, D. R., and Levy, O. 2019. Innate immunity. In *Clinical immunology* (pp. 39–53). Elsevier.

Menzies, R. E., Zuccala, M., Sharpe, L., and Dar-Nimrod, I. 2021. Are anxiety disorders a pathway to obsessive-compulsive disorder? Different trajectories of OCD and the role of death anxiety. *Nordic Journal of Psychiatry*, *75*(3), 170–175.

Middleton, R., Wheaton, M. G., Kayser, R., and Simpson, H. B. 2019. Treatment resistance in obsessive-compulsive disorder. *Treatment Resistance in Psychiatry*, 165–177.

Millichap, J. 2000. PANDAS, tics and OCD, and immunotherapy. *Pediatric Neurology Briefs*, *14*(1).

Murphy, M. L., and Pichichero, M. E. 2002. Prospective identification and treatment of children with pediatric autoimmune neuropsychiatric disorder associated with group A streptococcal infection (PANDAS). *Archives of Pediatrics & Adolescent Medicine*, *156*(4), 356–361.

Murphy, T. K., Sajid, M. W., and Goodman, W. K. 2006. Immunology of obsessive-compulsive disorder. *Psychiatric Clinics*, *29*(2), 445–469.

Murphy, T. K., Storch, E. A., Lewin, A. B., Edge, P. J., and Goodman, W. K. 2012. Clinical factors associated with pediatric autoimmune neuropsychiatric disorders associated with streptococcal infections. *The Journal of Pediatrics*, *160*(2), 314–319.

Nicholson, T. R., Ferdinando, S., Krishnaiah, R. B., et al. 2012. Prevalence of anti-basal ganglia antibodies in adult obsessive–compulsive disorder: Cross-sectional study. *The British Journal of Psychiatry*, *200*(5), 381–386.

Novellino, F., Saccà, V., Donato, A., et al. 2020. Innate immunity: A common denominator between neurodegenerative and neuropsychiatric diseases. *International Journal of Molecular Sciences*, *21*(3), 1115.

Overduin, M. K., and Furnham, A. 2012. Assessing obsessive-compulsive disorder (OCD): A review of self-report measures. *Journal of Obsessive-Compulsive and Related Disorders*, *1*(4), 312–324.

Pauls, D. L., Abramovitch, A., Rauch, S. L., and Geller, D. A. 2014. Obsessive–compulsive disorder: An integrative genetic and neurobiological perspective. *Nature Reviews Neuroscience*, *15*(6), 410–424.

Paxos, C. 2022. Moving beyond first-line treatment options for OCD. *Mental Health Clinician*, *12*(5), 300–308.

Pearlman, D. M., Vora, H. S., Marquis, B. G., Najjar, S., and Dudley, L. A. 2014. Anti-basal ganglia antibodies in primary obsessive–compulsive disorder: Systematic review and meta-analysis. *The British Journal of Psychiatry*, *205*(1), 8–16.

Perlmutter, S. J., Leitman, S. F., Garvey, M. A., et al. 1999. Therapeutic plasma exchange and intravenous immunoglobulin for obsessive-compulsive disorder and tic disorders in childhood. *The Lancet*, *354*(9185), 1153–1158.

Rapaport, M. H., Clary, C., Fayyad, R., and Endicott, J. 2005. Quality-of-life impairment in depressive and anxiety disorders. *American Journal of Psychiatry*, *162*(6), 1171–1178.

Rapinesi, C., Kotzalidis, G. D., Ferracuti, S., Sani, G., Girardi, P., and Del Casale, A. 2019. Brain stimulation in obsessive-compulsive disorder (OCD): A systematic review. *Current Neuropharmacology*, *17*(8), 787–807.

Ravindran, A. V., Griffiths, J., Merali, Z., and Anisman, H. 1999. Circulating lymphocyte subsets in obsessive compulsive disorder, major depression and normal controls. *Journal of Affective Disorders*, *52*(1–3), 1–10.

Rehn, S., Eslick, G. D., and Brakoulias, V. 2018. A meta-analysis of the effectiveness of different cortical targets used in repetitive transcranial magnetic stimulation (rTMS) for the treatment of obsessive-compulsive disorder (OCD). *Psychiatric Quarterly*, *89*(3), 645–665.

Robbins, T. W., Vaghi, M. M., and Banca, P. 2019. Obsessive-compulsive disorder: Puzzles and prospects. *Neuron*, *102*(1), 27–47.

Kahana, D. S., Shelef, A., Weizman, A., et al. 2019. The effectiveness of high-dose escitalopram in the treatment of patients suffering from schizophrenia with comorbid obsessive-compulsive disorder: An open-label study. *International Clinical Psychopharmacology*, *34*(4), 179–183.

Ruscio, A. M., Stein, D. J., Chiu, W. T., and Kessler, R. C. 2010. The epidemiology of obsessive-compulsive disorder in the national comorbidity survey replication. *Molecular Psychiatry*, *15*(1), 53–63.

Sayyah, M., Boostani, H., Pakseresht, S., and Malayeri, A. 2011. A preliminary randomized double–blind clinical trial on the efficacy of celecoxib as an adjunct in the treatment of obsessive–compulsive disorder. *Psychiatry Research*, *189*(3), 403–406.

Schenten, D., and Medzhitov, R. 2011. The control of adaptive immune responses by the innate immune system. *Advances in Immunology*, *109*, 87–124.

Scherma, M., Giunti, E., Fratta, W., and Fadda, P. 2019. Gene knockout animal models of depression, anxiety and obsessive compulsive disorders. *Psychiatric Genetics*, *29*(5), 191–199.

Schiepers, O. J., Wichers, M. C., and Maes, M. 2005. Erratum to "Cytokines and major depression" [Prog. Neuropsychopharmacol. Biol. Psychiat. 29(2), 201–217]. *Progress in Neuropsychopharmacology & Biological Psychiatry*, *29*(4), 637–638.

Sigra, S., Hesselmark, E., and Bejerot, S. 2018. Treatment of PANDAS and PANS: A systematic review. *Neuroscience & Biobehavioral Reviews*, *86*, 51–65.

Swedo, S. E., Leonard, H. L., and Kiessling, L. S. 1994. Speculations on antineuronal antibody-mediated neuropsychiatric disorders of childhood. *Pediatrics*, *93*, 323.

Srivastava, V., Singh, A. K., Yadav, V., Singh, R., and Verma, N. K. 2022. Medicinal uses (wound healing) and pharmacological properties of *Chromolaena odorata*: A Review. *IASR Journal of Medical and Pharmaceutical Science*, *2*(6), 01–07.

Stein, D. J. 2002. Obsessive-compulsive disorder. *The Lancet*, *360*(9330), 397–405.

Stein, D. J., Costa, D. L., Lochner, C., et al. 2019. Obsessive–compulsive disorder. *Nature Reviews Disease Primers*, *5*(1), 1–21.

Swedo, S. E. 1995. Brown University Child and Adolescent Behavior Letter 11, 3.

Swedo, S. E., Leckman, J. F., and Rose, N. R. 2012. From research subgroup to clinical syndrome: Modifying the PANDAS criteria to describe PANS (pediatric acute-onset neuropsychiatric syndrome). *Pediatric Therapeutic*, *2*(2), 113.

Swedo, S. E., Rapoport, J. L., Cheslow, D. L., et al. 1989, Feb. High prevalence of obsessive-compulsive symptoms in patients with Sydenham's chorea. *The American Journal of Psychiatry*, *146*(2), 246–249.

Tadayonnejad, R., Wilson, A. C., Corlier, J., et al. 2020. Sequential multi-locus transcranial magnetic stimulation for treatment of obsessive-compulsive disorder with comorbid major depression: A case series. *Brain Stimulation: Basic, Translational, and Clinical Research in Neuromodulation*, *13*(6), 1600–1602.

Teixeira Júnior, A. L. 2003. Coréia de Sydenham: um modelo de transtorno neuropsiquiátrico autoimune? *Psiquiatria Biologica*, 133–135.

van Roessel, P., Grassi, G., Aboujaoude, E. N., Menchon, J. M., Van Ameringen, M., and Rodriguez, C. I. 2022. Treatment-resistant OCD: Pharmacotherapies in adults. *Comprehensive Psychiatry*, 152352.

Veale, D., and Roberts, A. 2014. Obsessive-compulsive disorder. *BMJ*, *348*, g2183.

Wahl, K., Kordon, A., Kuelz, K. A., Voderholzer, U., Hohagen, F., and Zurowski, B. 2010. Obsessive-compulsive disorder (OCD) is still an unrecognised disorder: A study on the recognition of OCD in psychiatric outpatients. *European Psychiatry*, *25*(7), 374–377.

Walitza, Susanne, Melfsen, Siebke, Jans, Thomas, Zellmann, Henrike, Wewetzer, Christoph, Warnke, Andreas. 2011. Obsessive-compulsive disorder in children and adolescents. *Deutsches Arzteblatt International*, *108*(11), 173–179

Williams, K. A., Swedo, S. E., Farmer, C. A., et al. 2016. Randomized, controlled trial of intravenous immunoglobulin for pediatric autoimmune neuropsychiatric disorders associated with streptococcal infections. *Journal of the American Academy of Child & Adolescent Psychiatry*, *55*(10), 860–867.

Wong, P. H., and White, K. M. 2016. Impact of immunoglobulin therapy in pediatric disease: A review of immune mechanisms. *Clinical Reviews in Allergy & Immunology*, *51*(3), 303–314.

Yin, L., Han, F., Yu, Y., and Wang, Q. 2022. A computational network dynamical modeling for abnormal oscillation and deep brain stimulation control of obsessive–compulsive disorder. *Cognitive Neurodynamics*, 1–18.

Youssef, N. A., Phung, P., and Patel, R. S. 2021. Characteristics of patients who received deep brain stimulation in obsessive-compulsive disorder versus major depressive disorder. *Journal of Psychiatric Research*, *136*, 384–387.

Index

H

I

L

M

N

O

P

Q

R

Made in the USA
Las Vegas, NV
19 September 2026

308d0dad-1167-4468-903e-a6ccd00850fbR01